Mathematical Modeling
in the
Environment

Part 1
An Interdisciplinary Introduction to Selected Problems in Ground Water, Air Pollution, and Hazardous Materials

Part 2
Further Development of Modeling Concepts

© 1998 by
The Mathematical Association of America (Incorporated)
Library of Congress Catalog Card Number 98-85695

ISBN 0-88385-709-X

Printed in the United States of America

Current Printing (last digit):
10   9   8   7   6   5   4   3   2   1

Mathematical Modeling
in the
Environment

Part 1
**An Interdisciplinary Introduction to Selected Problems
in Ground Water, Air Pollution, and Hazardous Materials**

Part 2
Further Development of Modeling Concepts

Charles R. Hadlock

Published by

THE MATHEMATICAL ASSOCIATION OF AMERICA

CLASSROOM RESOURCE MATERIALS

Classroom Resource Materials is intended to provide supplementary classroom material for students—laboratory exercises, projects, historical information, textbooks with unusual approaches for presenting mathematical ideas, career information, etc.

MAA Service Center
P. O. Box 91112
Washington, DC 20090-1112
1-800-331-1622   fax: 1-301-206-9789

*This book is dedicated to Joanne,
my wife, my best friend, and my greatest source
of encouragement and support.*

Preface

I wrote this book because I felt it just had to be written.

Too few mathematics professors have an opportunity to leave the academic world to pursue a different career, in my case thirteen years in environmental consulting, and then return to academia to tell about it. I left academia in 1977 with a craving to learn more about "real world" applications. Even though I was certified by my graduate degree as an applied mathematician, I had never really delved deeply into a single, real applied problem, and that was not unusual. Within a short time, I found myself totally immersed in the issues of Love Canal, Bhopal, Three Mile Island, and many other major environmental cases of the day. I finally left the full-time consulting environment in 1990 because I had learned much, wanted to share it in the classroom, and wanted to think about it unfettered from the daily pressures of the "bottom line." I think that there is a great deal in this field that is likely to be of interest to mathematics teachers and students at all levels, as well as to other students and professionals.

This book is primarily intended for use in college-level mathematics or interdisciplinary courses. However, there are several quite distinct ways in which one might use it:

- as a text for a lower level, interdisciplinary course in an inherently interesting subject area, one with a strong problem solving component and with natural opportunities to support other important skills such as information research and computer use;
- as a text for a first or later course in applied mathematics, requiring minimal prerequisites but able to make use of more advanced background, if available, in selected sections and exercises;
- as a source of individual topics that could be incorporated into other courses, such as calculus;
- as fairly easy "bedtime reading" for faculty and graduate students, who may well find that the discussion here puts many traditionally more abstract concepts in a more concrete perspective, and who may also pick up more of an appreciation of how modeling is used in the real world;
- as a source of some well motivated derivations and exercises that could enrich many courses or independent study programs.

For many years, the mathematics community placed its highest value, at least implicitly, on "pure mathematics," and it continued to educate the bulk of both undergraduate and graduate students with scant attention to mathematical modeling of real world problems. Even courses in applied mathematics were more survey courses in mathematical techniques than courses intended to give the students a real taste of the very different world in which mathematical modeling for applications is usually carried out. Many of today's professors are the products of those years.

The grim job prospects for many recent PhDs in mathematics are now causing this to change; but, more generally, the need for math graduates at all levels to be able to compete for jobs with students of engineering, physics, economics, and other quantitative fields encourages a reexamination of our educational programs. To put it more positively, I think that many of the traditional aspects of mathematics education, such as the logical analysis of issues and the search for simple, basic principles of understanding, can prepare mathematics students with unique strengths that will let them make excellent contributions in real world applications and give them valuable advantages that others may not have — as long as they can learn to reach outside the limits of their own discipline.

But beyond serious and committed students of mathematics, there is also a larger group of students who have practically no appreciation for the degree to which mathematical modeling permeates the operations of our modern society. They range from beginning freshmen who may be trying to decide what field of study to major in to that somewhat disaffected group of college students who want to study the very least mathematics that their schools will permit. If one is to give these students an experience that will cause them to explore their options more fully or to reexamine their assumptions, one must get to them early, for their first college math course may be their last. There is no time to teach a plethora of techniques or abstract concepts. Nor should one want to. Let's take these students right out into the real world and show them the societal problems that modelers are grappling with day in and day out. You don't need a lot of mathematical machinery to get involved in some of these problems. Furthermore, most students are already well-disposed to studying environmental issues; they only have to get over the shock of seeing a mathematician in front of the classroom instead of a biologist or someone else!

The biggest challenge in developing this book was in offering material to more than one audience. I wanted to provide something engaging to the quite different groups that are characterized above. The solution was the division of the book into two parts. The first part (Chapters 1–4) uses no calculus; the closest it comes is the use of the exponential function e^x, with a self-contained introduction even to that. The second part (Chapters 5–7) further develops each of the three environmental themes from earlier, making various uses of one-dimensional calculus and multidimensional calculus in well defined subsections for which the prerequisites are spelled out. This part also touches upon other fields, either to reinforce previous study (e.g., linear algebra) when the student has had that material, or to try to provide an enticing introduction to some new ideas (e.g., numerical analysis, probability).

For a more elementary class, the first part is enough for a semester's course, especially if the nonmathematical aspects of the material are explored as well. Even if such aspects are not treated as fully as I myself might like to do, there are additional topics in part 2, labeled as such, that do not use any calculus, as well as others that use only one-dimensional calculus. General education courses usually require minimal prerequisites and are sometimes offered even in the freshman year. I believe that this material is particularly well suited for such courses, as it uses no calculus, is interdisciplinary (math, science, societal issues, computers, information research, and field-trip opportunities), and deals with a subject about which students are motivated to learn. It also lends itself to group work in class and homework projects. A number of high schools have also incorporated material from the first part as enrichment units in their junior and senior level math courses.

A more advanced class could easily become quite involved in the mathematical issues raised in the second part of the book, and some of the exercises will be quite challenging even to the professional. Either a traditional course or a seminar structure could be used in such a class. The approach of this book is not to focus on mastering a wide range of techniques nor bringing students to the current state-of-the-art, however. Rather, it first immerses the students in an applied area with engaging nonmathematical assignments (e.g., reading news accounts of local environmental emergencies, web browsing), and then gives them experience with a number of important math topics (e.g., Laplace's equation, heat flow, diffusion, numerical methods, Gaussian distribution, physical modeling, fluid mechanics, and others) that structure the modeling approach to these issues.

Two key components found throughout are:

- constant interplay between physical reasoning and mathematical modeling;
- "learning by doing," as the material is presented via a combination of text and interspersed exercises, the latter suitable (if desired) for group work, even in class, or student presentation and discussion. (The questions are sometimes open-ended and invite disagreement and discussion.)

Computers are a natural tool for mathematical modeling, and I believe it would be difficult and less than optimal to try to teach about modeling without making use of them. There is a natural progression in the text. Roughly speaking, Chapter 2 is hand calculation, although some problem types could be simplified and organized with something like a spreadsheet program. Chapter 3 would benefit further from some programming device, such as a programmable calculator or computer spreadsheet. Chapter 4 is based on using one of several commercial packages, although one free one that has minimal computer hardware requirements is available on a single diskette. On the non-math side, on-line computer resources are encouraged for some of the information research questions.

A teaching challenge is that the teacher has to build up enough self-confidence to talk about various scientific concepts, such as soil, molecules, vapor pressure, and fire. Nothing is needed beyond what is in the book, but we in the mathematics profession simply are not used to teaching concepts that don't fit the nice structure of an equation or theorem. The students seem to handle it very well, even very weak students who always hated science in high school. Dealing with some of these concepts in response to computer menu requests seems to make it more motivating for the students to think about what they really mean. And a class field trip to a fire department (or a class speaker), where uniformed personnel throw these same concepts around constantly, makes them seem so much less academic that the students are no longer intimidated. The *Supplementary Material and Solutions Manual* is intended to help teachers over these obstacles, as well as to offer complete solutions to every single exercise, mathematical and nonmathematical, in the text.

Organizing this material into the above framework was quite a challenge. I started with many more topics in my initial teaching efforts, but gradually decided to provide a more thorough treatment of three themes that had a universal presence and many good local examples.

I sincerely hope that the reader will discover new interests in this material and that this work will make at least some small contribution to the improvement of our corps of citizens, leaders, and professionals as we approach the important environmental decisions of the future.

Charles R. Hadlock
Waltham, Massachusetts

Acknowledgments

I am deeply indebted to five institutions and to numerous individuals for their help in bringing this project to fruition.

Bentley College, in offering me a professorship in 1990, gave me the opportunity to realize my hopes of bringing into the classroom so many of the new viewpoints and interesting ideas I had encountered while working in the environmental industry since 1977. Dean Lee Schlorff and all my colleagues in the Mathematical Sciences Department have been strongly supportive of this curricular innovation. Barbara Nevils offered to be the first "guinea pig" in trying to learn and teach this material without previous environmental background, and her success at this convinced me of the viability of trying to share it with the greater mathematical community. David Carhart, Norman Josephy, Erl Sorenson and others have offered valuable comments and suggestions on the manuscript. Alex Schuh verified my solutions to almost every problem in the book and recommended valuable improvements. Alice Laye provided superb typing and production assistance.

The National Science Foundation provided generous support over the course of three years, and it was their acceptance of my proposal that represented the "alea iacta est" of the project. I must single out my project manager, Myra Smith, from Cornell University but on a temporary assignment at NSF, whose enthusiasm and confidence in this project were an inspiration. My NSF project advisory board, consisting of Ira Gessel (Brandeis), William Keane (Boston College), and James Ward (Bowdoin), were a constant source of good advice and encouragement.

The Massachusetts Institute of Technology provided the perfect working environment for much of the writing by offering me a visiting professorship for the academic year 1996–97. Not only did I have within the Department of Earth, Atmospheric, and Planetary Sciences a stimulating group of colleagues and excellent resources, but I also had an opportunity to teach from the manuscript to a group of lively and encouraging students from a wide range of disciplines. Daniel Burns, Timothy Grove, and Thomas Jordan made all this possible.

The Mathematical Association of America, in the personages of Andrew Sterrett and Don Albers, expressed strong interest in this project right from the start. During production, it has been a pleasure to work with Elaine Pedreira Sullivan and Beverly Ruedi.

Last but not least among the institutional groups, the many professional opportunities I received by virtue of my position at Arthur D. Little, Inc., really provided the underlying substance for the whole project. I value greatly my many stimulating associations with colleagues and clients, especially the late Dan Egan (EPA) who was most supportive in my early years of working in this field.

During the preparation of this work, I have also had the benefit of many valuable discussions with professionals in the field and with fellow mathematicians and educators, not listed above, some of whom provided detailed reviews of portions of the manuscript. I would particularly like to acknowledge contributions from Ben Fusaro (Florida State University), Anita Goldner (Framingham State College), Jack Guswa (HSI/GeoTrans), John Hagopian (Hazmat America), Don Miller (St. Mary's College), Steven Pennell (UMass Lowell), and Lou Rossi (UMass Lowell).

I would also like to recognize the key contribution made by a number of my former students at Bentley and at MIT, whose sharp intellect, attention to detail, and interest in the subject were both an encouragement to follow through on the project and a constant check against ambiguity and error. These include Matthew Berman, Wendy Brown, Ellen Clark, Diane Curry, Jennifer Forman, Hau Hwang, Caroline Martorano, Walter VanBuskirk, and many others.

In listing all the above contributors, I must myself take responsibility for any residual errors, and I would appreciate having them called to my attention for correction in later printings.

Contents

Part 1

*An Interdisciplinary Introduction
to Selected Problems in
Ground Water, Air Pollution, and
Hazardous Materials*

1

Introduction

This book has two principal objectives:

1. To provide an in-depth introduction to several very important environmental issues.
2. To show how mathematical models are used routinely to help analyze these issues.

These objectives are closely interrelated. On the one hand, essentially every environmental problem has key quantitative aspects. For example, it is not enough simply to know that there is pollution somewhere, for, unfortunately, there is pollution almost everywhere. One needs to know how serious the pollution is, if it is getting better or worse, and how much it might be improved by implementing various possible remedial actions. These are all quantitative questions for which one would expect quantitative methods to be useful. On the other hand, by trying to make precise quantitative statements about environmental problems, one will actually develop a more precise and refined understanding of the underlying problems themselves.

The level of environmental awareness of society (both in the US and in many other countries) has grown immensely in the last decade or two. For example, the following are all practices that were routinely accepted a relatively short time ago, and, while still widely practiced, are now under much closer scrutiny and control:

- The disposal of garbage and other wastes in municipal dumps or "landfills," often located in swampy areas less desirable for construction and habitation. Now we recognize the migration potential of contaminants contained in waste materials, and we further understand that wetland areas are key entry points to surface and groundwater pathways through which such contaminants can easily move. Furthermore, the filling in of wetlands by waste disposal and other human activities eliminates a vital source of stability in the hydrologic cycle and can have far-reaching ecological implications.
- The use of buried underground storage tanks, often made of ordinary steel, for fuel oil and other chemicals. We now better recognize the "obvious" fact that such tanks will eventually corrode and leak, and that the resulting chemical contamination may persist for long periods of time and travel long distances before it is detected. By then it can be very hard and prohibitively expensive to clean up the problem.
- The disposal of sewage in such a way that it can enter lakes, rivers, and harbors in relatively untreated form. The scale of this problem ranges from large cities, where untreated sewage has often been allowed to flow into a harbor, to small bungalows and cottages on the

shores of lakes, where inadequate cesspools or septic systems contribute substantially to water pollution.

- The construction of very tall smoke stacks to disperse pollutants from power plants and other facilities where fuels or materials are burned. We now better understand that such pollutants do not "disappear" or necessarily reach harmless concentrations when injected into the atmosphere; they may just fall into someone else's "backyard" at a greater distance from the source, often in a modified form such as acid rain, or they may have other harmful effects.

- The consumption of natural resources (e.g., forests, minerals, fuels, fish stocks) at a very high rate. More people are finally coming to question our right to use the earth's resources at a rate that cannot be sustained for the benefit of future generations, but we hardly know how to change past habits without unacceptable economic reverberations.

- The wide use of chemicals that may have devastating implications for the earth's atmosphere or its ecosystems. Examples include pesticides, such as DDT, which have brought certain species of birds to near extinction, as well as the chlorofluorocarbons (CFCs) used to provide the pressure inside aerosol cans. These latter have been found to damage the earth's ozone layer, with resulting effects on health (e.g., skin cancer and cataracts of the eye), agriculture, and natural ecosystems.

With the exception of the last item in this list, where some of the effects took years of scientific research to uncover, we can look back today on these practices and ask whether anyone "back then", meaning in 1950, 1960, or 1970, ever really thought about the damage we were doing to our environment as we encouraged industrial development, international competitiveness, and growth in our apparent living standards. The answer is that of course some people were thinking about these aspects, but they were not dominant themes in our decision making, and they were too easy to downplay or postpone in the face of other priorities. We were engaged for many years in a dangerous Cold War, and much of our economic growth was fueled by the arms race with the Soviet Union. For a considerable length of time, the war in Vietnam also stole much of our national attention and energy.

It took the occurrence of a number of major environmental disasters in the post-Vietnam era to bring the attention of political and business leaders both in the US and abroad to the kinds of environmental concerns that had been raised from time to time by citizens groups, scientists, and others. Examples of such incidents include several major petroleum tanker accidents at sea, including the Amoco Cadiz (1978) and the Exxon Valdez (1989), the circa-1978 official recognition of extensive underground contamination at Love Canal in upstate New York, leading to the evacuation of a neighborhood, the 1979 Three Mile Island nuclear accident in Pennsylvania, involving a partial core meltdown, the 1984 accident at a Union Carbide chemical plant in Bhopal, India, in which a toxic gas release caused thousands of deaths and injuries, the 1986 explosion at the Chernobyl nuclear power plant in the Ukraine, which caused many deaths and other health effects in the local area and also sent airborne radioactive contamination around the globe, and, also in 1986, the extensive contamination of the Rhine River along its entire length following a pesticide warehouse fire near Basel, Switzerland. There are unfortunately many more similar examples, each of which has served further to heighten public awareness of the hazards associated with our increasingly technological society.

This heightened attention and concern has led to wholesale changes in the way individuals and businesses now interact with the environment. Government regulatory agencies have been very active in improving their systems of regulations so as to protect the environment, and businesses have changed their practices significantly, not only because of these new regulations, but also because of their fear of clean-up and liability costs associated with any environmental problems that they may create. For example, the financial effects of the Bhopal incident on Union Carbide were considerable, facilitating a hostile takeover attempt and contributing to the near dismantling of that major multinational corporation. The direct and indirect financial impacts of these kinds of incidents are positively frightening to corporate boards, CEOs, and shareholders, over and above the basic human and environmental impacts themselves.

Even though we now can see more clearly the kinds of environmental problems associated with past or present societal practices, finding good solutions is not necessarily easy, often involving the substitution of one environmental problem for another. If you try to reduce the volume of solid waste to be buried in a landfill, for example, by incinerating it instead, you may at the same time create a potential air pollution problem. If you excavate the contaminated ground around leaking gasoline tanks at a service station, you will need to process or dispose of that contaminated material in some other location, thereby creating a potential new contamination problem there. If you build new and better treatment plants for municipal sewage, you then have the problem of disposing of the waste materials from those plants, as well as the myriad environmental impacts associated with taking land for and constructing what are often regarded as somewhat undesirable facilities. People who think there are obvious and easy answers to environmental issues usually do not have a full understanding of the complexities involved. So while this book will give you an introduction to several of the basic areas of environmental concern, please be cautioned that making good environmental decisions for society is a very complex process that invariably involves tradeoffs between one kind of environmental impact and another. This kind of process also involves a wide range of participants, not just scientists and engineers, but also business people, lawyers, civic leaders, economists, mediators, journalists, information specialists, and ordinary citizens.

Table 1-1 gives an overview of the diversity of environmental issues of concern today. It would be good to spend a few minutes reading through this long list and gaining a realization of the wide range of environmental issues that require attention. To some people, environmental protection means protection of endangered species and habitat; to others, it means preventing the pollution of air and water; to others still, it means protection of key aspects of the earth's atmosphere related to global warming or the ozone layer. The point here is that the environmental field is very broad, and there are many areas that still require much more study and much difficult decision making by society.

While every single environmental issue in Table 1-1 has important quantitative components, in this book we will focus on some of the more "everyday" environmental issues that are likely to arise in the reader's own city or town. Thus we will be focusing on environmental issues with primarily local impacts, such as air pollution, water pollution, and the handling of hazardous materials within local communities. It is hoped that this introduction to the field of environmental management will provide a good foundation for the reader who is interested in pursuing some of the other major environmental issues of the day.

TABLE 1-1. Examples of the wide range of environmental issues and some typical aspects where models are applied

Issue	Overview	Typical aspects addressed by modeling
Air pollution	Local pollution (e.g., urban smog, toxic releases), acid rain, global issues (e.g., CFCs and related chemicals).	Concentrations, health impacts, trends, impact and cost/benefit analysis of proposed control measures.
Water pollution	Sewage, agricultural runoff, pesticide residues, acid rain, chemical discharges. Relatively few people in the world have access to clean water.	Transport and environmental fate of pollutants. Effectiveness of remedial measures.
Ground water	Availability of ground water and the investigation or remediation of contamination problems.	Safe yields from wells, direction and rate of movement, spread of contaminants, evaluation of remedial measures.
Water supply	There are many notable instances where surface or ground water is being used at a non-sustainable rate.	Determining the capacity of supplies. Effectiveness of policy and incentive programs.
Global warming	Greenhouse gases, such as CO_2, cause more of the Sun's energy to be retained by the Earth, leading to a rise in average temperature.	Interactive effects of greenhouse gases, sunlight reflection by aerosols and increased cloud cover, changes in vegetation, etc. Economic impacts (e.g., on agriculture).
Sea level changes	Melting of polar ice due to increase in global temperature could redefine coastlines, displace populations, and affect ocean currents and climate.	Feedback loops; e.g., larger ocean may lead to higher amount of evaporation, more water in cloud state, perhaps more reflection of solar energy, etc.
Forest destruction	Injects more greenhouse gases into atmosphere and reduces world photosynthetic capacity (which removes CO_2). Also leads to erosion, silting of waterways, and species loss.	Role of forests as carbon reservoirs in world climate models. Impacts on deforestation of trade policies or economic incentives.
Ozone depletion	Certain chemicals (e.g., CFCs) deplete ozone in the stratosphere. Such ozone shields the surface from ultraviolet light, which can harm humans and plants.	Short and long term ozone predictions. Quantifying physical and economic impacts on health and agriculture. Effects of CFC replacements.
Hazardous materials	Dangers to the public and the environment from both planned and unplanned releases.	Concentration, transport through the environment, population exposure, health effects, probability and effects of accidents.
Solid waste	Management and disposal of municipal solid waste, industrial waste, and miscellaneous wastes (e.g., mine tailings, dredging spoils).	Evaluation of both technical and cost/benefit aspects under uncertain future scenarios regarding quantity, characteristics, etc.
Hazardous wastes	Hazardous wastes are now variously incinerated, buried, reprocessed, or otherwise treated for disposal or reuse.	Environmental effects of various disposal systems. Effectiveness of alternative regulatory or economic incentive programs.
Ocean pollution	Years of routine dumping of waste material at sea by many nations make it difficult to control this practice, despite some treaties.	Environmental fate of materials disposed of at sea, both in the physical environment and in the food chain.

Issue	Overview	Typical aspects addressed by modeling
Environmental impact analysis	Evaluation of the total environmental impact of major development projects or industrial facilities.	Air, water, noise, traffic, thermal, ecological and other issues. Cost/benefit analysis of alternatives.
Population dynamics	Ecological effects on plant and animal populations of natural changes or perturbations by human action (e.g., overfishing, predator control, dams).	Retrospective or advance evaluation of projects or government policies by model simulation.
Environmental health impacts	Determining whether there is a connection between observed disease and some environmental cause or contributor.	Exposure modeling; statistical evaluation of suspected disease anomalies.
Biodiversity	Preservation of diverse species and gene pools for reasons ranging from aesthetics to ecology to development of medicines and other products.	Quantitative measurement of diversity and how it is affected by human actions or environmental changes.
Wetlands	Wetlands are important for many reasons (e.g., habitat, flood and erosion control, groundwater recharge).	Quantitative evaluation of value, either for advance decisions or setting charges for past damage.
Desert growth	In many parts of the world, human overpopulation combined with poor conservation practices has led to the growth of deserts in areas that were previously covered by vegetation.	Interactive effects of human activities and climatic episodes. Viability of restoration programs.
Ionizing radiation	Radiation exposure from normal operations (e.g., nuclear power plants) or accidents (e.g., Chernobyl). Disposal of radioactive wastes.	Environmental and health impacts of releases. Probability of major accidents. Suitability of waste disposal concepts.
Non-ionizing radiation	Radiation from electromagnetic sources, such as power lines and electronic equipment.	Estimation of population exposure at different levels. Statistical correlation with health data.
Indoor air pollution	Many substances in the indoor environment may be harmful (e.g., radon gas, household materials).	Movement of air pollutants through buildings and ventilation systems. Correlation of health impacts to population exposure data.
Noise pollution	An area of great importance to abutters of many projects (e.g., highways, airports, rest areas, construction activities, windmills, industrial plants).	Advance calculation of expected noise levels and evaluation of control strategies.
Aesthetic degradation	Human aesthetic response to the environment is recognized as a formal element of assessing environmental impact and degradation.	Sometimes evaluated using quantitative ranking schemes to measure preferences and responses.
Natural hazards	Natural environmental occurrences can cause great environmental damage (e.g., volcanoes, mud slides, forest fires, earthquakes, floods, etc.)	Prediction, risk of damage to engineered systems (e.g., water supply, power, hazardous facilities), cost effectiveness of investments to prevent or mitigate.

Issue	Overview	Typical aspects addressed by modeling
Land subsidence	Land subsidence caused by underground mining, withdrawal of ground water, underground dissolution of soluble rock formations, consolidation of sediments, or other factors.	Prediction of such effects and evaluation of control measures (e.g., rock mechanics modeling to change mining system).
Salt water incursion	Incursion of salt water from coastal areas into nearby aquifers due to overpumping.	Calculation of safe yields by ground-water modeling.

2

Ground Water

2.1 Background

You probably already know that if you start digging a hole almost anywhere—in your backyard, out in the woods, even in the middle of a big city—you will eventually hit water. It might happen just a few inches or feet below the surface, or it might take hundreds of feet of digging or drilling; maybe you would even have to break through solid rock. But the point is: you would eventually hit the so-called *saturated zone,* where all the spaces between the soil particles or all the cracks in the rock are full of water. The water in this zone is called *ground water.**

Now, that ground water isn't just sitting there. It's actually moving. It's not rushing along like a river on the surface; and, in fact, it might just be creeping along at only a few inches or feet a day or even less. Whether it's moving fast or slow, it *is* moving, which is the critical point, because if it ever gets contaminated by some kind of toxic material, then that contamination is going to move along with it, possibly leaving in its wake a very large underground zone where the ground water may no longer be fit for consumption or other uses.

Imagine, for example, an underground storage tank, such as a gasoline tank at a service station, that is leaking very slowly (so that the leak is not detected by the service station operator). The gasoline moves downward vertically through the soil until it reaches the ground-water zone, and then it begins to migrate with the ground water. Even if the ground water is moving relatively slowly, let's say ten feet per day, in the course of a whole year the contamination would have reached a distance of over 3,600 feet, which is almost 3/4 of a mile. Along the way it could have contaminated private drinking water wells or town wells, or it could have reduced the value of future home sites because wells might no longer be feasible on the contaminated property. How could you even begin to clean up such a large area of underground contamination? If the leak continued undetected for a longer period of time, say five years, the zone of contamination would have extended to over three miles from the original site.

---

* Note that the spellings "ground water" and "ground-water" are both used in this text. When being used as a noun, the preferred usage is as the two words "ground water." As an adjective, the hyphenated form is used, as in "ground-water contamination." The spelling "groundwater" is also gaining acceptance in the literature.

To take this example one step further, think of all the problems that would emerge once this contamination had been detected. On the one hand, there is the technical problem of how much the contamination could be cleaned up at all. Maybe you could dig up the contaminated soil right around the underground tank at the service station, but you certainly couldn't dig up the wide zone of contaminated subsurface along which the gasoline has moved. Think of the lawsuits that might be brought against the service station operator, through whose negligence some of the neighbors may have been exposed to health risks by drinking contaminated water or whose properties may have decreased in value because of this problem. Think of the actions that could be brought by governmental agencies, who would likely impose substantial fines on the operator for causing this environmental pollution. Then there might be insurance issues involved. The service station operator is probably carrying some form of liability insurance. Does it cover this kind of long-term process of leakage, or is it just limited to isolated accidents? What if the service station operator is forced into bankruptcy because of the overwhelming costs associated with cleaning up this problem? Might this cause him to default on his mortgage and thereby let the property be transferred into the hands of the bank that holds the mortgage? But now the bank would own a piece of contaminated property, and it might have inherited many environmental problems that it had never anticipated having and that it is ill-equipped to deal with. As you can see from this example, the problems can get more and more complicated. But these are real problems, experienced every day by citizens, companies, and governmental agencies.

Now take the previous example and multiply its degree of severity a thousandfold, say. Imagine a chemical company operating on the outskirts of a large city that also happens to be a major tourist destination because of a famous waterfall located nearby. Imagine that for year after year the chemical company kept dumping all kinds of highly toxic chemical wastes in an old canal that was no longer used for navigation. Think of these chemicals as being in standard 55-gallon steel drums, which rust away fairly quickly when exposed to the ground and to the elements. Imagine that when the entire canal area is full of toxic materials, the top is covered over with dirt, and the land is donated to the city for public recreational purposes. A school and playground are built in the area, and eventually even houses are built on the site—all this while the chemical drums are corroding and leaking, and the chemicals are starting to migrate through the soil from the original dump-site to the surrounding area. The ground water is becoming heavily contaminated in the vicinity. After quite a few years pass, people notice funny smells in their houses, and peculiar chemicals are even found oozing out of some basement walls.

Now take this situation and multiply *it* a thousandfold or more to indicate in a qualitative way that the same kind of scenario has been occurring to a greater or lesser extent in all parts of the country and in almost every state. This is the situation the United States found itself in during the 1970s when the famous case of "Love Canal" described above began to make headlines, first in its home city of Niagara, New York, and then nationally. As scientists, engineers, government regulators, citizens, and business leaders started to look around, they began to realize the disaster that had been allowed to develop and fester during most of this century. Highly toxic chemicals had been disposed of in "dumps" often legally and sometimes illegally. No immediate problems from this practice had usually been detected. But ground water is like the fabled tortoise: it may move slowly, but it moves steadily. If you give it enough time, it can cover many miles.

People became very concerned by developing stories of contaminated sites throughout the country, not just associated with toxic or hazardous waste dumps, but also from more mundane situations such as municipal landfills, the oiling of gravel roads to reduce dust, underground storage tanks (at service stations, homes, schools, and almost every other kind of facility), leaking chemical pipelines, and so on. One started to notice more and more articles in the newspaper recounting the contamination of town and city drinking-water supplies.

While there had always been an environmental movement within the country, it now went "mainstream." The Congress enacted a famous law known as the "Superfund Act" setting up a whole system of procedures for prioritizing and cleaning up such contaminated sites. A key portion of the act and the one that got the greatest attention from the business community was the assignment of "joint and several liability" to any person or organization that had participated in the ownership or contamination of such a site. This meant, for example, that if you were one of a hundred companies that disposed of wastes legally (at that time) in a site that had since led to ground-water contamination, then you could be held *totally liable* for the *total* costs of cleaning up that site, such as if the other companies had gone out of business, gone bankrupt, or did not have the financial resources to contribute substantially to the cost to cleanup. It also meant that if you now owned a site that had been contaminated before you had any interest in the site, you could also be held totally responsible for the cost of cleanup. This law gave the Environmental Protection Agency (EPA) and the Department of Justice tremendous power in trying to recover the costs of cleanup from a wide variety of sources, although they certainly have tried to use principles of fairness and moderation in using this power. It also made companies extremely wary about any of their ongoing operations that could possibly lead to similar contamination problems in the future.

So you can imagine how important the study of ground water is to the evaluation of environmental risk from various kinds of industrial and municipal operations. Not only scientists and engineers, but government officials, lawyers, bankers, real estate developers, and many others all began during this period to learn more about ground water and ground-water contamination.

Since much of what goes on in the realm of ground water is beneath the surface of the earth, where it cannot be seen or directly observed, mathematical models have come to be a key tool for ground-water investigators to use. Based on a few sampling points where wells are drilled or the ground water is directly observed or sampled, we create mathematical models to make deductions about the direction in which the ground water is flowing, the rate at which it is moving, and the locations at which it may have picked up contamination. This subject therefore provides a perfect introduction to how mathematical models can be used to help us understand environmental problems.

Exercises for Section 2.1

For research questions such as those below, one way to proceed is to use an on-line computer search program, available in many libraries or on many computer networks. Another approach is to look through the telephone directory for relevant government agencies or other possible sources of information. Library reference departments are also excellent sources of help.

1. Identify at least one government-recognized site with contaminated ground water in either your home town or a town that borders it, or your home country (if not the US). In one page, summarize its status in terms of being investigated and/or cleaned up. You should identify the contaminants of interest there, how the contamination arose, and other items of interest. Almost every community has such sites.

2. Compile a list of all government-recognized contaminated underground sites in your home city or town. (You will probably be quite surprised at this number.) Not all of these sites necessarily have contaminated ground water, because many of them are sites where there have been leaky underground storage tanks which have since been removed and where some contamination was found in the soil around the tanks. It is not always the case that this contamination has moved downward to the water table where it actually enters the ground-water system.

3. Compile a list of all National Priority List (NPL) sites identified by the EPA within your home state, and for three such sites provide a brief summary of the nature of the contamination.

4. Find out how many sites on the National Priority List (NPL) published by the EPA have actually been completely cleaned up to the government's satisfaction or have at least completed the process of construction of facilities for long-term remedial activity. For one NPL site in your home state (every state has at least one), determine the actual or projected cost of cleanup.

2.2 Physical Principles

As mentioned in the previous section, if you start to dig down into the ground essentially anywhere on earth, you will eventually come to water. This is because the soil and rock below the surface all have a certain amount of open spaces, and over time water that seeps into the ground from surface water bodies or from precipitation eventually moves downward by gravity, thereby filling up all these open spaces up to a certain height. This height is called the *water table* and is depicted in a general way in Figure 2-1. It can also be seen from this figure that a surface body of water, such as a pond or a river, is simply an extension of the water table where the latter actually extends above the surface of the land.

The water below the water table is called ground water, and any portion of the underground through which it can move relatively easily is called an *aquifer*. Assuming that the portion of the subsurface just below the water table is an aquifer, it is called the water table aquifer, to distinguish it from deeper zones of soil or rock that may also behave as aquifers.

Returning to Figure 2-1, it can be seen in this case that the upland pond actually serves as a source of water seeping into the water table aquifer, and thus it is considered as a source

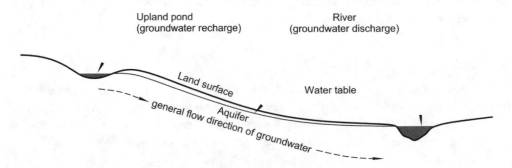

FIGURE 2-1
General schematic of the water table

of ground-water recharge. It is not the only source of ground-water *recharge,* of course, since a substantial portion of any precipitation landing on the surface would also be expected to seep into the ground and eventually enter the water table aquifer. (The rest of the precipitation might either run off the land into small streams or other surface water bodies, or it might return to the air by evaporation or plant transpiration.)

Figure 2-1 shows in addition that a surface water body, in this case the river shown at the right, can also serve as a *discharge* zone for ground water since in this case water from the aquifer may return to the surface through the stream bank below the surface of the river.

These possible relationships between the water table aquifer and surface bodies of water have important implications for the spread of contamination. On the one hand, you can imagine from Figure 2-1 that if the upland pond were to become contaminated from one source or another, then that contamination might spread out in the subsurface along the entire course of the aquifer. On the other hand, if the aquifer were to become contaminated, such as from rainwater percolating through a municipal landfill or leakage from an underground fuel storage tank, then that contamination might even show up some distance away at the surface as a result of the aquifer's discharging into a surface body of water. For this reason, it can sometimes be very difficult to understand why a river or stream is contaminated with chemicals even though there are no identifiable sources along the shores. The contamination may be entering via the ground-water system, and there may be a cumulative effect from various sources along the watercourse.

Even though our primary modeling of ground-water systems will be based on the relatively simple setting described in Figure 2-1, it is important to be aware of the kinds of complexities that often arise in real life applications. These complexities generally result from the geological structures that can exist underground, some of which are depicted in Figures 2-2 and 2-3.

Figure 2-2 is a very simple schematic of the typical underground situation, where, if you were to dig or drill down beginning at the surface, you would generally first encounter various

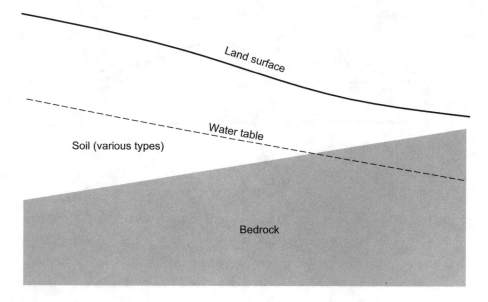

FIGURE 2-2
General structure of the subsurface

kinds of unconsolidated materials, called soils, and then you would eventually reach solid rock, known as bedrock. Examples of soils include gravel, consisting of relatively coarse particles, sand, silt, and clay, in order of decreasing individual particle size. Within any general soil classification such as this, there are also many additional differences based on particle shape, chemical composition, mixtures of particle sizes, presence of organic matter, and other factors. In fact, you might even encounter half a dozen distinct soil types even in the first 10 feet. Geology is the science that investigates the long-term processes that have led to the presence of these distinct underground zones, whether they be soil or rock.

After passing through the soil column, you will eventually reach solid rock, but even in this case its internal structure is more complicated than you might first expect. This is suggested in Figure 2-3, which shows a combination of typical sedimentary rock types on the left (meaning that these rocks formed as compacted sediments laid down by ancient bodies of water), bordered on the right by a body of igneous rock (formed by heating and melting deep below the earth's surface) that is now located close to the surface and that in fact "outcrops" at the surface just to the right of the body of surface water.

You may not think of ground water as flowing through such bedrock bodies, but it often does. This is either because they may have an inherently porous structure, such as is often the case with sandstone, or because they may have extensive fracture zones that are sufficiently interconnected to allow the movement of fluids. This is often the case with limestone, and it is likely to be the case with the granite shown in the figure, at least in the vicinity of the fault zone at the left and the weathered zone near the surface.

When we discuss the movement of ground water within aquifers, these aquifers might be portions of either the soil column or the bedrock. The most likely aquifers among the soils would be sands and gravels, and within the bedrock they might be sandstones, certain kinds

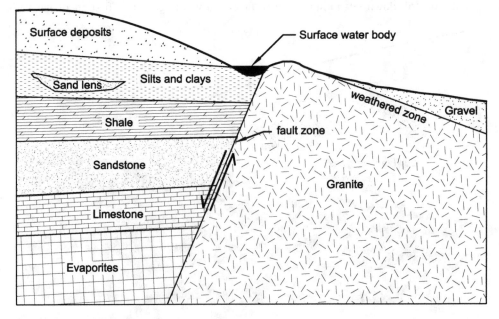

FIGURE 2-3
Potential bedrock complexity in the subsurface

of limestones, and certain disturbed portions of other rock bodies. On the other hand, other portions of the subsurface are generally quite resistant to the flow of water and are classified as *aquicludes* (blocking the flow of water) or *aquitards* (substantially resisting the flow of water). Typical examples in the soil column would be clays and fine silts, and in the rock column relatively plastic rock, such as salt, or relatively unfractured formations of shale, granite, or many other kinds of rocks. (For this discussion we are mentioning only a small sample of the wide variety of underground geologic formations that are commonly found.)

While the water table aquifer most often follows the general shape of the topographic surface of the land, as suggested by Figure 2-1, it is quite possible that there might be a number of distinct aquifers at different depths and that these might be sufficiently insulated from each other by aquicludes that their flow patterns may be quite different. There are a number of potential examples of this in Figure 2-3, for example, where the surface soils, sand, sandstone, and limestone may all be sufficiently different in properties and isolated from each other so that what happens in one may not be a good indicator of what happens in another. Deeper aquifers, in particular, may have quite complex flow regimes because they may be influenced by elevated recharge zones many miles distant and they may be affected by pressures of overlying rock bodies and by intervening structures such as fault zones (where one rock mass has slid along another). Further discussion of these complicating features would go beyond the scope of this brief introduction.

Exercises for Section 2.2

For the exercises below that pertain to your own community, good local sources of information are your municipality's public works department, local science teachers, civil engineers who work in the area, or environmental agencies or organizations. Environmental science texts and environmental reference books are good sources for more general information.

1. For your own home residence, determine where the drinking water comes from. If it comes from ground water, determine the general nature of the aquifer formation being used.

2. Determine the general allocation of water supplies in the US to various distinct purposes. (Consumption is only a small fraction.)

3. Determine the extent to which ground water is used in your own home community for various purposes. To what extent is it treated or tested before domestic use?

2.3 Typical Quantitative Issues

In the next section we will investigate how some very simple physical principles and mathematical equations enable us to investigate cases of ground-water contamination. But let's begin here by making up a "wish list" of the kinds of information you would really want those equations to tell you. The best way to do this might be to imagine yourself in the role of someone who has to deal with such a ground-water contamination problem. Your first reaction may be that putting yourself in such a role on the basis of your limited knowledge of ground water, based perhaps entirely on the previous section, would be unreasonable. But it isn't. As was discussed earlier, the range of people generally being asked to understand and deal with ground-water contamination problems includes not only scientists and engineers, but also lawyers, real estate agents, bankers, town officials, members of citizens groups and others. People could well find themselves in one of these roles with even less background than you already have!

Here is the first situation:

Situation 1. You are a member of the Board of Health of your town or city, a part-time volunteer position. You have been called by some acquaintances in town and told that they have noticed the underground gasoline tanks being dug up at a local service station. Since they live about a tenth of a mile from the service station and rely on a drinking-water well on their property for water, they are wondering whether there is any health risk they should be aware of. See Figure 2-4 for the general layout of this situation.

What should you do? Well, first you should see if any other town official has been involved in this situation, perhaps the town engineer or building inspector. If not, then someone should discuss the situation with the service station operator to determine why the tanks are being replaced. The operator is likely to respond that a "small leak" was found in one of the tanks and so to be on the safe side the gasoline supplier, a major oil company, decided to replace all the tanks with brand new ones. The operator assures you that the leak just developed and that if it had been occurring for any length of time they would have found it earlier by their tracking of inventory. (They may actually believe that, but you shouldn't. Keep on going.)

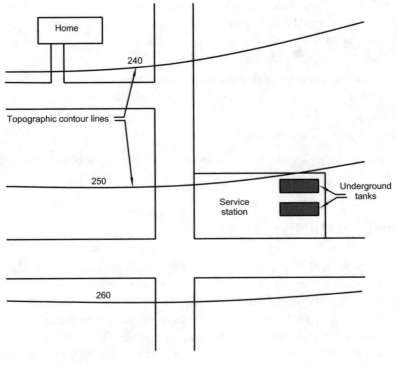

FIGURE 2-4
Schematic representation of "Situation 1"

Logic tells you that if there were a leak, the surrounding soil would have been contaminated to some extent. Any rainwater percolating down through the soil would have carried some of the contamination down into the water table aquifer. Once the contamination reached the aquifer, it would move in the direction of general ground-water movement. So now you want to get quantitative: How much gasoline leaked? How long has it been leaking? What direction is the ground water going in? How fast is it going? And how sure are we about the answers to all these questions?

You talk to the town engineer and you decide to insist that some underground sampling be done to determine the extent of contamination. Underground sampling in these kinds of cases is usually done by drilling wells. You may generally think of a well as a hole you drill into an aquifer in order to pump water for consumption, but in fact among the most common and useful kinds of wells are simple *test wells* or *monitoring wells*. These are holes sunk down through one or more layers of the subsurface and lined with pipe that is perforated to allow water to enter from the portion of the subsurface in which you have particular interest.

But wells can be quite expensive to install, and even the equipment for drilling them can disfigure the location where it is used. (The equipment consists of truck-mounted or track-mounted drill rigs with perhaps some support vehicles.) You can't require the operator to put holes all over the place both because of the expense and because of the environmental impact of that operation itself. If you put the holes in the wrong place, you may miss the migration pathway of contamination and be led to the false conclusion that there is no environmental problem. So what you really need to do first is form a preliminary concept of what might be going on underground. For example, you might use reference material on the geology and hydrology of the area to begin this process of analysis. Then you might insist on a first round of sampling, on the basis of which you would refine your conceptual framework to see whether a second more refined round of sampling may be necessary. At this point, some very rough mathematical calculations might be useful.

This kind of situation occurs all over the country all the time and is the main line of work for many civil and environmental engineering firms. While the engineers and hydrologists do much of the analysis, the people who ultimately need to understand the results and make the decisions are usually not engineers at all. As stated before, they might be town officials, nearby residents, or business owners. You should be able to imagine yourself in a number of roles in this kind of situation.

Let's consider a second situation:

Situation 2.   You are a homeowner with a private drinking water well, but you are located in a town which also has a public water supply that about half the residents are connected to. Because of recent developments in the town, it has been decided to augment the town's existing water supply wells by developing an additional well sited as shown in Figure 2-5. For many years you have been concerned about your location, which is relatively close to a formerly used town landfill, but you have been assured by the town engineer that the landfill has been properly "capped" with clay and topsoil and that the contaminants are not subject to migration.

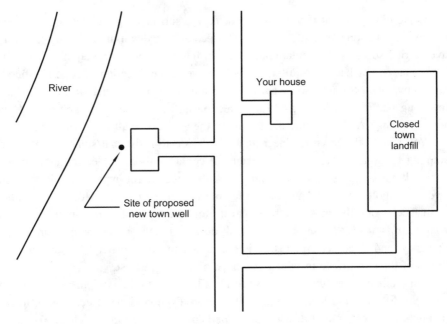

FIGURE 2-5
Schematic representation of "Situation 2"

But now you begin to think logically based on your general understanding of how ground water moves. Isn't it likely, you ask, that the new town well will be pulling or drawing ground water into it from essentially all directions around it? While some of this water will likely be river water drawn into the well through the surrounding sediments, won't the well also be drawing water from the direction of your house and the old town landfill? On the one hand, might this lower the water level in your own well and thereby imperil your own water supply? On the other hand, might it do something even worse, namely, draw contaminated water from under the landfill site in the direction of your house and well? When the landfill was shut down, the engineering studies probably proved that everything was stable under the conditions existing *at that time*. But now the conditions are changing as a result of the new town well, and this may put your water supply at risk. This phenomenon is suggested in Figure 2-6, where you can see how the withdrawal of water from a well by pumping has a tendency to depress the water table in the vicinity of the well and to draw water in from a wider and wider radius.

You would clearly be within your rights asking the town to investigate the impact of the new well on your water supply and on the migration of potential contaminants under the closed town landfill. But unlike Situation 1, where you were trying to investigate something that occurred in the past, in this situation you are asking them to make a prediction about the *future behavior of the ground water in the aquifer,* even before the well is built and operated. How do you make such predictions? You collect the relevant data from the site, and you use mathematical models.

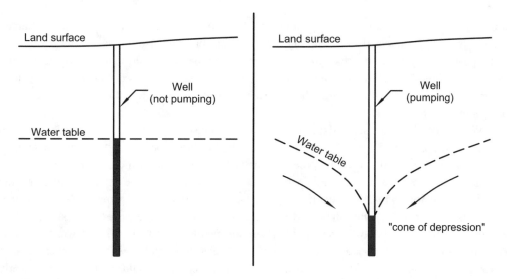

FIGURE 2-6
The effect of pumping on the water table in the vicinity of a well

To generalize, the questions that someone might ask about flow within a ground-water system usually boil down to two basic aspects:

1. How much ground water is flowing through a portion of an aquifer?

This is important because if you can also estimate how fast some pollutant is seeping into the ground water, you could then estimate its concentration in the aquifer. You would then have a better idea of the degree of risk as well as the kinds of levels you should be trying to detect in any samples that you take. The answer to this question usually depends on an equation called *Darcy's law,* which as you can see looks very simple:

$$Q = KiA.$$

You will have to wait just a little while for the individual variables to be defined precisely.

2. How fast is the water moving?

This would certainly give you an idea of how far the contamination might have spread or how much time you have to act to keep it from spreading much farther. The answer to this question usually depends on a simple equation called the *interstitial velocity equation* which, as you can see, also looks quite simple:

$$v = \frac{Ki}{\eta}.$$

Once again, the variables will be defined shortly.

So now we move on to the details of Darcy's law and the interstitial velocity equation.

2.4 Darcy's Law

In developing any kind of mathematical model, you need to figure out what the key physical variables are that control the situation of interest. We are interested here in the flow of ground water through some kind of porous geologic medium. In about 1850 a French engineer named Henri Darcy was interested in essentially the same question because he was trying to set up a system for filtering water in the city of Dijon, France, by passing it through beds of clean sand. (Such sand filters are still commonly used today.) The question he faced was really how much water could move at what rate through what size sand filter. To answer this question, he set up some simple experiments.

While not quite the same experimental layout as used by Darcy, Figure 2-7 shows an experimental apparatus in which water can be pushed by a piston through a geologic sample, such as sand, gravel, silt, or even some kind of porous or fractured rock. Look at the picture and make sure you understand the general layout because this will be fundamental to all the discussion that follows. Obviously if the geologic sample is of some material through which water passes quite easily, then it would not take much force on the piston to push the water through. On the other hand, if the sample does not readily permit the flow of water, such as a silt or a clay, then the same amount of force on the piston would result in a much smaller water flow rate through the sample. In this latter situation, you would expect that by greatly increasing the force, you might be able to bring the flow rate back up to that of the earlier case.

Because we will be referring to this experimental apparatus over and over again in our discussions, a simplified version in two dimensions is shown in Figure 2-8. This is basically just a side view of the apparatus shown in Figure 2-7. Let us suppose that we select a specific geologic material to use in all the experiments. We will refer to the situation in Figure 2-8 as the "baseline experiment." For a given force on the piston we will obtain a certain flow rate.

Now look at Figure 2-9. The only difference between Figure 2-9 and Figure 2-8 is that in Figure 2-9 we are applying more force to the piston so that the water pressure to the left of the sample has been doubled to a new pressure $2P$. In this situation, *what do you think would happen to the rate at which water would be forced through the sample?* Think about this and formulate your answer before reading further.

The most reasonable answer to the previous question would probably be that if you applied twice the pressure to force the water through the sample, then you should expect to roughly double the flow rate. This is a very reasonable answer, and it also happens to be true. How do

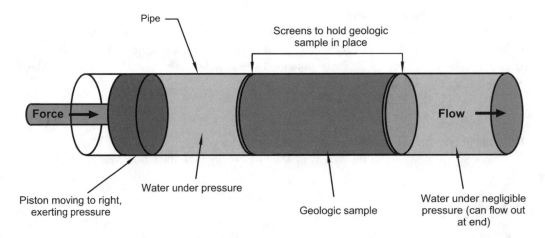

FIGURE 2-7

Hypothetical experimental apparatus for pushing water through a geologic sample

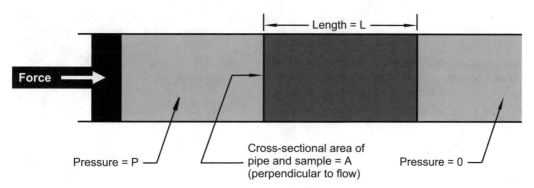

FIGURE 2-8

Two-dimensional simplification of experimental apparatus shown in previous figure

we know it is true? Well, this was exactly the point of the experiments conducted by Darcy, who set up a mechanism similar to ours and actually verified by repeated experiments under various conditions that the results were as expected.

Let's try this same kind of exercise again. First, remind yourself of the baseline situation described in Figure 2-8. Now look at the modified experimental setup in Figure 2-10. In this case the inlet pressure is just the same value, P, as it was in the baseline case, but we have put twice as long a geologic sample in the flow pathway so that the water has to be pushed through twice as much of this material. *What do you think would happen to the flow rate in such an experimental setup?* Once again, formulate your answer before reading further.

This time you certainly should have expected the flow rate to decrease because we have put more resistance in the way of the flow. That is, water has to pass through a longer pathway of this porous material, so there is naturally going to be more resistance. With more resistance, the flow will decrease. In fact, by doubling the length of the sample, it would appear that we are doubling the total resistance to the flow, not adding any more pressure to help the water get through it, and so we would expect the flow rate to be cut in half. Once again, this intuitively believable hypothesis was also verified by the experiments of Darcy.

Now you will have a chance to try your intuition on a more complex variation on the baseline case. First look back briefly at the baseline case shown in Figure 2-8 and get it firmly

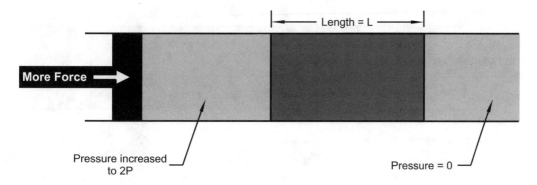

FIGURE 2-9

Schematic diagram showing a doubling of the pressure pushing the fluid through the sample

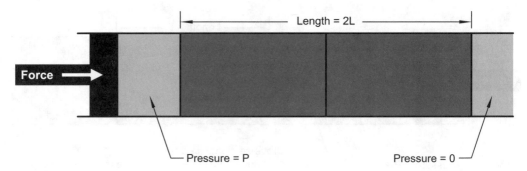

FIGURE 2-10
Schematic diagram showing a doubling of the length of the sample

set in your mind. Now consider the modified situation in Figure 2-11, where two modifications have been made to the baseline case. First we have doubled the input pressure by increasing the force on the piston. Second we have increased the resistance of the pathway by a factor of two by increasing the length of the sample. *What would you expect the net effect on the flow rate to be?* Think about this for a moment before proceeding further.

If you answered that you would expect the flow rate to be identical in the two cases, then you would seem to have a good understanding of the key factors affecting the movement of ground water. After all, doubling the length of the flow pathway doubles the amount of the resistance to flow. At the same time, we have compensated for this by doubling the driving pressure. The first process would have cut the flow rate in half, except that the second process exactly compensates for this. So the bottom line is that the flow rate is identical to the baseline case. We should make one additional observation about this case, as indicated in Figure 2-11, namely, that if the pressure at the left end of the sample is $2P$ and the pressure at the right end is 0, then obviously the pressure gradually decreases throughout the sample, and at the midpoint marked in the diagram you would find that it would be exactly P.

Now we need to consider one additional experiment. Refresh your memory of the baseline case shown in Figure 2-8 and then look at the revised experimental setup shown in Figure 2-12. Figure 2-12 is meant to depict two copies of the original experimental apparatus coupled

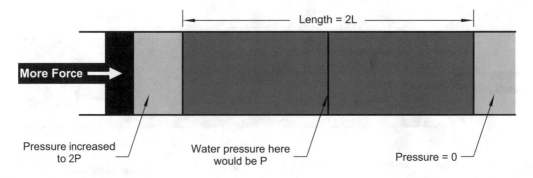

FIGURE 2-11
Schematic diagram showing both a doubling of the pressure and a doubling of the length of the sample

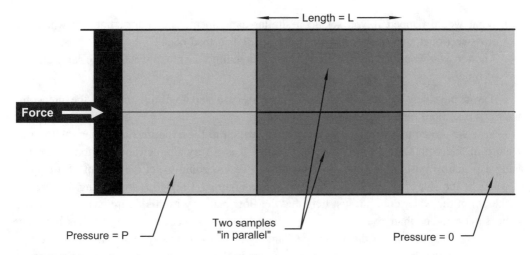

FIGURE 2-12

Schematic diagram showing an effective doubling of the cross-sectional area of the sample

together along their length. The pressure on the left is still the value P, as in the baseline case, and the length of the sample is still L. But our question is the following: how does the *total amount of water* being forced through the composite apparatus in Figure 2-12 compare with the *total amount of water* being forced through the apparatus in Figure 2-8? As usual, think about your answer before proceeding.

In the modified experimental setup shown in Figure 2-12, you should have guessed that twice the amount of water would be flowing through the double sample in a given period of time, compared to the baseline case given earlier. Your first reaction to this may be "so what?"; but in a few moments you will see the significance of this simple observation. For now, just note that what we really did in this experimental set-up was to *double the cross-sectional area* of the flow pathway, keeping all other conditions the same as in the original experiment. We could even have just used a larger diameter pipe (by a factor of $\sqrt{2}$) so as to have a single apparatus with this larger cross section, and the resulting flow would still be twice the original.

Looking back on the preceding series of experiments, it should be fairly obvious that the use of a factor of two to modify some of the input conditions was chosen only for simplicity. The underlying conclusions that you drew would apply to any similar modifications by other factors. For example, go back and look at Figure 2-8 one more time and try to answer the following questions.

a) What would happen if you tripled the length of the sample, keeping all other variables the same?

Answer: you would decrease the flow rate to a third of the original because you have tripled the resistance to flow.

b) What would be the effect of tripling the input pressure P, keeping all of the factors the same as in the baseline case?

Answer: you would expect to triple the flow rate because the resistance would remain the same but you have three times more pressure trying to push the water through.

c) What would happen if you set up a similar experimental apparatus, but one where the cross-sectional area of the flow pathway was $3A$ instead of A?

Answer: you would be able to push through three times as much water in the same amount of time.

If you have any doubts about these answers, you should think through the preceding diagrams one more time before proceeding further.

Now we are going to make one more modification to the experimental apparatus to bring it more in line with Darcy's original experiments. It is a very simple modification. Instead of using a piston to push water through the sample, we are going to let the weight of a column of water be the source of the driving pressure. See Figure 2-13. Before discussing the issue of the height of the water column, just note that the flow pathway in this figure has been redrawn to be square rather than round in cross section. The cross-sectional area still has the same value A as during the earlier discussions. Changing this flow pathway to a square cross section only simplifies the drawing and requires no change in the basic principles that we discussed in connection with the previous figures.

Applying your intuition to Figure 2-13, it should be obvious that if you keep pouring water in at the top of the left column, it will continue to flow through the system until it eventually overflows at the top of the right column, which is lower. If the heights of the two columns

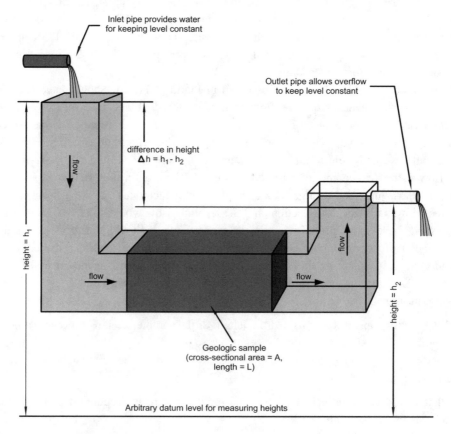

FIGURE 2-13

Revision of the experimental apparatus shown in Figure 2-7

were identical, no water would flow through the system. But as the left column is made higher and higher, there is more and more of an imbalance in pressure between the left side and the right side, therefore leading to an increase of flow through the system. Thus the difference in the heights of the two columns, represented on the figure by the quantity Δh, given by

$$\Delta h = h_1 - h_2,$$

is really a measure of the "net" pressure tending to force water through the sample. If you double Δh, you are providing twice as much pressure, and you would expect to double the rate of flow through the sample. Note that the individual values of the heights of the columns, h_1 and h_2, are not really the important factors. It is only the difference between them that affects the rate of water flow. Therefore h_1 and h_2, which in the figure are shown measured with respect to a line below the experimented apparatus, could be measured with respect to any common elevation point.

If we put all the above observations together in mathematical language, we will in fact have arrived at Darcy's law! In the terminology of mathematical proportionality, we can make the following statements:

1. The flow rate is proportional to the net driving pressure, represented by Δh.
2. The flow rate is inversely proportional to the length L of the sample.
3. The flow rate is proportional to the cross-sectional area A of the flow pathway.
4. The "constant of proportionality" for the above relationships will naturally depend on the specific geologic medium under consideration. (See below for further elaboration on this concept.)

Remember that when you say that one quantity is proportional to another, it simply means that the first quantity can be written as a constant times the second quantity, and this constant is called the *constant of proportionality*. Therefore, putting these principles together in the form of a mathematical equation, we can write:

$$Q = K \times \Delta h \times \frac{1}{L} \times A$$

where:

Q = the total flow rate through the sample measured in some units that represent volume per unit time (e.g., gallons per minute or cubic feet per day); and

K = the corresponding constant of proportionality.

K is the place in the equation where the properties of the geologic material enter. Materials that water can easily be transmitted through would have a large value for K, whereas those that provide significant resistance to the flow of water would have a small value of K. This constant K has a special name: *hydraulic conductivity*. Reference books contain values of hydraulic conductivity for different kinds of geologic materials, and a representative table of such values is shown in Table 2-1. It should be noted that the use of the term "hydraulic conductivity" presumes that the geologic material behaves as a porous medium with respect to fluid flow. In the case of fractured geologic media, this assumption applies best when the flow is predominantly through an interconnected network of many fractures and joints, rather than through only a limited number of very large ones.

TABLE 2-1

Typical ranges of values for porosity and hydraulic conductivity for selected geologic media

Geologic medium	Porosity, η		Hydraulic conductivity, K (ft/day)	
	%	decimal	decimal	exponential
Gravel	25–40	.25–.40	100–100,000	10^2–10^5
Sand	25–50	.25–.50	0.01–1000	10^{-2}–10^3
Silt	35–50	.35–.50	0.0001–0.1	10^{-4}–10^{-1}
Clay	40–70	.40–.70	0.0000001–0.001	10^{-7}–10^{-3}
Sandstone	5–30	.05–.30	0.00001–0.1	10^{-5}–10^{-1}
Limestone	0.1–20	0.001–.20	0.0001–0.1	10^{-4}–10^{-1}
Granite (weathered or fractured)	0.01–10	0.0001–.10	0.0001–10	10^{-4}–10^1

The equation for Q discussed in the previous paragraph is very close to the form of Darcy's law that was stated earlier. In particular, if we make the definition

$$i = \frac{\Delta h}{L},$$

then the above equation does indeed reduce to the standard form of Darcy's law,

$$Q = KiA.$$

The quantity i is called the *hydraulic gradient*. It represents the combined effect of the pressure difference between the two ends of the geologic medium and the length of the flow pathway through that medium. So, for example, if you were to double the value of Δh (similar to what we did in Figure 2-9) and also double the length of the flow pathway (similar to what we did in Figure 2-10), the net effect would be to keep the i value the same as the starting value and thus to maintain the same level of flow. This was the observation we made in connection with the discussion of Figure 2-11, which you should review to verify.

Now we are ready to apply these concepts not only to the experimental apparatus described earlier, but to a typical sketch of a ground-water system. Figure 2-14 presents such a side or "section" view of a portion of the subsurface along with information on two monitoring wells that have been drilled into the aquifer in order to measure the height of the water table.

This figure introduces one new term that was not present in our previous discussions. In particular, it refers to the height of the water level at any point as the *head,* also known as *hydraulic head.* This standard usage among hydrologists is meant to accommodate more complex situations than we are discussing here—situations in which there are forces other than simply the difference in the heights of the water level at two points that may also have an important influence on the rate of movement of such water. However, for our purposes, we will use the word "head" to mean the height of the water level, and we will continue to use the letter h to refer to it. Similarly, the expression Δh will now be called the *head loss* or the *change in head* as we move from one location to another.

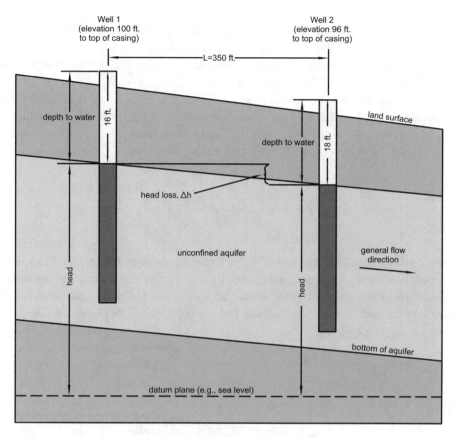

FIGURE 2-14
Section view of unconfined aquifer with two monitoring wells (not drawn to scale)

Does Figure 2-14 give you sufficient information to determine the amount of water flowing through the indicated aquifer? Look carefully and see if you can find everything you would need to apply Darcy's law.

You certainly have the head values for each of the two wells. For example, for Well 1, a surveyor has apparently determined that the top of the well casing (the pipe that lines the hole) is at an elevation of 100 feet above sea level. By putting a measuring tape down into the hole until it reaches water, someone has determined that the depth of the water table is 16 feet, so that the difference would be the elevation of the water table, namely,

$$h_1 = 100 \text{ ft} - 16 \text{ ft}$$

$$= 84 \text{ ft.}$$

Similarly, you should be able to verify that the value of h_2, the head at Well 2, is 78 feet. But if you wanted to apply Darcy's law to determine the flow through the aquifer, you would need to know the value of K for the kind of soil or rock the aquifer consists of. Such a K has not been specified in Figure 2-14. In addition, in order to determine the amount of flow, you would also need to know the cross-sectional area, A, but in this case you do not know the extent of the aquifer in the direction perpendicular to the cross-section, nor, in fact, are you even given the thickness of the aquifer in the vertical direction.

Figure 2-15 provides a more complete description of a typical situation, showing a buried sand and gravel deposit (such as an old stream bed). Let us assume that the hydraulic conductivity for this particular formation averages 100 ft/day. In this case, it is now fully possible to exercise Darcy's law. The problem is to determine the total flow rate, sometimes called the *volumetric flow rate,* within this given portion of the aquifer. Pay careful attention to the following calculation and be sure that you understand where each of the components comes from:

$$Q = K \times i \times A$$
$$= 100 \text{ ft/day} \times \frac{65 \text{ ft} - 54 \text{ ft}}{1200 \text{ ft}} \times (40 \text{ ft} \times 200 \text{ ft})$$
$$= 7{,}333 \text{ ft}^3/\text{day}.$$

That's obviously quite a large amount of water (equal to about 55,000 gal/day) and is a good indication of the substantial quantities of ground water moving around beneath the surface.

Calculations of the type illustrated above can also be used as part of the estimation of the concentration of a contaminant in an aquifer. The key requirement for this estimation process is

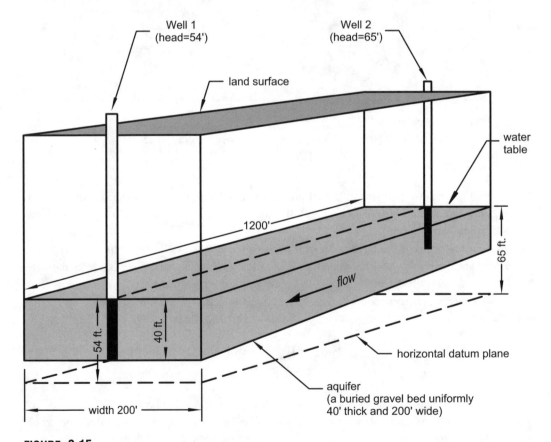

FIGURE 2-15

Block diagram of idealized aquifer showing two monitoring wells along a flow line and aquifer dimensions (not drawn to scale)

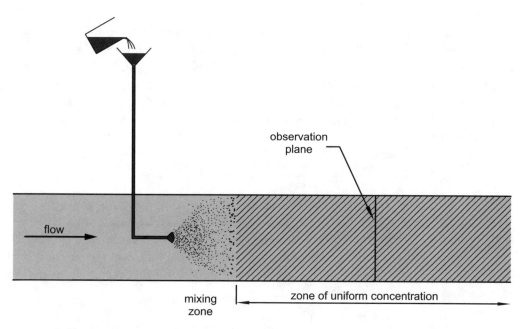

FIGURE 2-16

Schematic representation of small contaminant influx to aquifer

that you have some reasonable idea of the rate at which the contaminant has been leaking into the aquifer. This is illustrated schematically in Figure 2-16, where the aquifer is represented by a large pipe with water flowing from left to right, and the contaminant source is represented by a much smaller tube that is being used to inject material into the flow. As the contaminant enters the main pipe through the small diffuser nozzle, it essentially immediately starts to move downstream at the same speed as the water itself. It also spreads out laterally so that after passing through a brief "mixing zone," it may be regarded as being uniformly distributed throughout the water.

To give a concrete example of how to calculate the resulting uniform or average concentration, suppose that the flow rate through the aquifer is 1200 ft^3/day and that the contaminant is leaking in at a rate of 2 lb/day. Since everything is moving downstream at the same rate, this means that in the course of one day, 1200 ft^3 of water will have passed through the imaginary "observation plane" shown on the figure, and 2 lb of contaminant will also have passed by. Therefore there would have to be 2 lb of contaminant per 1200 ft^3 of water. In mathematical form, this simply says that the concentration C in that water must be:

$$C = \frac{2 \text{ lb}}{1200 \text{ ft}^3} = 0.00167 \text{ lb/ft}^3.$$

This kind of calculation is really very simple: you essentially divide the contaminant inflow rate by the water flow rate in the aquifer, which is the way the calculation is rewritten below:

$$C = \frac{2 \text{ lb/day}}{1200 \text{ ft}^3/\text{day}} = 0.00167 \text{ lb/ft}^3.$$

You can see that the unit "day" cancels out between numerator and denominator. In fact, this idea of manipulating or canceling out units as though they were actually numbers is a very powerful method used for many purposes in both elementary and advanced calculations. It is called "dimensional analysis." It is very useful in many environmental calculations because it is so frequently necessary to convert from one system of units to another. This is a good place to elaborate on it, so here are two closely related examples.

Suppose you wanted to convert the above value, 0.00167 lb/ft^3, to grams per cubic centimeter, g/cm^3. From a dictionary, handbook, or standard conversion computer program, it is easy to determine that 1 pound = 453.6 grams, and that 1 inch = 2.54 centimeters. Therefore you can consider the quotients

$$\frac{1 \text{ lb}}{453.6 \text{ gm}} \quad \text{and} \quad \frac{1 \text{ inch}}{2.54 \text{ cm}}$$

simply to be unusual ways of writing the number "1" or unity, since the numerators and denominators in each case are really exactly the same thing. And you also know that when you multiply anything by unity, you do not change its value. Therefore, we can take the original value and multiply by various combinations of these and similar fractions, chosen in such a way that when we cancel out all the units possible, we are left with an answer in precisely the set of units we wanted. For example:

$$\frac{0.00167 \text{ lb}}{\text{ft}^3} \times \frac{453.6 \text{ g}}{1 \text{ lb}} \times \left(\frac{1 \text{ ft}}{12 \text{ in}}\right)^3 \times \left(\frac{1 \text{ in}}{2.54 \text{ cm}}\right)^3 = 2.675 \times 10^{-5} \text{g/cm}^3.$$

Note how it was necessary to cube the factors in parentheses in order to have enough "ft" and then "in" units to cancel out with the ft^3 in the first fraction. So the 12 and the 2.54 also had to be cubed in the calculation.

For the second example, let us suppose that we also begin with the original value, 0.00167 lb/ft^3, but now we want to convert it to "parts per million" or "ppm" units. For ground-water contamination problems, ppm units refer to mass or weight.* To illustrate, if we had 6 pounds of contaminant in 2 million pounds of water, the concentration could be expressed as 3 ppm, since there would be 3 parts (by weight) of contaminant for every million parts (also by weight) of water. It would be the same if we had 6 grams of contaminant in 2 million grams of water. Making use of one additional fairly standard conversion factor, namely, that 1 cubic foot of water weighs 62.4 pounds, we can use the same kind of calculation as above:

$$\frac{0.00167 \text{ lb (con.)}}{\text{ft}^3(\text{water})} \times \frac{1 \text{ ft}^3(\text{water})}{62.4 \text{ lb (water)}} \times \frac{10^6 \text{ lb(water)}}{1 \text{ million lb (water)}} = \frac{26.8 \text{ lb (con.)}}{\text{million lb (water)}} = 26.8 \text{ ppm.}$$

Note that in the final conversion factor above, we are treating "million lb" as a single composite unit itself.

Exercises for Section 2.4

1. Imagine an underground sand aquifer that is 30 feet thick, 400 feet wide, and has a hydraulic conductivity of 30 ft/day. Two test wells 550 feet apart have been drilled into the aquifer along the axis of flow, and the measured head values at these wells were found to be 115 and 103 feet respectively.

---

* For concentrations of pollutants in the air, ppm units refer to volume fractions rather than weight fractions. Such units will be used in Chapter 3, so it is important to remember that the ground-water and the air cases are quite different.

a) Draw a really clear diagram of this situation.

b) Find the total flow rate in a given cross section of this aquifer.

2. This problem refers to the situation of Figure 2-14. In addition to the information provided in that figure, suppose that the aquifer is composed of a coarse porous sand. What would be the volumetric flow rate through any single imaginary planar surface of area one square foot, oriented perpendicular to the direction of flow? Describe how such a surface would be oriented with respect to the plane represented by Figure 2-14.

3. Consider the aquifer shown in Figure 2-15 and for which a volumetric flow calculation was carried out in the text. If an engineer calculates that over the past five years, an old landfill above the aquifer has caused about 100 pounds per year of a certain contaminant to leach into the aquifer, what would this imply for the value of the resulting long-term average concentration in the aquifer? Express your answer in the following three sets of units: lb/ft^3, g/cm^3, and ppm.

4. Consider the aquifer discussed in Exercise 1, above. Suppose it is discovered that a chemical pipeline passing over the aquifer has apparently had a small leak for a long period of time. Furthermore, by comparing chemical inventory records at both ends of the pipeline, it is determined that about 10 pounds per week have probably been lost through this leak. What would be the resulting long-term average concentration of the chemical in the aquifer? Express your answer in the following three sets of units: lb/ft^3, ppm, and kg/m^3. (Note: "kg" refers to kilograms, or one thousand grams; "m" refers to meters.)

2.5 Interstitial Velocity Equation

As you may recall from earlier in this section, we set out to investigate two general issues:

- the quantity of water flowing through an aquifer, and
- its corresponding velocity.

We have now finished the first and are about to embark on the second. Thus we are now going to discuss the basis for the "interstitial velocity equation," which was mentioned briefly in an earlier section.

Suppose water is flowing through an aquifer according to the principle described in Darcy's law and suppose for a moment that we focus our attention on a single square foot of the aquifer's cross-section, as suggested in Figure 2-17. In fact, to make matters very simple, assume at the outset that this pathway is actually an "open channel," that is, an actual hole or an underground opening that presents absolutely no resistance to the flow of water at all. (Obviously, this is a very uncommon and extreme case, but it is very useful for the purposes of this discussion.) The right side of Figure 2-17 presents a three-dimensional sketch of this portion of the flow pathway. The left side presents an "ends on" view where you should be thinking of the ground water as flowing directly out of that end towards you as you look at the paper. Make sure you understand this picture before proceeding.

Now suppose someone tells you that *seven cubic feet of water per minute* are flowing out of the end of that section of the aquifer. Again, that means that seven cubic feet per minute of water will be flowing out of that square on the left towards you. *How fast* do you suppose the water in that section of the aquifer would have to be flowing, speed-wise or velocity-wise, in

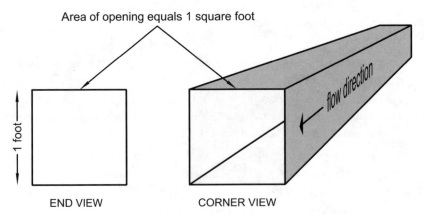

FIGURE 2-17

Flow through open channel with unit cross-sectional area

order to yield seven cubic feet of water per minute pouring out of the end? Think about this for a moment and then proceed to read after you have formulated a potential answer.

One way to answer the above question is to picture the situation as follows. In one minute, seven cubic feet of water come out the end of the aquifer section. Since the cross-sectional area is exactly one square foot, this would mean that all the water in the last seven feet of the aquifer section would need to be expelled in the given minute. In other words, the water from as far upstream as seven feet would have to make it to the end by the end of this one-minute time interval. The only way for that to happen would be if the velocity of the water in the aquifer were exactly seven feet per minute. If the flow were at a rate less than seven feet per minute, then fewer than seven cubic feet of water would make it out the end within one minute. Similarly, if the flow were at a rate faster than seven feet per minute, then more than seven cubic feet would come out the end in the given one minute of time. Be sure you fully understand this before reading on.

Next consider the situation shown in Figure 2-18, which is simply a modification of the previous figure. Here you have another corner view and end view of the same aquifer, but you should imagine that half of the original flow pathway has been sealed up with concrete or some other solid filler material. Now suppose you are also told in this case that seven cubic feet of water are pouring out the end of this channel every minute. Can you figure out what the fluid velocity in the channel would need to be in order to provide this level of flow at the end?

In this case, since a given length of the aquifer pathway contains only half as much volume of water as in the previous diagram, you would need to be drawing water from twice as far in the same amount of time. Therefore, if you wanted to get seven cubic feet of water out of the end, you would need to be drawing water from as far away as 14 feet. Therefore, in order to achieve the same volumetric flow rate out the end, the actual fluid velocity within the channel would need to be 14 feet per minute.

The same principle would also apply even if the half of the channel that was sealed off were not all against a single side. For example, consider Figure 2-19. This shows a section of an aquifer made up of a geologic material that has a "porosity" of 0.5, also often referred to as 50%. The *porosity* of a geologic material is the fraction of the bulk volume that is actually open space or voids, such as spaces between particles in a soil matrix, or open fractures in a

Effective area of opening now equals one half square foot
(the other half having been blocked off completely)

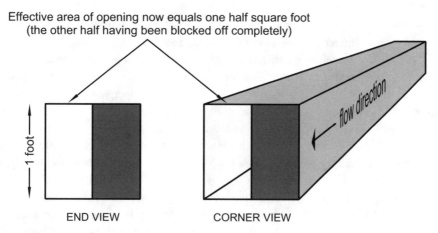

END VIEW   CORNER VIEW

FIGURE 2-18
Flow through open channel of previous figure, but with half the cross-sectional area blocked off

solid rock material. You can imagine a geologic medium with a porosity of 50% as behaving just like the partially sealed-off channel shown earlier in Figure 2-18, except that the material sealing off the channel is spread rather evenly throughout the channel rather than being all concentrated on a single side. But the logic is exactly the same as before. If we knew that seven cubic feet per minute were pouring out of the end of the channel, then that would require that water from as far away as 14 feet upstream would need to make it to the end in the time of our one-minute experiment. Therefore, the water velocity would be 14 feet per minute, just as in the previous case.

This last statement requires one important clarification that the reader may already have anticipated. There is actually an important difference between the situations depicted in Figures 2-18 and 2-19. In Figure 2-18, half the pathway is blocked off by an impervious material in a straight line along the entire length. Therefore, in the situation of Figure 2-18, the water continues to flow in a straight line along the axis of the pathway. However, in Figure 2-19, when

Effective area of opening equals one-half square foot
(although it is distributed throughout)

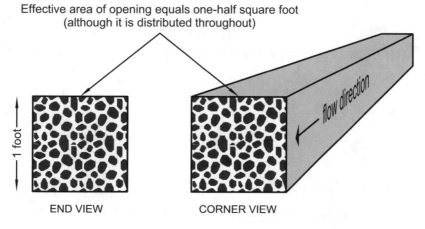

END VIEW   CORNER VIEW

FIGURE 2-19
Flow through channel filled with porous material with porosity $\eta = 0.5$

you fill the pathway with a porous material, the path followed by individual droplets of water will necessarily be somewhat more circuitous, since they will need to wind their way among the soil particles, say, therefore moving to the left and right, at the same time as they gradually migrate in the predominant flow direction shown on the diagram. The argument given in the previous paragraph actually still demonstrates that the net or effective fluid-flow velocity in the flow direction indicated must be 14 feet per minute. This is inescapable since it is still necessary that all the water from up to 14 feet away make its way down through the cross-section by the conclusion of the one-minute experiment. This net velocity along the axis of predominant flow is what we will consistently refer to in this book as the interstitial velocity or, simply, velocity of the ground water. This net velocity is the measure that we would generally be interested in because it will tell us how far from their source any dissolved contaminant might migrate over a given period of time. Nevertheless, it is good to keep in mind that the actual velocity of individual water droplets as they move along the flow pathway could be higher than this number because they must, in addition to covering the downstream distance, follow numerous detours as they move around solid particles and follow the special geometric arrangement of the pore spaces of the geologic material through which they are flowing.

There is nothing special about the specific values chosen for the foregoing discussion. For example, if you knew that the total volumetric flow rate of water through the end of the channel was 5 cubic feet per minute (instead of seven) and if the porosity was still 50%, then you could calculate that the actual fluid velocity would be 10 feet per minute. To take one more example, if you knew that the actual volumetric flow rate through the end of your channel was three cubic feet per minute and in this case you were told that the porosity was only 10%, then the velocity of the water would have to be 30 feet per minute in order to yield the required amount of fluid during the one-minute experiment.

Summarizing the above calculations, if you want to determine the actual fluid velocity in an aquifer, all you need to do is look at the volumetric fluid flow through a *unit* cross-sectional area (one square foot in the examples above) and then divide this number by the porosity of the aquifer. To put this in mathematical terms, first let us calculate the volumetric fluid flow q through a *unit cross-sectional area* of the aquifer. We obtain this by setting $A = 1$ in Darcy's law:

$$q = Ki(1)$$
$$= Ki.$$

Then we divide by the porosity η to obtain the *interstitial velocity equation*:

$$v = \frac{Ki}{\eta}$$

where, to review the parameters all together,

$v =$ the effective fluid velocity along the axis of the flow pathway;

$K =$ the usual hydraulic conductivity;

$i =$ the usual hydraulic gradient;

$h =$ the porosity of the medium (this is the Greek letter "eta").

Some typical porosity values are shown in Table 2-1. This fluid flow through a unit cross-sectional area has a special name: it is called the *flux* and is denoted by the lower case letter q.

So we can also write the interstitial velocity equation in the form

$$v = \frac{q}{\eta}.$$

Note that to calculate the fluid velocity, you do not need to know anything about the thickness of the aquifer or its lateral extent. You only need information on two of its inherent material properties—hydraulic conductivity K and porosity η—as well as the hydraulic gradient i over the portion of the aquifer you are studying.

As an example of the use of the interstitial velocity equation, let us return to the situation described in Figure 2-15. In the previous numerical example based on this figure, we calculated the volumetric flow rate in the given portion of the aquifer, and we found it to be approximately 7,333 cubic feet per day. Assume now that the porosity of this aquifer, which was not necessary for the previous calculations, is 0.3, a typical value for such an aquifer. We can now use the interstitial velocity equation to calculate the actual fluid velocity as follows:

$$v = \frac{Ki}{\eta}$$

$$= 100 \text{ ft/day} \times \frac{65 \text{ ft} - 54 \text{ ft}}{1200 \text{ ft}} \times \frac{1}{0.3}$$

$$= 100 \text{ ft/day} \times 0.0092 \times \frac{1}{0.3}$$

$$= 3.06 \text{ ft/day}.$$

You can also see how the units work out in this equation, leading to the expected velocity units of feet per day.

Looking at the final result of 3.06, you can see that the ground water in this aquifer is moving relatively slowly, only about three feet per day (or a little over an inch an hour) even though the total quantity of water flowing through the aquifer is quite substantial. This illustrates an earlier statement in the text to the effect that ground water is not usually racing along at a high rate of speed.

Once the velocity has been calculated, we can use its value to estimate the time it would take for the ground water to cover a certain amount of distance. To refresh your memory on how this general kind of calculation works, just think for a moment about driving a car. If your velocity is 60 miles per hour and you have to go 90 miles, the time it takes would be:

$$\text{time} = \frac{\text{distance}}{\text{velocity}} = \frac{90 \text{ miles}}{60 \text{ miles per hour}} = 1.5 \text{ hours}.$$

Applying this same elementary logic to the ground water case, suppose we wanted to find the time necessary for the ground water in this example to travel one mile (= 5,280 ft). We would proceed as follows:

$$t = \frac{d}{v} = \frac{5,280 \text{ ft}}{3.06 \text{ ft/day}} = 1,725 \text{ days}.$$

This result, 1,725 days, is just under five years. This travel time calculation gives another perspective on the potentially slow movement of ground water. This slow movement can have both advantages and even some disadvantages. On the plus side, there are certain kinds of contaminants which may degrade over time as they are transported with the ground water or that might enter into chemical reactions with the surrounding soil or rock matrix in such a way that their mobility is substantially reduced. On the minus side, you might think of a contaminated plume of ground water as an underground "time bomb" that can go very long periods of time without detection. When it is finally detected some distance from its source, the spatial extent of the problem and the total amount of contaminant that has been allowed to leak into the aquifer may be considerable, and remedial action might be extremely costly or perhaps not even feasible.

Exercises for Section 2.5

1. A homeowner notes a strange taste entering into his well water, which comes from a deep well in a bedrock aquifer over 250 feet deep. The nearest source of contamination he can think of is a gasoline station three-quarters of a mile away. He contacts an attorney, who in turn hires an engineering consultant to collect and analyze available data relevant to this situation. The data are as follows. The house and the gasoline station are underlain by a bedrock aquifer, and the hydraulic head according to published ground-water contour maps for the area is 38 feet at the house and 44 feet at the service station. The hydraulic conductivity of the aquifer has been estimated in a regional study to be about 3 ft/day, and the porosity (corresponding to the fracture system) to be about 0.001. The tanks at the gasoline station are pressure-tested once a year for leakage, and the last test, which still showed no leakage, took place 7 months before the contamination was noticed by the homeowner. What do you think the consultant's analysis showed? Should the homeowner pursue his case?

2. Referring to the situation of Figure 2-14 and its further elaboration in Exercise 2 of Section 2-4, calculate the actual ground-water velocity.

3. Suppose you were trying to pin down further the ground-water travel time in Exercise 1, above, and you had a fixed amount of money available to invest in technical services to refine your estimates of the hydrologic parameters for the locality. In general, do you think it would be more productive to direct the money towards improving your estimate of hydraulic conductivity K or of porosity η, if this were the choice you were given? (Assume that the money would be equally effective in narrowing the uncertainty band for each of these parameters.)

4. Discuss the following hypothesis: in general, lower porosity geologic media usually tend to have lower ground-water velocities.

2.6 Discussion of Parameters

Before proceeding to the next concept in ground-water modeling, it may be helpful to make a list of the various parameters that have been defined so far and to make a few additional comments concerning several of them. These parameters are listed below:

K hydraulic conductivity (units are length/time, such as feet/day; value depends of material in aquifer)

i hydraulic gradient (dimensionless; i.e., no units)

A cross-sectional area of the flow pathway or portion of aquifer under consideration

η porosity (dimensionless; represents fraction of "open space" in aquifer material)

Q volumetric flow rate (units are volume/time, such as cubic feet per day)

q flux, the volumetric flow rate through a cross-sectional area of one unit (units are length per time, such as feet per day)

v velocity of fluid (units are distance per time, such as feet per day)

h hydraulic head, or the height of the water level above a given reference point (units are distance, such as feet)

L length of flow pathway under consideration (units are distance, such as feet)

t travel time of ground water from one point to another.

Notice that in the above list two key parameters are dimensionless. Both the hydraulic gradient i and the porosity η are independent of the units in which any of the measurements are being carried out.

Note also that three of the parameters have units that make them look like velocities. This is a source of confusion to a number of new students of this subject. These parameters are K (hydraulic conductivity), q (flux), and v (interstitial velocity). In the units in which our examples have been carried out, each of these parameters would be expressed in feet per day. Returning to the previous numerical example, these parameters had the following values:

$$K = 100 \text{ ft/day}$$

$$q = 0.92 \text{ ft/day}$$

$$v = 3.06 \text{ ft/day}.$$

Here is the best way to think about these units:

When we say that $K = 100$ ft/day, the units given there should not be given any physical interpretation, but thought of only as the units in which the value of hydraulic conductivity K may conveniently be expressed.

When we say that $q = 0.92$ ft/day, a sometimes useful physical interpretation of this fact is to rewrite it as follows:

$$q = 0.92 \text{ ft/day} = 0.92 \text{ ft}^3 \text{ per ft}^2 \text{ per day.}$$

This says that 0.92 cubic feet of water flow through each square foot of cross section of the aquifer per day, which is exactly the way we previously described the meaning of the flux q.

When we say that $v = 3.06$ ft/day, we mean exactly that, namely, the average water velocity in the direction of movement is 3.06 feet per day. Naturally, some droplets of water might be moving faster than others, depending on the nature of the physical pathway through the porous material. Furthermore, because as the water makes its way through the material it is not really traveling in a straight line but winding its way around grains or through uneven fractures, the actual average velocity of an individual water droplet or molecule might be considerably faster than 3.06 feet per day. Nevertheless, the average net velocity of the water in the direction of overall movement is this number 3.06 feet per day, and this is the right number to look at when we are thinking about the potential rate at which a contaminated plume might be migrating downgradient with the ground water.

If you did not have available a table of hydraulic conductivity and porosity values for the geologic media that you were working with, you might be tempted to think that the media with the higher porosity values would tend also to have higher hydraulic conductivity values. After all, if there is more open space within the geologic medium, wouldn't it be easier for water to flow through it? While this is a reasonable initial hypothesis, it turns out not to be anywhere near reliable. For example, comparing the hydraulic conductivity and porosity values for similar media in Table 2-1, it is easy to see that certain media, such as clay, which have surprisingly high porosity values, are also likely to have quite low hydraulic conductivity values. In the case of clay, this is because the pore spaces, while occupying a large portion of the overall volume, tend to be so tiny that it is difficult for the flowing water to squeeze itself through them. To put it another way, the small size of the pores in clay provide a relatively high resistance to the flow of water.

Another interesting aspect of porosity is the so-called "fracture porosity" in rocks, or the total volume within fractures. For many kinds of rocks, this is the main porosity, and it is quite small. Even though ground-water movement through individual fractures doesn't really follow Darcy's law in the reliable way that flow through truly porous materials does, it has been found that, considered on a large scale, where the flow is not restricted to one or two fractures but rather works its way through a whole network of them, Darcy's law is still a reasonably accurate representation of the flow conditions.

It may be useful for you to spend a few moments comparing corresponding entries in Table 2-1 in order to get a feel for the range of hydraulic conductivity and porosity values for various geologic materials.

So now you know how to calculate volumetric flow rates and fluid velocities in aquifers given the necessary parameters such as head, hydraulic conductivity, cross-sectional area, and porosity.

2.7 Use of Head Contour Diagrams

When a hydrologist first begins to investigate the ground-water regime in an underground aquifer, he or she needs a convenient way to keep track of any data that may be available on the hydraulic heads in that aquifer. After all, the hydraulic head may be the most fundamental parameter governing ground-water flow because it determines both the direction of flow and the pressure difference that is driving that flow. Except for cases where the water table may be at equilibrium with a surface water body, as suggested earlier in Figure 2-1, the general method of determining ground-water head values in an aquifer is to drill a number of test wells, which are simply holes drilled into the aquifer from which you can measure the depths of the water table, and use that information, as we have done earlier in this chapter, to determine the water table elevation. Once the data from a number of such test wells are collected, the corresponding head information is usually put together in a *water table contour map,* like the one shown in Figure 2-20.

Incidentally, if you have ever used a topographic map to go on a hike, this is really the same thing. On a topographic map, the contour lines show points of equal elevation; on a water table contour map, the contour lines show points where the head values are the same, and their values are labeled on the map as they are in Figure 2-20. It should be emphasized that maps such as that shown in Figure 2-20 are based on a large amount of interpolation and extrapolation. You may have drilled only quite a limited number of wells into an aquifer, but on the basis of the values obtained at those wells, you have drawn lines to estimate where you think the locations of constant head values are. Often one takes into account the general

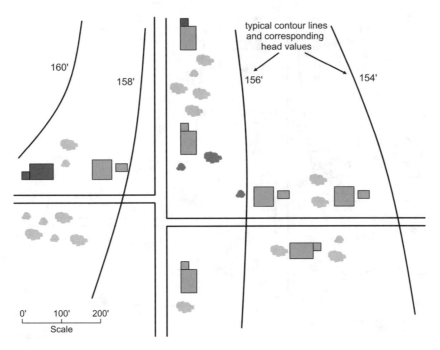

typical contour lines and corresponding head values

160' 158' 156' 154'

0' 100' 200'
Scale

FIGURE 2-20
Water table contour map for a residential subdivision

topography of the land in making these extrapolations because, as has been mentioned earlier, at least in a general way the water table often mirrors the slope of the land above.

Given a water table contour map such as Figure 2-20, an absolutely key fact is the following: *the ground water at any point almost always tends to flow in a direction that is perpendicular to the head contour line at that point.* This is suggested in Figure 2-21, where a number of ground-water "flow lines" have been sketched in on the basis of this principle.

The reason that ground water tends to flow in a direction perpendicular to the contour lines is analogous to the following situation. Think of a marble starting to roll slowly down the side of a hill. From any given point, it wants to head down the hill in a direction that is always as steep or directly downward as possible. In mathematical language, the marble would try to follow the "path of steepest descent." This path will always be perpendicular to the topographic contour lines because such a path is basically always following the most direct route to the "next lower" contour line. Ground water behaves the same way. Basically it also is responding to gravity and wants to follow the steepest or quickest path possible down the water table.

The only reason we said above that ground water "almost always" follows this path of steepest decent is that occasionally something might get in the way to "deflect" the water into an otherwise less than steepest direction. Just like the situation where the marble might want to roll directly down the hill but it runs into a tree and gets deflected sideways, the ground water might be trying to flow along the path of steepest descent in the head level, but it might encounter a zone where the particular shape of the fractures in the rock through which it is moving, or orientation of the pore spaces, preferentially deflect it into some other direction. However, from now on in this chapter we shall ignore this qualification because it is a rather unusual and specialized case. In all the cases to be treated from here on, except during a more

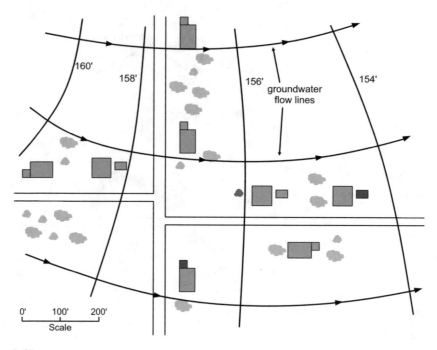

FIGURE 2-21
Water table contour map from previous figure, but with ground-water flow lines also shown (indicated by arrows)

detailed discussion of this issue in Chapter 5, *we shall assume as a general rule that the flow of ground water will be perpendicular to the head contours.* This is a very good assumption for almost every common situation.

We are now ready to use this principle to illustrate the final type of calculation to be addressed in this section. Consider the hydraulic head contour map shown in Figure 2-22. Assume that the aquifer under consideration has a hydraulic conductivity of 50 feet per day and a porosity of 25%. Assuming that a source of contamination exists at the point marked "X", we wish to determine which of the three wells W1, W2, or W3 will most likely be directly affected and how long it will take for the contaminated ground water to travel from point X to the affected well. This is a fairly typical type of problem, and its solution might proceed as follows.

Starting at point X, it is clear that the direction of ground-water movement would be generally towards the lower right, since that is in the direction of decreasing head and it is generally perpendicular to the corresponding head contour lines. Now start to draw a smooth flow line beginning at X so that each time it crosses a head contour line, it is perpendicular to it. Such a path has been drawn in Figure 2-23.

From this diagram, it can be seen that well W3 is the well most directly at risk, since it is located directly down a ground-water flow line from the point of contamination X. Once this pathway has been drawn, the travel time can be calculated as follows. Break the flow line up into segments at each one of the contour lines. For example, the first segment of the flow line would be from contamination source X to the 85-foot contour line; the second would be from the 85-foot contour line to the 80-foot contour line; and so on. For each of these individual

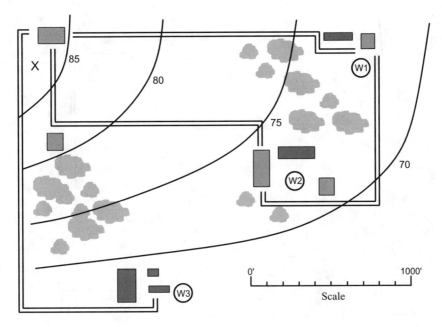

FIGURE 2-22
Water table contour map showing source of contaminant and three wells

segments, we should calculate its length (based on the map scale shown in the figure) and the hydraulic gradient (based on the head values at the two endpoints of the segment and on the length of the segment). The only segments for which this process has a slight complication are the first and last segments because the figure does not specify the head values at one of their endpoints (X for the first segment and W3 for the last segment). Therefore we shall postpone calculations for these two segments until after we finish the easier ones.

By combining the calculated hydraulic gradient with the K and η values for the aquifer, we can calculate the interstitial ground-water flow velocity in each segment and hence the corresponding travel time as well. Once the travel times for each of the five segments in this case have been calculated, they can be added up to give the total travel time from the source to the receptor, in this case well W3. These calculations are illustrated below, where we have:

$\eta = .25$

$K = 50$ ft/day

$d =$ length of each of five segments from point X to point W3 along
slightly curved flow line, estimated using the scale on the figure.

Segment 2:

$$d = 460 \text{ ft}$$
$$i = \frac{\Delta h}{L} = \frac{5}{460} = 0.011$$
$$v = \frac{Ki}{\eta} = \frac{50 \times 0.011}{0.25} = 2.2 \text{ ft/day}$$
$$t = \frac{d}{v} = \frac{460}{2.2} = 209 \text{ days.}$$

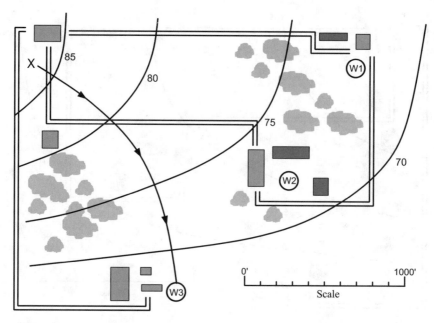

FIGURE 2-23

Contaminated ground-water flow line superimposed on water table contour map from previous figure

Segment 3:

$$d = 530 \text{ ft}$$

$$i = \frac{\Delta h}{L} = \frac{5}{530} = 0.009$$

$$v = \frac{Ki}{\eta} = \frac{50 \times 0.009}{0.25} = 1.8 \text{ ft/day}$$

$$t = \frac{d}{v} = \frac{530}{1.8} = 294 \text{days}.$$

Segment 4:

$$d = 440 \text{ ft}$$

$$i = \frac{\Delta h}{L} = \frac{5}{440} = 0.011$$

$$v = \frac{Ki}{\eta} = \frac{50 \times 0.011}{0.25} = 2.2 \text{ ft/day}$$

$$t = \frac{d}{v} = \frac{440}{2.2} = 200 \text{ days}$$

With respect to Segments 1 and 5, the only difference is that, given the lack of head information, we shall make the reasonable assumption that the gradient in these segments can be well approximated by the gradient in the nearest segment. Since values tend not to change drastically in ground-water situations, this is a sound approach. Thus we have:

Segment 1:

$$d = 170 \text{ ft (from the center of the X to the contour line)}$$

$$i = 0.011 \text{ (based on the value calculated above for Segment 2)}$$

$$v = \frac{Ki}{\eta} = \frac{50 \times 0.011}{0.25} = 2.2 \text{ ft/day}$$

$$t = \frac{d}{v} = \frac{170}{2.2} = 77 \text{ days}$$

Segment 5: $d = 270$ ft (from contour line to center of circle containing W3)

$i = 0.011$ (based on Segment 4)

$$v = \frac{Ki}{\eta} = \frac{50 \times 0.011}{0.25} = 2.2 \text{ ft/day}$$

$$t = \frac{d}{v} = \frac{270}{2.2} = 123 \text{ days}$$

Thus, adding up the five segments in their actual order, the total travel time from point X to well W3 would be about $77 + 209 + 294 + 200 + 123 = 903$ days, or roughly $2\frac{1}{2}$ years.

In studying the above calculation, people often ask whether it is really necessary to divide the flow pathway up into so many individual segments. For example, why not simply look at the pathway as one big segment from the source X to the well W3, calculate its length, estimate the head value at each end (by extrapolating from nearby values), and then apply the interstitial velocity equation to this whole segment? This is a very natural question, and, in fact, sometimes this simpler method yields reasonably accurate answers with much less work. Although it is too complex to explore the reasons here, the fact is that this abbreviated procedure will work well whenever the hydraulic gradient is fairly uniform along the flow path, but will diverge from the correct values in cases where the gradient is significantly different over different parts of the flow path. (This all assumes that the other hydrologic parameters—porosity and hydraulic conductivity—are constant throughout.)

For the problems in this book, we recommend that you continue to break up the flow pathway into individual segments extending from each head contour line to the next, or between start and end points and the nearest contour line. This is what was done in the previous example. However, to take the tedium out of the calculations, you may want to set up a small computer program or spreadsheet, so that all you have to do is input the basic data. (In fact, you might even want to let the program apply the scale to your measurements from the figures.) For example, a spreadsheet program for the previous example might take the form shown in Table 2-2,

TABLE 2-2

Ground-water travel time spreadsheet for example in text (Note: shaded cells denote input values.)

Conductivity =	50	ft/day							
Porosity =	0.25								
Segment	Length	h1	h2	i(calc)	i(extrap)	i	v(ft/day)	t(days)	t(years)
1	170				0.011	0.011	2.2	78.2	0.21
2	460	85	80	0.011		0.011	2.2	211.6	0.58
3	530	80	75	0.009		0.009	1.9	280.9	0.77
4	440	75	70	0.011		0.011	2.3	193.6	0.53
5	270				0.011	0.011	2.3	118.8	0.33
Total								883.1	2.42

where the scale conversion has been done manually in this case. Note that the values calculated by the spreadsheet are slightly different from those calculated above when the example was first worked out. This is because the spreadsheet carries many decimal places through the intermediate calculations even if the display only shows a limited number. In any case, the differences between the two sets of calculations are insignificant as a practical matter because all ground-water calculations are quite approximate, and one should not try to attach too much precision to the results.

Exercises for Section 2.7

The following three notes may be useful to keep in mind as you work the exercises below:

a) These problems generally involve the construction of flow lines for contaminated ground water, beginning with a given head contour diagram. You will want to try to draw a smooth curve for these flow lines, with the requirement that it always crosses the contour lines at a 90° angle. If you have difficulty telling whether the angle of crossing is close to 90°, as some people do, it may help to use the corner of a piece of paper as a template to compare against your sketch.

b) The term "boundary of the region" or similar term refers to any or all of the four sides of the rectangular boundary of the map given in the problem.

c) In some of these problems, you are asked to sketch in a "contaminated plume." Even though contamination from a single, specific "point source" will tend to spread out by diffusion, as discussed earlier in connection with Figure 2-16, in these cases we are looking at a more widespread "distributed source" of contamination. And even though there may be some additional diffusion out at the edges to widen the plume further as it moves downstream, for simplicity we shall restrict our attention to only that portion of the aquifer that is directly downstream from any portion of the original source. For concentration calculations, the water in this section may be regarded as the counterpart to the water in the pipe in Figure 2-16.

1. Consider the situation described by Figure 2-24. The contour lines represent estimated contours of hydraulic head in a shallow aquifer composed chiefly of coarse sand. If a major gasoline spill onto the ground occurs at point X, as shown in Figure 2-24, and some of the gasoline seeps down to the water table, indicate on the diagram its likely migration path with the ground water, and estimate how long it might take for the first traces of dissolved gasoline to reach the boundary of the region shown on the figure. Pick representative values of hydrologic parameters you need from Table 2-1. If you need to make any additional assumptions, be sure to explain what you do. (Note: the bulk of the gasoline would not be expected to mix with or dissolve in the ground water; it would accumulate near the top of the aquifer and move downgradient according to some complex processes. However, a small amount will actually dissolve in the water and have a noticeable effect on taste, even at very low concentrations.)

2. Consider the situation described by Figure 2-25. The contour lines represent estimated contours of hydraulic head in a shallow aquifer composed chiefly of a slightly silty medium sand. If a leak of hazardous plating wastes (which may contain heavy metals such as chromium and cadmium) occurs and seeps into the ground at point X, as shown on the figure, indicate on the diagram the likely migration path with the ground water, and estimate how long it might

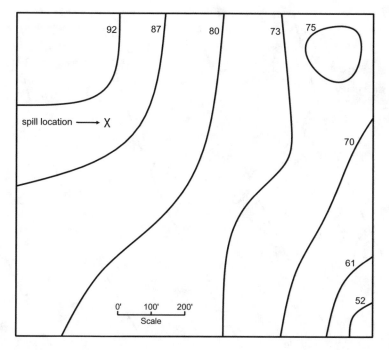

FIGURE 2-24
Hydraulic head contours in the vicinity of a major gasoline spill (Exercise 1)

take for the first traces of contamination to reach the boundary of the region shown on the figure. Use reference values from Table 2-1 to choose your parameters.

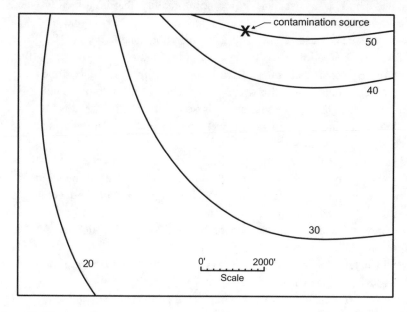

FIGURE 2-25
Hydraulic head contours in a shallow aquifer contaminated by plating wastes (Exercise 2)

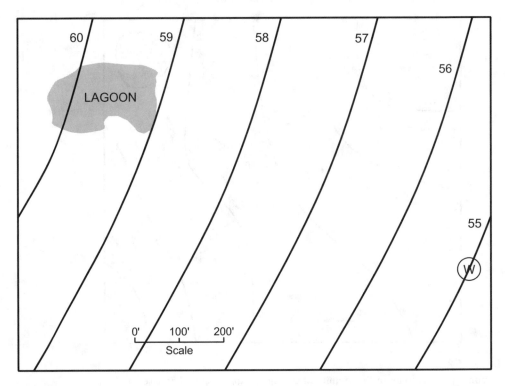

FIGURE 2-26
Ground-water regime downstream from a waste rock lagoon at an ore processing plant (Exercise 3)

3. The diagram in Figure 2-26 shows the hydraulic head contours in the vicinity of a waste rock lagoon or "slimes pond" at a secondary gold mining facility, that is, a facility where old waste rock from an earlier gold mine is being reprocessed to pull out more of the gold. Suppose that the aquifer under the site averages 20 feet thick, its hydraulic conductivity is 190 ft/day, and its porosity is 0.25. Cyanide solution (used to separate gold) is known to be seeping from the lagoon into this aquifer. By considering rates of precipitation, evaporation, and slimes disposal, a consulting engineer has estimated that this seepage is occurring at a rate of roughly 50 cubic feet per day uniformly under the lagoon site, and this solution actually contains about 0.01% cyanide by weight. You are interested in both travel time to and concentration at a planned drinking water well at W.

As discussed in Note c at the beginning of these exercises, you may ignore diffusion of contaminated material out the sides of the plume, assuming instead that all the contamination stays in the portion of the aquifer that passes under at least some part of the lagoon, and that the concentration is uniform therein. Express your concentration results in ppm units (for which you may wish to review the material at the end of Section 2.4). The following sequence of individual questions may help you analyze this problem:

a) How long will it take for the contamination to reach W? Analyze the flow path from the lagoon to W as you have done earlier.

b) What is the cross-sectional area of the contaminated portion of the aquifer? (This refers to the cross section relevant to Darcy's law calculations.)

c) What is the volumetric flow rate in ft³/day for this portion of the aquifer?

d) If roughly 50 cubic feet per day of cyanide solution are seeping into this aquifer, how much is this in units of pounds/day? (Assume normal water density for the solution as well.)

e) Now, this is a 0.01% solution, so how many pounds/day of cyanide are entering the aquifer?

f) What is the concentration of cyanide, expressed in ppm, in the ground water at the time it leaves the lagoon site or reaches W?

4. Suppose that a shallow aquifer averaging about 15 feet thick runs under the locale sketched in Figure 2-27. Its hydraulic conductivity is 50 ft/day, and its porosity is 30%. The contour lines show head values (in feet) in this aquifer. Also shown is an equipment repair facility ("plant") which, over the years, has been careless in its disposal of waste solvents and cleaning agents used to clean parts and equipment. In fact, the residue was generally dumped on the ground "to evaporate." Unfortunately, a substantial amount seeped into the ground before it could evaporate. Based on records of chemical purchases over the years, a consulting engineering firm determines that a long-term average of about two gallons per week were dumped out to evaporate and that about a quarter of this is estimated to have actually seeped into the ground. This has been taking place for at least the past 30 years. To test their understanding of the situation, they decide to sink test wells into the aquifer directly along the centerline of the plume along both the 65- and 75-foot contours.

a) What concentrations should they expect to find in these two test wells (expressed in ppm)?

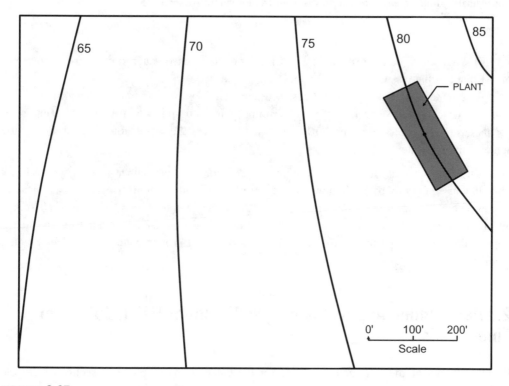

FIGURE 2-27
Hydraulic head contours in the vicinity of an equipment repair facility (Exercise 4)

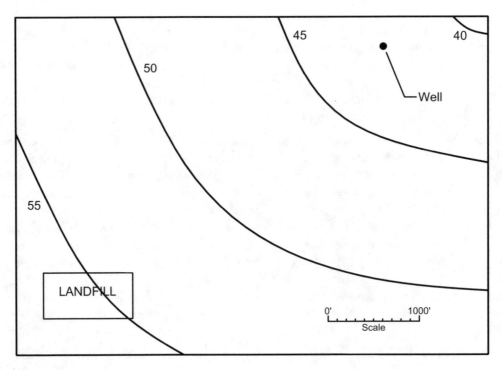

FIGURE 2-28
Conceptual hydrologic environment downstream from a landfill (Exercise 5)

b) What would be the expected travel time for contaminated water to move from the center of the plant to the test well on the 65-foot contour line?

Assume that the materials of interest here are water-soluble and that they just move along with the water at the same rate. This is a simplifying assumption that applies to some such agents but not to others. Assume further that the density of the waste liquids is essentially the same as that of water.

5. Consider the situation described in Figure 2-28, which shows a lined landfill with a leaky liner that is underlain by a shallow sand aquifer with head contours (in feet) as indicated. Suppose that the aquifer is uniformly 30 feet thick, its hydraulic conductivity is 10 ft/day, and its porosity is 0.4. Why is this situation, as described here, self-contradictory and hence impossible? (Hint: begin by carefully sketching the plume of ground water contaminated by the landfill.)

2.8 Determining Approximate Flow Directions Using Data from Three Wells

Recall that two of the key factors needed for the calculations in previous sections were the hydraulic gradient, i, and the hydraulic head contour lines, representing locations of constant hydraulic head. These latter were used in combination with the general principle that the flow

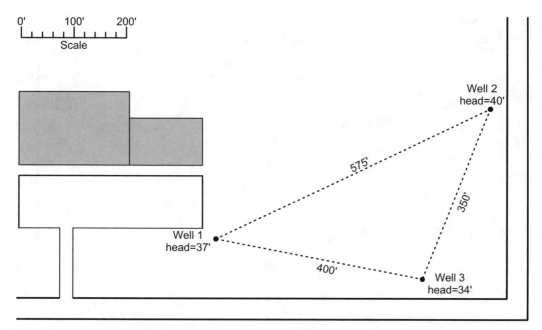

FIGURE 2-29
Data from three wells being used to investigate the ground-water regime adjacent to a commercial site

direction is always perpendicular to the contour lines (at least in the range of situations discussed so far), to enable us to sketch in the actual flow lines.

But how are these important model inputs actually determined from field data? For example, consider the situation shown in Figure 2-29. The figure shows a commercial building with three wells drilled into the ground nearby. One of the wells is just to the right of the parking lot, and the other two wells are at the sides of nearby roads. Now, without reading ahead in the text yet, could you look at the data shown in this figure and estimate approximately which direction the ground water is flowing in and roughly what the hydraulic gradient might be in that direction?

Now let's see how your own thoughts compare with a typical approach to answering that question. We can break down our solution into the following logical steps, which you can see illustrated in Figure 2-30:

1.   Identify the well that has a head value that is intermediate between the other two.

In this case you can see that Well 1 has the intermediate value of 37 feet.

2.   Since the head at Well 1 is intermediate between the head at Well 2 and Well 3, there must also be some point along the line segment from Well 2 to Well 3 at which the head value is the same as at Well 1.

This is fairly obvious because the head must gradually change from a higher value at Well 2 to a lower value at Well 3. So it must hit the intermediate value 37 somewhere in between.

3.   Find the location of this intermediate point between Well 1 and Well 3.

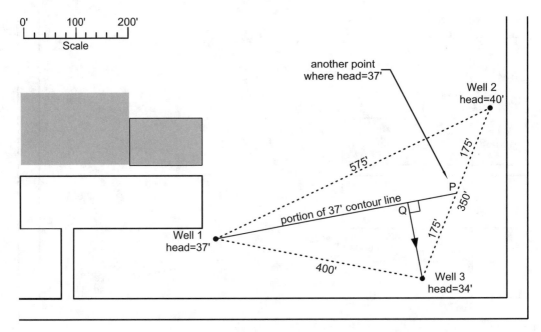

FIGURE 2-30
Further elaboration of the previous figure, leading to estimated contour line and flow direction

You can estimate this exact point by using the simple principle of proportions. Look at it this way: if you were looking for the point where the head value is halfway between the Well 2 value and the Well 3 value, you would expect it to be halfway along the 350-foot distance between the two wells. If you were looking for the point where the head value is 2/3 of the way from the Well 2 value to the Well 3 value, you would go 2/3 of the distance from Well 2 to Well 3. So all you have to do to get started is determine how far the number 37 is from the Well 2 value of 40 to the Well 3 value of 34. The answer is that it is exactly halfway between them because:

$$\frac{37-34}{40-34} = \frac{3}{6} = \frac{1}{2}.$$

Therefore, we should go half the 350-foot distance to get to the point we are looking for. The resulting point is 175 feet from Well 2 along the line from Well 2 to Well 3. Remember that we have calculated it as the point where we estimate the hydraulic head should also be 37 feet, the same value as at Well 1. Let us call this new point P, as shown on the figure.

4. Now construct your best estimate of the 37-foot head contour line.

We now have two points that belong to the 37-foot contour line: the original Well 1 location and the new point P we have constructed on the line between Well 2 and Well 3. There are many possible paths that the head contour line could take so as to connect these two points, but in the absence of any further information, standard practice is to draw the simplest kind of line connecting these two points, which would of course be a straight line. Therefore the best estimate on the basis of the data currently available for the 37-foot contour line is simply a straight line passing through these two points. This line is shown in Figure 2-30.

5. Estimate the general ground-water flow direction based on the contour line calculated earlier.

Since the flow direction is perpendicular to the contour lines, the flow should be in the general direction suggested by the flow line emanating from the new point Q, shown on the figure.

6. Calculate the corresponding hydraulic gradient in the direction of flow.

In order to calculate the actual hydraulic gradient we need to have two points along a flow line at which we know the values of hydraulic head. In this case, focus your attention on the particular flow line drawn in Figure 2-30 that connects the point Q on the 37-foot contour line to the point identified as Well 3. (In fact, Q was chosen so that this line would go through Well 3.) The head at Q must be 37 ft, since it is on the 37-foot contour line. The head at Well 3 is known to be 34 ft. Therefore we can calculate the difference in head between these two points.

The only additional information we need in order to calculate the hydraulic gradient is the distance between these two points or, equivalently, the length of the perpendicular segment from Well 3 to the contour line at Q. There are two ways to determine this, one being simply to estimate it from the figure, using the scale provided there, and the other being to use trigonometry. Careful use of the scale would yield a value of about 150 ft. Using this value, we estimate the gradient as follows:

$$i = \frac{\Delta h}{L} = \frac{37 - 34}{150} = \frac{3}{150} = 0.02.$$

In summary, we have used the above sequence of steps to take data from three given wells in order to calculate the general direction of the hydraulic head contour lines, the corresponding flow lines, and the actual hydraulic gradient.

It is actually quite expensive to drill wells because it requires heavy equipment and skilled operators to do so. Furthermore, it is often desirable to minimize the number of wells drilled because the process may be disruptive to the properties where it takes place. For both of these reasons, hydrologists often do carry out simple calculations such as described above in order to maximize the information they can obtain from a very limited number of wells.

Exercises for Section 2.8

1. Use trigonometry to determine the exact length of the arrow discussed under step 6, above. This is the distance from the 20.9-foot contour line to Well 3. Use this new value to recalculate the hydraulic gradient.

2. Consider the situation shown in Figure 2-31, which is analogous to the example treated above. Determine the general direction of flow and the corresponding hydraulic gradient.

2.9 Guide to Further Information

The following is a list of suggested additional subjects and activities readers may wish to undertake in order to round out their understanding of the topics treated in this chapter. Some

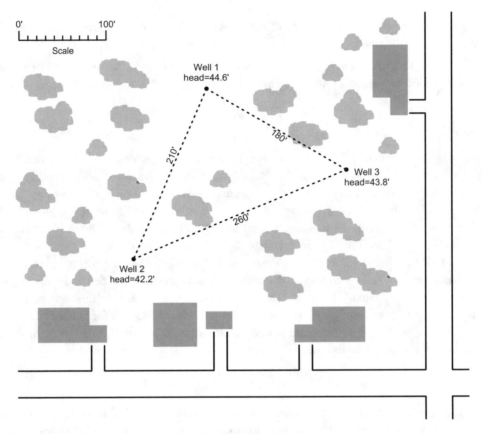

FIGURE 2-31
Data from three wells in a residential subdivision

of these activities are better suited to a group framework, and others are more appropriate as
individual projects.

1. From your initial review of contaminated sites in your own or surrounding communities,
identify the ones that seem to be the most serious. For example, if you have a so-called NPL
site in your area, this corresponds to a site on the "national priorities list" published by the
EPA in connection with the Superfund Program. Such sites are usually the object of extensive
investigation, and it is recommended that you try to arrange a field trip and talk with people who
are carrying out these investigations. Since such sites are almost always under the supervision
of a government agency, even though private companies or private contractors may be carrying
out the work, a call to the EPA or to the state department of environmental protection will put
you on the track to finding a person who would be willing to organize a productive field trip
for you or for your group.

2. Independent of the issue of identifying sites with actual ground-water contamination, also
consider the possibility of simply arranging a field trip to any site that has been investigated
extensively from the point of view of the ground-water system. For example, some sites are
studied as potential sources of drinking water wells or wells for industrial water needs. As has

been discussed earlier, most of the modeling connected with evaluating even contaminated sites is simply modeling of the ground-water flow regime itself independent of the contamination. It should be relatively easy to identify sites and contacts by considering calls to the following: the geology department in a local college or university, the EPA, the USGS (the United States Geological Survey), the state department responsible for the environment, or even the Yellow Pages under listings for hydrologists or for environmental consultants. You will probably be surprised at the degree to which many of these professionals welcome the opportunity to discuss their work with students and faculty.

3. Arrange a visit to campus by a person involved in the field of ground-water modeling, preferably to discuss a particular case study on which he or she could bring slides, photographs, or diagrams, which would certainly make such presentation much more interesting. Frameworks for this could be either a class in mathematical modeling or in environmental studies, as well as a meeting of a student organization such as a math club or environmental association.

4. Do some additional reading in connection with some of the topics that were treated only relatively briefly in this introductory chapter. If you like this material, general reading that might be of interest to you could range from a basic book on geology or hydrology to a survey book on current environmental issues, in which ground water is likely to receive extensive discussion. You would learn much more about the practical aspects of how government and industry are pursuing the cleanup of extensive ground-water problems created in the past.

5. Considerable free or low-cost information on the topics treated in this chapter is available from government agencies who often have as part of their charge education of the public about the activities they are carrying out. For any large government programs, it is easy, once you identify the lead agency, to get your name on mailing lists for newsletters and miscellaneous general and technical reports. A particularly rich source of information in this regard is the national Nuclear Waste Program, which served as a major driving force for many years in the development of new techniques for ground-water modeling in complex situations. This program is under the aegis of the US Department of Energy, which, as with many government organizations, can be contacted easily through the Internet. One of the beneficial aspects of information obtained from these programs is that it often can be obtained for free. It is also often written for a more general audience than specialists in mathematical modeling, so it tends to be more comprehensible.

3

Air Quality Modeling

3.1 Background*

This chapter deals with air pollution, and a good way to begin would be to consider the following question: What do you think of when you hear the term "air pollution"? Think about this for a few moments before proceeding further in the text.

Different people will respond to the above question in different ways, no doubt conditioned to some extent by the kind of environment in which they live. For example, if you live in a large city, you have probably had the experience of standing by the curb near a bus stop and getting bathed in black smoke as a bus starts back down the street. Whereas this used to be a relatively normal state of affairs, in most cities now the control of vehicle exhausts is much better and the black clouds of exhaust smoke are less common.

Others may picture large smoke stacks when they think of air pollution, and, in fact, if you live in any kind of developed area, you can probably look out your window and see several such smoke stacks within view. When you see smoke coming out, you might think of that as pollution, and if you don't see any smoke coming out, you might think that either the facility is not operating or the exhaust is clean. These assumptions might not be exactly correct. For example, much of the smoke that comes out of smoke stacks is just water vapor that condenses into small water droplets upon contacting the cooler surroundings, for water is a natural chemical by-product of the combustion of hydrocarbon fuels. On the other hand, smoke stacks that are emitting no visible gases might well be emitting large quantities of relatively invisible gases that could present real problems either for people in the local area or perhaps even for others hundreds of miles away.

Maybe you have had the experience of driving to or flying into a city where there was a noticeable gray or brown haze in the air. In that case you have experienced smog, and, if so, you might well think of this image in response to a query about air pollution.

---

* The numerical calculations in this chapter will require, at a minimum, access to a calculator or computer that can raise numbers to powers, and preferably one that has the exponential function e^x built in. A programmable calculator would be more convenient, and a computer with a spreadsheet program or other programming capability would be ideal in terms of convenience. The mathematics in this chapter does not require any background in calculus. The exponential function e^x is used, but the chapter does include a self-contained introduction.

But how many of the following items passed through your mind in response to the original question:

- The smell of drying paint, stain, or varnish, whether you have just painted a piece of furniture, or a room in your house, or the outside of your house. These smells depend, of course, on the type of paint being used, but often derive from so-called *volatile organic compounds* (VOCs) which have quite a life of their own once they enter the atmosphere. For example, they undergo a whole chain of chemical reactions that contribute to the production of smog. The use of such materials is being phased down significantly in favor of water-based paints and coatings.

- Other odors, such as those from a garbage dump or waste transfer station, a factory, the spreading of manure on a farmer's field, or those found in a brand new automobile or when you install new carpeting. You know by the smell that you are inhaling chemicals that are not ordinarily there. Could they be harmful?

- Cigarette smoke, universally recognized as a major health risk for smokers themselves, but now also understood to represent a risk of health damage, not just annoyance, to others in the vicinity.

- Radon gas, a radioactive gas produced naturally within the ground, from where it can seep into houses through their foundations and cause health risks to the occupants. You cannot see it or smell it, but can detect it only by using special instruments or detection devices.

- CFCs, or chlorofluorocarbons, used as propellants to pressurize spray cans of household products, and also used in many other ways such as in refrigeration and air-conditioning systems and in fire extinguishers. Some of these uses are being phased out under international agreements because such compounds, once they enter the atmosphere, are involved in a sequence of chemical reactions that lead to a reduction in the ozone in the stratosphere (the upper portion of the atmosphere). When this ozone is depleted, we lose what is effectively a shield from a substantial amount of the ultraviolet radiation coming from the sun, with a consequent increase in certain illnesses (e.g., skin cancer and cataracts of the eye) and negative effects on agricultural output.

- Acid rain, which is essentially the "fallout" of certain pollutants in the atmosphere as they return to the earth's surface in raindrops that can be quite acidic. This can change the chemistry of bodies of surface water, such as lakes and streams, so much so that important species of fish and other aquatic organisms can no longer survive. There are also readily observable effects on forests, soils, historic stone monuments, painted surfaces, etc. Because the deposition of acid rain can occur at large distances from the original source of pollution, the analysis of such relationships can be quite complex and can result in inter-regional and international disputes.

- Nitrogen oxides (NO_x), which are a very interesting pollutant because their main source is not the fuel being burned or any other chemical being used, but rather the ordinary nitrogen and oxygen found in the air. The key point is that they are chemically combined with each other in the presence of the high temperatures achieved when fuels are burned in boilers, furnaces, vehicle engines, and other facilities or equipment. Not only do such nitrogen oxides contribute to acid rain, but they also contribute to urban smog and have a direct negative effect on the respiratory system.

- Carbon monoxide (CO) and carbon dioxide (CO_2), which are byproducts of the combustion of hydrocarbons. Carbon monoxide, as the reader is probably aware, is a highly dangerous gas because it combines with the hemoglobin in the blood in such a way as to reduce the ability of the blood to carry oxygen from the lungs to the other organs. Carbon dioxide, while relatively harmless in terms of direct exposure, is one of the key "greenhouse gases," meaning that its effect on the overall atmosphere is such as to hold in heat from the surface of the earth (just as the glass in the greenhouse keeps the heat in the greenhouse from re-radiating to the outside), with the likely effect of a gradual increase in global temperature.

- Naturally occurring materials in the air, such as hydrocarbons released from trees and other plants, methane from the decay or digestion of organic material, and pollen, to which many people have allergies.

There are many additional examples of air pollution aside from those listed above, but those given at least indicate that the subject covers a very wide range, perhaps wider than one might have originally thought. The basic conclusion is that when we breathe in the air, we are actually breathing in a soupy mixture of a wide variety of chemicals, some perhaps detected by the senses and some not, some which are key to basic life functions (especially oxygen), and others which may cause a variety of harmful health effects either through short-term or long-term exposures.

Considerable progress has been made in the last few decades to improve the air pollution situation in the United States. Keep in mind that the emphasis during the first half of the twentieth century was on developing our industrial economy at a rapid rate, spurred on, of course, by the need to support our forces in two world wars. Much, if not most, of this development took place without serious evaluation of environmental impacts. At best, smoke stacks were built higher simply to reduce the fallout for the people in their immediate vicinity. The first version of the Clean Air Act was passed in 1963, followed by important rounds of more stringent amendments over the ensuing years, and the Environmental Protection Agency (EPA) was established in 1970. In that era, the country began to focus more of its resources on cleaning up some of the environmental problems that had been created in the past and on preventing further ones from developing in the future. Air pollution received prompt attention. For example, there had been a number of incidents in urban and industrial areas where peculiar weather conditions combined with air pollution to provide stifling breathing conditions that led to significant numbers of cases of death and illness.

But even as the nation tried to control its traditional major air pollution sources, such as factories and power plants, other sources of pollution became more severe. For example, in the 30 years from 1950 to 1980, the number of passenger cars in this country tripled from 40 million to 120 million. And although the engines gradually became more efficient, the vehicles also became larger and heavier, the net result being that the number of miles per gallon achieved by vehicles during this period did not change substantially (averaging about 15 mpg). The result of this is that the transportation sector became an increasingly important source of pollution. It is not surprising therefore that California, marked by one of the classic cases of urban smog in Los Angeles, largely caused by automobiles, has led the way in developing and enforcing stricter regulations on vehicles and fuels. Other states have often followed their example.

Aside from the aggravation of recognizable problems, such as pollution from automobiles, recent developments in earth-monitoring by satellite and in related scientific fields such as atmospheric chemistry and physics have made it possible to understand the state of the earth's atmosphere on a more global scale. This has enabled us to recognize with greater clarity issues such as stratospheric ozone depletion* and the trend towards global warming due to the accumulation of "greenhouse gases" in the atmosphere.

Mathematical modeling is used to analyze the full range of air pollution issues. However, because the issues are so diverse, many different kinds of mathematical models need to be employed depending on the specific problem one is concerned with. Since the emphasis in this book is on explaining the basic principles of mathematical modeling of environmental problems, we will restrict ourselves to some of the simpler problems such as the release of air pollutants from individual, fixed sources.

For such fixed facilities, the general regulatory process followed in the United States and most other countries is that the operator must obtain a permit from the environmental regulatory authority in order to build and operate a plant which will release pollutants into the atmosphere. The application for such a permit generally involves detailed descriptions of the underlying industrial process as well as estimates of the amount of pollutants in various categories that are likely to be released during operation. If acceptable to the regulatory authority, which operates within the general guidelines of the actual written regulations, the permit will be issued and will include specified limits on releases, perhaps over different periods of time, requirements for monitoring such releases, and requirements for notification of the regulatory authorities in case there are unexpected excursions above the limits. Not only are such permits issued for large smoke stacks, such as you might see along the skyline of a large city, but even for much smaller stacks and vents, such as are quite common throughout individual plants. So, for example, a typical industrial plant, hospital, or large facility might have tens or hundreds of individual environmental permits for each of its individual sources of air pollution, or at least some comprehensive permit that references each of the individual sources.

Operating plants must control their processes in order to be sure to avoid exceeding their permitted levels of release. For example, trash incinerators, which are quite common in certain parts of the country, generally use an overhead crane to mix the various loads of trash being brought in by waste haulers so as to minimize the concentration of particular items that may contribute harmful pollutants to the stack releases. For example, tires and gypsum wallboard ("sheetrock") contain substantial quantities of sulfur, which can cause such a plant to exceed its regulatory release limits for sulfur compounds, even if only on a relatively short-term basis. Yard wastes, such as brush and leaves, contribute a substantial quantity of nitrogen, and hence additional NO_x, to the exhaust gases when they are burned, and this can also cause an excursion above the regulatory limit. This is why such materials are sometimes banned or controlled in some way in municipal waste operations.

The regulatory limit applied by the environmental authority often takes into account the current ambient levels of various pollutants. Therefore, more stringent requirements might be imposed on facilities operating in areas where the air quality is already substandard. In fact, in some areas the regulators apply the so-called "bubble concept" under which they look at the overall air quality in a well defined region and refuse to allow any new sources of pollution

---

* Ozone, which is simply a molecule composed of three oxygen atoms (O_3), is a curious substance. It is desirable to have it in the upper portions of the atmosphere, where it shields us from the ultraviolet radiation of the sun, but undesirable to have it in the lower portions of the atmosphere near the ground, where it is a principal component of smog. In the upper atmosphere it is easily destroyed through the catalytic action of CFCs, as has been mentioned earlier. In the lower portions of the atmosphere, it is produced by chemical reactions involving released VOCs. Thus the two issues involving ozone are really two entirely distinct environmental issues.

in that region without counterbalancing improvements in the release of pollutants from other sources. This has led to the interesting situation where some operators who wish to build plants wind up making investments in additional pollution control equipment for unrelated neighboring companies, or even the situation where companies now can "trade pollution rights" along the lines of a commodities or stock market. While these ideas may seem curious at first, what they really do is help to capture more effectively the economic costs associated with pollution control, and they target the pollution control investment in such a way as to maximize the environmental rate of return.

Exercises for Section 3.1

[Note: several of these exercises ask you to target your response to a specific type of imaginary audience. You are being asked to play a specific role, and you should be sure to put yourself in this role as you prepare your response. Remember that just collecting the appropriate information is not enough; you must be able to communicate it effectively and efficiently to whatever audience you are dealing with.]

1. You recently got a job at Ace Consulting. They have expanded their services to include air quality analysis. The last set of major changes to the Clean Air Act has their clients confused and calling daily for some help in understanding exactly what they need to do to comply. Your boss tells you that you are to participate in a workshop for these clients. Your assignment will be to introduce the program by giving a very brief synopsis of the Clean Air Act. You'll need to include its basic structure (e.g., the classes of pollutants being addressed) as well as how this law has been updated since its original enactment. You're told you need to submit a two-page summary of your introduction to be included with the papers given by the other associates. Here's your chance to shine. You need to show good judgment in picking information that will give both the "big picture" as well as key specifics. Target your audience and write the summary.

2. After college you wanted some international experience, and so you took a position in the international division of Hirosaki Industries in Japan. The company's business is building and operating trash incinerators that also usually generate electricity. The Board of Directors has suggested expanding from current markets in Asia and the Pacific into the US, and your Vice-President sends you to the US for a couple of weeks to find out what problems have occurred in permitting or operating trash incinerators there in the last ten years. You wind up the field part of your assignment and are about to take a leisurely trip back with a stopover for some much needed R&R in Hawaii when you get an urgent call from his secretary asking you to be back to meet with him in his office at 9 a.m. the next day, bringing along a memo summarizing your observations. He has apparently been asked to give an interim briefing to the board that evening. Nothing new, just another "crunch" moment you can expect in any new job! (You're the low person on the ladder, and you're used to them.) You race to airport for an all night flight, and along the way you type your memo on your notebook computer. You know that the boss likes facts, he likes graphic organization, and he likes brevity. One page maximum. Write the memo. (Note: assume that the US visit focuses either on the region where you live or on the region where your school is located, the term "region" referring to several states. Make good use of your library's computerized databases and other sources. Computerized newspaper files

can also be an excellent source of information, especially when there is controversy over some issue.)

3. Your state environmental agency probably has separate divisions responsible for different aspects of the environment, such as air, ground water, surface water, etc. As part of an ongoing election campaign, a prominent candidate promises that she will be able to reduce taxes by cutting back on unnecessary government operations. One of the agencies that she has proposed to cut back is the state environmental agency, with particular emphasis on the large contingent of staff responsible for air pollution work. Her position is that air pollution problems have been brought largely under control and that there is no longer nearly the same need for a large staff that there might have been a decade or two ago. You are working as a summer intern with a nonprofit citizens organization that is deciding which candidates to endorse for election, and you have been asked to prepare a briefing paper for the board of directors concerning the validity of this candidate's claims about the level of staffing. By contacting the agency and/or using publicly available reference material, find and present clearly and succinctly the following information: a general organization chart of the portion of the environmental agency that deals with air pollution, a breakdown of the staffing of this branch and its overall budget level, and the three most important activities of this part of the agency within the last five years.

4. This problem applies to the particular school or other institution with which you are associated. Find out who within the organization is responsible for environmental permits, and determine whether there are any such permits related to air pollution. If there are, compile a general list of emission sources for which such permits exist, and examine one individual permit, preparing a detailed summary of its content.

5. The text of this section referred to the occurrence of a number of incidents in the United States during which deaths and illnesses were caused by air pollution. Find an example of one such incident and provide a brief summary of what happened.

6. For both your own hometown and your institution's location, determine whether they are located in a "non-attainment area" for ozone, meaning that the national air quality standard for this pollutant is not being met. If so, determine the degree of severity of non-attainment according to the EPA's classification scheme. (Hint: this can be determined on-line by accessing and searching EPA's resources, or it can be determined by review of published reports or by making inquiries to environmental organizations or agencies.)

3.2 Physical Principles

As mentioned in the previous section, we will concentrate on individual "point sources" of air pollution, rather than on sources that have more complex geometries, such as a fleet of motor vehicles moving along a network of highways. We will also assume for the time being that the point sources we are interested in are operating at so-called "steady state," meaning that the amount of material being released stays essentially the same over the period of time we are interested in. Thus the best image for you to have in your mind as you read through this section is that of a typical smoke stack releasing a steady amount of smoke or gases at its top.

Think in some detail about the image of such a smoke stack, which you have no doubt seen many times in your life. It is essentially the same whether it is the large kind of smoke stack generally associated with a power plant or major factory, or whether it is a smaller smoke stack or chimney such as you might find on a school, apartment building, or even an individual house. You have probably seen smoke coming from such stacks on many occasions, and you have an image of how it disperses in the air as it comes out. The purpose of this section is to try to sort out the various physical processes that affect the way that smoke disperses from the moment of release.

As you work through this chapter on air pollution, it might be instructive for you to identify a number of such stacks that you can observe from your home, school, or office, and try to remember to observe them from one day to the next to see if there are any variations in the pattern of the exhaust emissions. Remember, as mentioned in the previous section, that sometimes the exhaust emissions are not visible even though the stack is emitting gases at even a relatively high rate. Nevertheless, you will have to confine your observations to stacks from which you can see some kind of emissions plume.

What mental image do you have of the emissions plume coming from such a stack? Does it go directly upward or does it turn to one side? Does it spread out in the vertical direction, or does it stay confined to a single relatively limited level? Does it climb higher and higher, or does some of it even work its way down to ground level? Think about what you have seen in the past and try to answer these questions, but also try to become more observant as you look at stacks while working on this section. There is much to be learned by simple physical observation.

If you observe many stacks over a long period of time, your conclusion would likely be that the emitted exhaust can form quite a variety of patterns. An obvious factor that is certainly going to affect the pattern of emissions is the wind. But there are others as well, as we shall shortly see. In general, the various patterns of exhaust plumes suggested in Figure 3-1 are only specific examples for which a whole continuum of possibilities exist. What causes these different patterns, and how can we predict what is going to happen at a plant that has not even been built yet? The answer is simple: we need to use mathematical models to make such predictions, taking into account the nature of the proposed operating plant, the characteristics of the exhaust gases, and the whole range of weather patterns historically experienced at the site. This is exactly what the people proposing the plant would be expected to do, and the same mathematical analysis and predictions are what the regulatory agencies would study before deciding whether the expected impact of the exhaust gases on the environment would be acceptable or not. So mathematics is at the center of these key issues.

Now we need to identify the key physical processes that control the way pollutants disperse in the atmosphere. These may be classified roughly into the following categories:

1. The propensity of gaseous materials to move from areas of higher concentration to areas of lower concentration by a simple diffusion process.
2. The patterns of movement of the surrounding air, including both horizontal movement (wind) and vertical movement.
3. The physical characteristics of the exhaust gases themselves, such as temperature, momentum, and perhaps other properties.
4 Other aspects of the surroundings, such as special topographic features or changes in elevation.

Let us discuss these topics in greater detail. If you poured a little perfume on a tissue placed in the center of a room, it would probably not be long before you could catch the scent of the

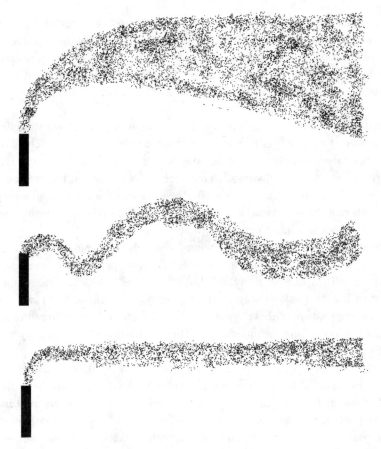

FIGURE 3-1
Representative examples of variability in plume behavior

perfume throughout the room. That is because the gas molecules generated by the evaporation of the perfume initially start out in a very high concentration right around the tissue, and then they tend to spread out by diffusion until they permeate the entire room. The same would be true if you spilled some gasoline on the floor of a garage. The gasoline would evaporate, and the corresponding vapors would gradually migrate from an area of higher concentration just over the spill to areas of lower concentration in other portions of the garage. This is what is meant by the "propensity of gaseous materials to move from areas of higher concentration to areas of lower concentration." Note that there does not have to be any kind of wind or air movement in the room or in the garage for this to happen. The molecules do it on their own. The driving force is actually the kinetic energy present in each of the individual molecules, which causes them to bounce around in a relatively random pattern until they are more evenly distributed. This is not unlike the process of shuffling a deck of cards. For example, if you start with a deck of cards with all the hearts together in one place and then you shuffle them a number of times, they will gradually be mixed up by this random process so that the final result is usually much more uniform and spread out.

The diffusion process is not limited to gaseous material or to an air environment. If you place a spoonful of sugar in the bottom of a cup of hot coffee, it will not only dissolve, but the

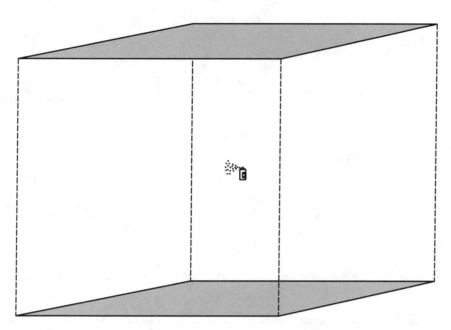

FIGURE 3-2
A typical example of a three-dimensional diffusion situation

dissolved sugar molecules will gradually spread out through the coffee even without any stirring (although it might take some time). This is another diffusion process. You could doubtless think of many other diffusion processes with which you are familiar.

Since diffusion is such a critical driving force in the dispersion of air pollutants, we need to discuss it in a finer level detail. For example, consider the situation depicted in Figure 3-2, which shows a momentary release of ordinary household insect spray in the center of a room in a house. Initially, all of the material is concentrated near the center, but gradually it will spread out in all three principal directions. This would be classified as a three-dimensional (or 3-D) diffusion situation because the diffusing material is going to be moving in all three spatial dimensions. If we were to graph the concentration of such material in the air in the room at a relatively early time t_1 and at later times t_2 and t_3, the resulting graphs would have the general shapes shown in Figure 3-3. You can see in these graphs what you would already anticipate from your own intuition, namely that the material tends to spread out as time goes on. (Once again, the molecules are getting "shuffled" as a result of their own random motions, driven by the kinetic energy which they have when they are released in the room.)

You might be tempted to think that three-dimensional diffusion is the only kind there is, but that would not be correct. For example, think of this experiment. Take a nice white handkerchief, wet it with water, rinsing out the excess, stretch it out, and then put a few drops of ink in the middle of it. (Just do this experiment mentally!) You will notice that the ink blotch starts to spread out at its edges because the ink is actually diffusing through the water that permeates the handkerchief. Now this is obviously two-dimensional diffusion because there are only two dimensions to the handkerchief. (Of course, the ink might not totally spread out evenly throughout the handkerchief, even if you wait a very long time, because of some physical-chemical reactions between the ink and the handkerchief itself.)

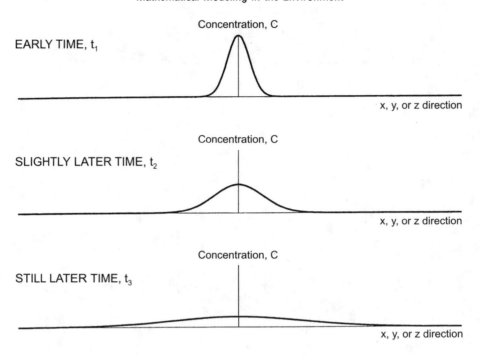

FIGURE 3-3
Conceptual shapes of the graphs of concentration of diffusing material in each of the three spatial dimensions and for early and later points in time

One way to think about two-dimensional diffusion in air would be along the lines suggested in the top part of Figure 3-4, which is similar to the three-dimensional diffusion case shown in Figure 3-2 except that the space into which the insecticide is sprayed has now been compressed down in the vertical direction, so that there are effectively only two real dimensions available for diffusion of the material.

Both this case and the handkerchief experiment may seem somewhat strange and unnatural to you, but they are a good way to introduce the really important case of two-dimensional diffusion, which is suggested schematically in the bottom half of Figure 3-4. Here the hypothetical situation begins as an ordinary three-dimensional problem, but the source of diffusing material is actually along an entire line running in the vertical direction, rather than at one individual point. You may think that this makes the problem more complicated, but actually it makes it easier. The key is the following observation: *since every single vertical level is essentially identical to the levels above and below it, there should never be a vertical difference in the concentration of the diffused material between two points that differ only in their height.* Since there is no difference in concentration from one vertical level to the next, there should be no diffusion in this direction, since diffusion is driven by a difference in concentration. Therefore the only diffusion that should take place in this case is diffusion in the two horizontal directions, and thus this is actually a two-dimensional diffusion problem!

You may have to read the above explanation over and over again and stare at the diagram in the bottom part of Figure 3-4 before you finally get the critical insight, but once you do, you will have made a very important step in understanding how a two-dimensional diffusion

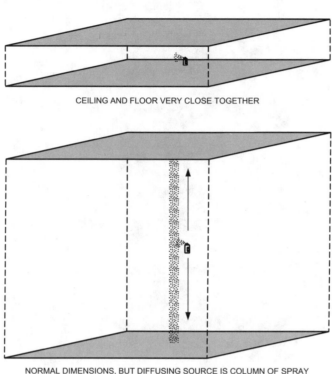

CEILING AND FLOOR VERY CLOSE TOGETHER

NORMAL DIMENSIONS, BUT DIFFUSING SOURCE IS COLUMN OF SPRAY
FROM CEILING TO FLOOR

FIGURE 3-4
Two possible frameworks for a two-dimensional diffusion problem

process can really be found in a three-dimensional setting. And two-dimensional problems are much easier to solve than three-dimensional ones!

Here are two further comments about this two-dimensional diffusion situation. First, we could refer to the situation shown on the bottom part of Figure 3-4 as a diffusion problem associated with a "line source." Second, if we were to graph the concentration of diffused material in each of the two directions of diffusion, that is, the x- and y-directions, the results would have the general shape shown earlier in Figure 3-3.

We might also occasionally encounter a one-dimensional diffusion situation. Once again, this could result from a physical situation in which the diffusion is actually physically restricted to one possible dimension, as shown in the top part of Figure 3-5, or to a situation where, because of the nature of the source, one expects a concentration difference to exist only in one dimension, so that there is only one dimension in which diffusion is an important process for moving material. This latter situation is shown in the bottom part of Figure 3-5, and as can be seen from this figure, the key component is the representation of the source of diffusing material as a two-dimensional plane. Because of symmetries among the points, meaning that every point on the source looks like every other point on the source, the only direction in which material is going to have a concentration difference that could drive the diffusion process would be perpendicular to the source plane. Thus this is really essentially a one-dimensional diffusion problem. The graph of the expected concentration of diffused material as a function of the x variable alone would once again have the shape seen earlier in Figure 3-3. Here x is, of course,

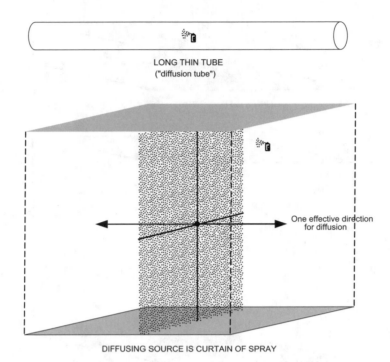

LONG THIN TUBE
("diffusion tube")

One effective direction
for diffusion

DIFFUSING SOURCE IS CURTAIN OF SPRAY

FIGURE 3-5
Two possible frameworks for a one-dimensional diffusion problem

the perpendicular distance from the source plane. One might also refer to this situation as one with a "planar source."

All of the previous discussion had to do with developing an understanding of the diffusion process, which is only one of the physical factors causing air pollutants to spread out when they are released from a stack or other source. The second factor identified earlier was the movement of the air itself, and obviously this is also very significant. Once again, the objective of this section is to help you develop a physical intuition about this process, so let us begin by posing the following question: suppose you live one-half mile from a power plant and the wind is blowing directly from the power plant toward your home. Surely you will have some concentration of pollutants in the vicinity of your home, even though this concentration might be relatively small. Now imagine the situation where everything is the same except the wind is blowing twice as hard. Do you think you will get a higher concentration at your home or a lower concentration? Think about this before proceeding further.

There are a number of different approaches you might take to answer the previous question. For example, you might argue that if the wind is blowing faster, it will bring the pollutants to your house sooner and hence with less time for them to diffuse and become diluted en route. This would lead to a higher concentration at your home. On the other hand, you might argue that when the wind is blowing harder, the air is being stirred up more, which might increase the actual rate of diffusion, thereby cutting back on the amount of pollution at your home. You will have to wait a while to see how this situation actually turns out.

There are a number of air movement factors that can affect the degree at which pollutants from a stack disperse. First, as shown in Figure 3-6, the wind speed itself is not constant at all

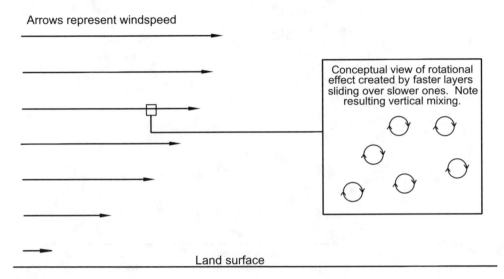

FIGURE 3-6
Wind-speed profile and effect on vertical mixing

heights above the ground, but rather, because of the friction of the air mass along the ground, its actual velocity there is very small and gradually increases as you move upward. This range of wind-speed values over a vertical profile is indicated by the varying length of the arrows shown in this figure. Thus we have a "wind-speed gradient" as we move vertically, and this has a tendency to encourage mixing between the air at one level and the air at those levels immediately above and below it. Therefore, even though the wind is blowing in a horizontal direction, variation in wind speed over a vertical profile can lead to vertical transport of pollutants.

Another important factor that affects vertical movement within the air mass is that of solar "insolation" or sunlight. Such insolation is absorbed by the ground, thereby heating it up. When the ground temperature rises, it heats the air immediately adjacent to the ground, and this air, once heated, expands, becomes less dense, and rises to higher levels. This is similar to the way a hot air balloon works, as shown in Figure 3-7, except that in the case of the balloon the heating is provided by a burner attached to the balloon itself, whereas in the case of the open atmosphere, the heating is provided by the sun-warmed ground.

Certainly, if some air is rising as a result of being heated, other air must come in to fill its place down at the lowest levels. Some of this air may come in from the sides and some may come from above, but this movement is also important because it is another mechanism for vertical mixing of the atmosphere and hence for the dispersion of pollutants both upward and downward.

Note that in the previous paragraph we used the word "dispersion" instead of "diffusion." The word "diffusion" is sometimes restricted to a situation that is purely a "molecular diffusion" process, meaning that the movement of material is driven solely by a concentration gradient and the random movement of the involved molecules. When the forces are somewhat more complicated and when there may be a number of different factors at work, it is more common to use the term "dispersion," although we will also use the word "diffusion" in this more general context.

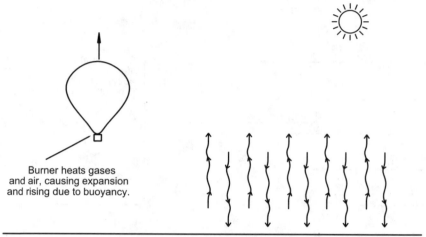

FIGURE 3-7

Comparison between the rise of a hot air balloon and the rise of air warmed by the ground

Another class of factors that can affect the distribution of pollutants once they emerge from the stack are the characteristics of the exhaust gases themselves. First, they are generally going to be hot. Therefore, just as hot air tends to rise, so do hot exhaust gases. In addition, the exhaust gases have generally been moving up the stack with considerable velocity, so that when they exit at the top, it is like a wind blowing out of the stack. Since the momentum of the exiting gases is therefore in the upward direction, they will continue to move upward vertically for some distance until this momentum is dissipated by mixing and integration with the surrounding air mass. The combined effect of these two factors—temperature and momentum—is to create an imaginary extension of the stack beyond its physical top, through which region in space the predominant movement of the gases is still vertical. This additional vertical height may be up to two to ten times the height of the physical stack itself. This concept is illustrated in Figure 3-8,

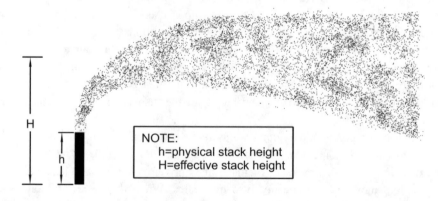

FIGURE 3-8

Illustration of the concept of "effective stack height" resulting from temperature and momentum effects within the exiting exhaust gases

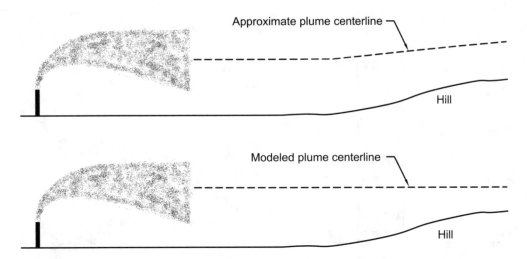

FIGURE 3-9
Physical behavior of plume and air mass approaching a topographic high and a conservative approach for modeling this situation

which suggests that for a physical stack height h, there is actually an "effective stack height" H which we could well think of as the effective height at which the gases are ready for horizontal transport and dispersion according to some of the factors we have considered previously.

The last category of factors to be considered in the transport of an exhaust plume is that of the characteristics of the earth's surface itself. One such factor will obviously be topography itself, referring to changes in the elevation of the land along the path of the plume. For example, if the prevailing wind and the entrained plume approach a hill, there will be some upward movement as the air moves up and over the hill. This is illustrated on the top part of Figure 3-9. On the other hand, since this upward movement can be somewhat difficult to predict with accuracy, it is not uncommon in air pollution calculations to assume that no such deflection of a plume takes place and that therefore an increase in elevation effectively places a receptor (such as a residence or individual members of the population) higher in the plume, where the concentration would naturally be expected to be higher. This latter approach is known as a "conservative" assumption, meaning that it might tend to overestimate the risk to people living downwind, and thus we would be being conservative by taking it into account in our planning.

Another characteristic of the earth's surface that can be quite important for certain kinds of air pollution calculations is that of "surface roughness." This refers to the fact that depending on whether the surface consists of open land, or heavily forested land, or buildings of various geometries and heights, there may be varying amounts of "friction" as the air mass moves over that land. This can affect the vertical velocity gradient (mentioned earlier in connection with Figure 3-6) as well as lead to various local phenomena such as eddies (which the reader might have encountered more commonly in connection with the flow in streams or rivers), downwash effects, and other interruptions of the normal flow scheme. Some of these effects are suggested in Figure 3-10.

In summary, the two dominant factors in the dispersion of air pollutants are their inherent tendency to diffuse as a result of concentration gradients and the effect on this process of

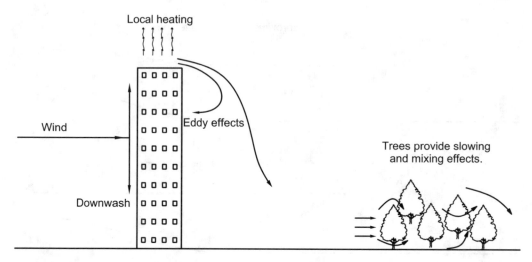

FIGURE 3-10
Conceptual representation of various ground effects on the movement of an air mass

movement of the air itself. That is, if there is more movement or mixing of the air itself, this would be expected to accelerate the rate of diffusion or dispersion. The additional effects of exhaust gas and ground characteristics are certainly important, but it will turn out that they can be incorporated in the calculations in a simpler and less fundamental fashion. So the one topic that needs some additional background discussion is that of air movements within the atmosphere.

We will use the term *stable* to refer to an air mass in which there is relatively little vertical mixing per unit of horizontal distance traveled. On the other hand, we will call an air mass *unstable* if the amount of vertical mixing is much higher per unit of horizontal travel. (There is a more technical scientific definition, but we do not need it for our purposes.) In the case of a stable air mass, you would naturally expect a plume of exhaust gases to disperse less during horizontal transport with the wind than you would if the atmospheric conditions were unstable. Thus a stable air mass can also be thought of as corresponding to a stable plume; similarly an unstable air mass corresponds to a relatively unstable plume with a higher degree of dispersion.

These concepts are illustrated pictorially in Figure 3-11. We have already discussed above a number of different characteristics of the atmosphere that might affect its level of stability. For example, a high level of solar insolation would be one factor that could encourage greater vertical mixing and hence greater instability. Meteorologists have studied various combinations of atmospheric conditions and have developed a relatively simple classification scheme for characterizing the degree of stability. This scheme is summarized in Table 3-1. Stability category A corresponds to the lowest level of atmospheric stability, whereas stability category F corresponds to the highest level of stability. Thus you would expect a plume to hold together more cohesively (and hence for a greater distance) under stability class F than you would for stability class A.

Air quality models based on this scheme have become widely accepted in industrial and regulatory practice, and the stability categories are sometimes referred to as the Pasquill-Gifford

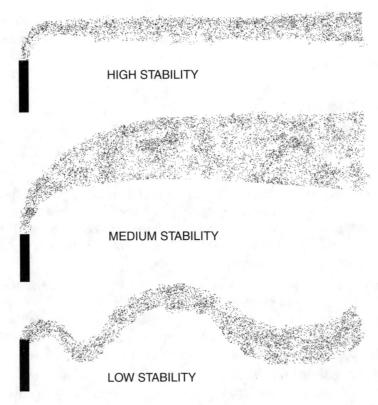

HIGH STABILITY

MEDIUM STABILITY

LOW STABILITY

FIGURE 3-11
Typical plume geometry as a function of atmospheric stability

TABLE 3-1
Definitions of atmospheric stability classes

Surface wind speed		Day			Night	
miles per hour	meters per second	Incoming solar radiation			Thinly overcast or ≤ 4/8 cloud cover	≤ 3/8 cloud cover
		strong	moderate	slight		
< 4.5	< 2	A	A–B	B		
4.5–6.7	2–3	A–B	B	C	E	F
6.7–11.2	3–5	B	B–C	C	D	E
11.2–13.4	5–6	C	C–D	D	D	D
> 13.4	> 6	C	D	D	D	D

Notes:
1. The neutral class, D, should be assumed for heavy overcast conditions during day or night.
2. Wind speed refers to values measured 10 meters above the ground.
3. For intermediate categories, such as A-B, both cases should be considered as possibilities.
4. Night refers to the period from one hour before sunset to one hour after sunrise.
5. Strong incoming solar radiation corresponds to the sun being more than 60 degrees above the horizon with clear skies; slight insolation corresponds to an angle of the sun between 15 and 35 degrees with clear skies. Incoming radiation that would be strong with clear skies can be expected to be reduced to moderate with broken middle level clouds or slight with broken low clouds.
6. Empty boxes are very rare conditions; use values for boxes below them.

stability categories. This entire classification scheme is somewhat more reliable over open country or rural areas than over highly developed areas, where there can be more important localized effects.

Exercises for Section 3.2

1. Give an example of a diffusion process with which you are familiar but other than those discussed in the text. Identify whether this situation involves one-, two-, or three-dimensional diffusion.

2. In the three-dimensional example of the diffusion of evaporating perfume through a room, imagine that you modify the situation by turning on a small fan in one corner of the room. The fan is not aimed at the tissue containing the perfume, but rather it is aimed along the wall on one side of the room. Would this increase or decrease the rate at which perfume disperses throughout the room? Explain your answer.

3. Suppose that exhaust gases are leaving a vertical smoke stack at a constant rate while the wind at the top of the stack is blowing at 10 miles per hour. There will be a certain effective height of release H somewhat larger than the actual physical height of the stack h, as discussed in the text. Now suppose that the wind speed is increased to 20 miles per hour. What effect do you expect this to have on the effective stack height H? Explain your answer.

4. As you know, gases from an exhaust plume can certainly disperse both horizontally and vertically. For gases that are dispersing in a downward direction, what do you suppose happens to them when they reach the actual ground level? For example—keeping in mind that for diffusion processes such as this, the net movement of materials is always from an area of higher concentration to lower concentration—could there ever be a layer just above the ground where the pollution would build up to a higher concentration than that just above it? Be very precise in your explanation.

5. Suppose you were trying to calculate the concentration of a pollutant at your residence, which is located exactly one mile downwind and in a direct line from a smoke stack that is 200 feet high. Here you know the value of h but you do not know how to calculate the effective stack height H. Suppose you used the value of h instead of H for your analysis. Would you be overestimating or underestimating the amount of pollution at your residence? Explain your answer carefully.

6. [Library research question.] Find and explain the precise scientific definition of atmospheric stability. (Hint: Look in a meteorology or air pollution text, or in some more general science or reference source, or consult on-line computer sources.)

3.3 Typical Quantitative Issues

Begin by putting yourself in the following typical, but imaginary situation. You work as a staff assistant to a city councilman in a large city. Up until now, the city's garbage has been shipped for burial in a landfill in a nearby suburb, but the capacity of that landfill is running out, and, in addition, environmental problems have been found there which are leading the city to worry about increased liability by continuing to ship more garbage to it. As part of its planning process, the city hired an independent consultant to sketch out various alternative technologies for trash management, and some time ago this consultant submitted a report in which three leading alternatives were proposed for evaluation: building a new landfill in an unspecified location, building a trash incinerator within the bounds of the city, and building a trash incinerator at an unspecified location outside the city. Based on this report, the city invited potential operating companies to submit detailed bids for the construction and operation of such facilities. All of these proposals were to incorporate a specified level of recycling of various materials, but they also require either the burial or incineration of substantial amounts of trash.

The city council member for whom you work has asked you to focus your attention on the trash incinerator being proposed for construction within the city. One of the bidders has submitted a detailed proposal on this alternative, and the documents submitted include roughly 800 pages devoted to environmental aspects, many pertaining to air pollution issues, including exhaust gas composition, control technologies to be employed, extensive tables and graphs of calculated concentrations under a wide variety of conditions, and a discussion of health impacts of different kinds of emissions.

Very few people would probably have the time or technical expertise to labor through all the details of this analysis. However, many people certainly would need to be well informed about the basic conclusions or "bottom line" associated with the air pollution impacts of this technology. In particular, the key questions which you would probably want to have in mind as you looked through the detailed report would certainly include the following:

1. What would be the "worst case" level of air pollution expected to be experienced by city residents as a result of this facility?
2. Where would this worst case pollution condition be expected to occur?
3. What would be the average level of air pollution expected to be encountered by residents in various parts of the city as a result of this facility?
4. Might any further development within the city have a significant impact on the amount of pollution caused by this facility? (For example, if other tall buildings were built in the vicinity of the waste incinerator, could this sufficiently modify the air currents so as to cause more pollution to wash downward upon the resident population?)
5. What types of weather conditions would lead to the highest pollution levels from this facility, and how likely are they to occur?

These are of course questions that would not only be of interest to you in your role as a staff assistant to a city council member, but they are questions likely to be asked by many persons at all levels of the decision-making process for this project. Note that they are not just questions for debate among scientists and engineers, but rather they would need to be addressed by politicians, city agencies, potential abutters of the new facility, members of the public, environmental organizations, and many others. While few of these parties would be likely to have the technical expertise to understand and critically review every single step in the analysis put forward by the bidders, certainly they would not be willing to lend their support to this project without being confident that they understood and believed the final results.

As has been stated in other chapters, the objective of this book is to enable you to develop a firm understanding of the basic principles of some of these key environmental issues, such as air pollution, and to introduce you to the main thought processes and mathematical techniques that are used in their analysis. As a brief preview of what is to be covered later in this chapter, a main objective is for you to develop a high degree of comfort and understanding of the following rather intimidating-looking equation:

$$C = \frac{Q}{2\pi\sigma_y\sigma_z u} \left[e^{-\frac{y^2}{2\sigma_y^2}} \right] \left[e^{-\frac{(z-H)^2}{2\sigma_z^2}} + e^{-\frac{(z+H)^2}{2\sigma_z^2}} \right]$$

This equation gives the concentration C of pollutant within a plume at essentially any location downwind and at any elevation. Even if you are well accustomed to the exponential function, you might still find this equation quite intimidating. If you are not at all familiar with the exponential function, you might feel like closing this book and studying something else instead. Don't be alarmed; we will work our way into this equation very gradually, and you will be able to understand and use it perfectly well very soon. In fact, just in case you never studied or remember very little about the exponential function, the next section provides a simple review of everything you need to know to proceed further.

Exercises for Section 3.3

1. As has been mentioned earlier, when a large project is proposed that may have significant air emissions of one kind or another, the regulatory authorities generally require that extensive air pollution calculations be carried out to predict the levels of concentrations of various pollutants in the vicinity of the facility. These environmental documents become matters of public record, and they are generally available for review by members of the public in the offices of the regulatory authority as well as quite often in town or city offices or local libraries in the vicinity of the proposed facility.

Identify one such project that has been analyzed in this way in your region or state. Find the location of available copies of the environmental studies, review these documents, and identify the key issues involving air pollution that needed to be resolved prior to making the decision on the acceptability of the project. Use secondary sources, such as newspaper accounts, if you do not have ready access to the primary sources. (However, you will learn much more about the context within which air pollution analysis is applied if you can gain access to the original documents, which is usually not difficult.) Examples of the kinds of projects for which such analysis would generally be required include: power plants (both nuclear and fossil fuel plants), chemical manufacturing facilities, incinerators, some large landfills, radioactive or hazardous waste management facilities, cement plants, and major transportation projects. You may be able to use computerized or paper resources in your library to identify such projects, or consult on-line sources. Telephone calls to environmental agencies and local governmental authorities are also a good way to get started on a project such as this.

2. Identify the role of quantitative air pollution analysis in the development of at least one environmental regulation by the US Environmental Protection Agency (EPA). Note that when an environmental agency intends to promulgate new regulations, it goes through a number of rounds of draft regulations for public review and comment. Furthermore, draft and final regulations are usually preceded by a carefully constructed preamble that summarizes the kinds of studies that were carried out in support of the new regulations. It will not be difficult to find extensive information of the type sought here from many different sources, such as accessing the EPA on-line, reviewing copies of the Federal Register (the official federal government document in which such new regulations or draft regulations are published), probably located in your library, or by using a variety of search tools, including newspaper indexes, available in your library.

3. Based on a review of newspaper articles or other information available in your library or on-line, identify what you would regard as:
    a) the city in the United States that has the worst air pollution situation;
    b) the city in the world that has the worst air pollution situation.
Describe briefly each of these situations and explain the reason why you would characterize it as the worst in its class.

4. [This problem is intended for automobile enthusiasts.] The problem of combustion-generated nitrogen oxides has been described in the text. What changes are being investigated for future automobile and truck engines that may decrease such emissions? (Hint: potential sources of information would include newspaper articles, popular magazines related to science or technology, automotive magazines, on-line searches or discussion groups, or discussions with people familiar with this technology.)

3.4 Brief Primer on the Exponential Function e^x

No doubt you have encountered the concept of a mathematical function before. It is no more than a rule that enables you to go from an "input" value to an "output" value. Such a rule is usually specified in the form of an equation. For example, when we write the equation

$$y = x^2,$$

we are specifying a rule for going from an input value x to an output value y by the process of squaring. Given an x value, say, 3, you can calculate the corresponding y value simply by plugging into the equation as follows:

$$y = 3^2 = 9.$$

In fact, sometimes functions have more than a single input value. For example, in the equation

$$z = x^2 + 3y^2,$$

we have two input values, x and y, that we need in order to calculate the corresponding value of the output value variable, z. For example, at the particular point where $x = 1$ and $y = -3$, we could calculate the output value z as follows:

$$z = 1^2 + 3(-3)^2 = 1 + 3 \times 9 = 1 + 27 = 28.$$

Sometimes the input values can show up in rather peculiar locations within the equation. For example, the following is a perfectly acceptable equation for defining a function:

$$y = 2^x.$$

Here, the input value x gets used as the exponent of the number 2 in the equation. For example, when the input value is 3, we can calculate the corresponding output value as follows:

$$y = 2^3 = 8.$$

In fact, we could even have two input values located up in the exponent of one of these functions, as follows:

$$z = 5^{x-3y}.$$

This is still a perfectly good function, for we can still take two input values and plug them into the equation in order to find the value of the corresponding output value. For example, when $x = 2$ and $y = 1$, we would find the value of the output z as follows:

$$z = 5^{2-3\times1} = 5^{2-3} = 5^{-1} = \frac{1}{5} = 0.2.$$

Note that in this last calculation there is a reminder of the fact that when you have a negative exponent, you can handle it by changing it to a positive exponent and putting the entire expression in the denominator, as we did in this previous equation in changing from 5^{-1} to $\frac{1}{5}$.

The last two examples of functions given above fall into a very special category called *exponential functions*. In those examples, the numbers 2 and 5 would be referred to as the *base* of the exponential functions. They are numbers that you have to raise to various powers to evaluate the functions.

You might think that 2 or 5 or other nice whole numbers would be the kinds of numbers we would like to have as the bases of any exponential functions we might want to deal with. For example, it would be much more complicated to deal with the exponential function

$$y = 1.162731^{x}$$

than it would be to deal with the nice simple exponential function given by

$$y = 2^{x}.$$

Now we come to a real surprise. The most common and most useful exponential function for use in mathematical modeling, not just for air pollution problems, but for all kinds of everyday situations such as bank interest, population growth, economic development, etc., is an exponential function that has a very strange base, namely, the famous number e. Maybe you have heard of e and maybe you have not. You have certainly heard of π, and you know that it is an infinite non-repeating decimal that begins with the digits $3.1415\ldots$. This number e is very similar to π in many respects. It too can be represented only by an infinite non-repeating decimal, the first digits of which are:

$$e = 2.718\ldots.$$

Even though we cannot write down its exact value as a fraction or a finite decimal, it is still a perfectly good number, and we can perform the ordinary operations of arithmetic on it, such as squaring it, cubing it, taking its square root, etc. As long as we can do these things, we can refer to *the* exponential function

$$y = e^{x}.$$

Now, if you were given an input value such as $x = 2$, you might wonder how you could actually calculate the output value for y. That's a good question. You might try this along the following lines:

$$y = e^2 \approx (2.718)^2 = 7.387524.$$

Here we have used an approximate value for e based on its first three decimal places, and then we have simply squared that result using ordinary multiplication (preferably on a calculator). In the above equation, the wavy lines in the middle ($\approx$) are used to indicate "approximately equal."

But there is an even easier way to perform calculations using this exponential function as long as you have a calculator or a computer available. They generally have built into them a much more elaborate way to approximate this function that gives accuracy to a larger number of decimal places. If you have a calculator handy, take a look at it and see if it has the exponential function built into it. This is usually indicated by a key that is marked either e^x or `exp`. Perhaps you have also seen this function in some computer program, where it is very often written either as `EXP( )` or `@EXP( )`. The latter terminology is often used in spreadsheet programs. If you are not very familiar with this function, it would be a good idea to take a few minutes to develop some comfort in carrying out calculations preferably using both a calculator and a computer program, such as a spreadsheet program. Here are some reference values for you to check your calculations against:

$$e^1 = \exp(1) = 2.7183$$

$$e^2 = \exp(2) = 7.3891$$

$$e^{-1.5} = \exp(-1.5) = 0.2231$$

There are a couple of other useful facts that would be good to call to your attention for the exponential function. The first is the value of the function when the input value is 0:

$$y = e^0 = 1.$$

This is from one of the basic laws of exponents in algebra which says that when you raise any positive number to the 0 power, you get 1. The other important fact has to do with the general behavior of this function for values of the input that are either very large positive numbers or very large negative numbers (meaning 'very negative'). In particular, suppose the input value is $x = 1,000$; the output value would be

$$y = e^{1000},$$

and if you either think about this logically or try to calculate it on your calculator, you will see that it is an extremely large positive number, probably too large to show up on your calculator. That is because this number would be the number e, or roughly 2.7, multiplied by itself a thousand times. Each time you multiply it, you are more than doubling its previous value, so by the time you multiply it together a thousand times, you would have doubled its value so many times that the result would be astronomical. Similarly when the input value is a very large negative number, such as $-1,000$, the corresponding output value would be

$$y = e^{-1000} = \frac{1}{e^{1000}}$$

where we have again used the fact that with a negative exponent, you can turn it into a positive exponent by putting the expression in the denominator. But now you can see that the result is a fraction whose numerator is one and whose denominator is astronomical. Obviously, this would be a fraction whose value would be very close to 0, and we might on occasion refer to it as being "negligibly small." To sum up, when e is raised to larger and larger powers, the results become astronomical very fast. Similarly, when e is raised to more and more negative numbers, the results approach zero very fast.

There is one last observation that we need to make in connection with the kinds of functions to be encountered in this chapter. Sometimes, just as in the above examples, they yield numerical results that are either very large or very small. For example, let us consider the value of the exponential function when the input value is 20:

$$y = e^{20} = 485165195.4$$
$$= 4.8517 \times 10^8$$
$$= 4.85 \times 10^8$$
$$= 4.85(E08)$$
$$= 4.85(08)$$

Here the answer has been written in several different forms. The first form gives the normal decimal representation of the number to as many digits as are available on the calculator. But for a number this large, almost 500 million, the last few decimal places are not likely to be very important, the emphasis usually being on the general order of magnitude of the number. Therefore it has become customary to use a different kind of notation for numbers that are very large and very small, as indicated in the subsequent parts of the above equation. These expressions all represent an approximate value of this number using what is called "scientific notation." Ignoring the rounding off of some of the decimal points for the moment, the objective is to write the number as a "normal sized number," generally with one place to the left of the decimal point, times the appropriate power of 10 to bring it up to its correct size. Remember that each time you multiply a number by 10, that has the effect of moving the decimal place one place to the right. Therefore, multiplying the number by 10^8 has the effect of moving the decimal place eight places to the right, and you can verify that that brings the numbers 4.8517 or 4.85 in line with the full decimal version written on the first line, although not with the same decimal precision given there. Because computer printouts are not generally set up to deal with writing exponents, the expressions shown on the last two lines of the above equation are standard abbreviations used by various computers and calculators to indicate the power of 10 by which a number should be multiplied to give its full value. Try to repeat this calculation on your own calculator and/or computer to see how the results would be indicated there. On many such machines, you have control over the form of the output, and so you should find out what you need to do in order to put output values in scientific form and to adjust the number of places shown to the right of the decimal point.

Similarly, when the value of the input is a negative number, the same kind of representation can be used, but you will find that the exponent of 10 involved will generally be negative. For example, let us calculate the exponential function when the input value is -15:

$$y = e^{-15} = 0.000000306$$
$$= 3.059 \times 10^{-7}$$
$$= 3.059(E - 07)$$
$$= 3.059(-07)$$

Here, of course, the negative exponent of 7 on the number 10 means that the decimal point needs to be moved seven places to the *left* to give the correct value. It is interesting to note here that the scientific notation actually allows for a more precise representation of the answer, because in displaying the number in fixed decimal format at the outset, the calculator needed to round off the last places. It did not need to save as much space by doing this in expressing the number in scientific notation. The actual number of decimal places represented when using scientific notation can vary and is usually something the user can select on the calculator or computer.

Exercises for Section 3.4

For Exercises 1–4, use your calculator or computer to determine sufficient values of the indicated function so that you can draw a reasonable sketch of its graph on the interval from -2 to 2. Pick a single common vertical scale for all the graphs that will be sufficient to cover enough of the corresponding y interval so that the differences among the graphs can be seen easily. (Feel free to use the computer to draw all the graphs as well if you know how.)

1. $y = e^x$

2. $y = e^{-x}$

3. $y = e^{x^2}$

4. $y = e^{-x^2}$

5. Carry out sufficient calculations to enable you to develop rough sketches of the graphs of $y = e^{-x^2}$ and $y = e^{-x^2/5}$ in a format that will enable you to compare them.

6. Can you use the laws of exponents from algebra to simplify the expression: $e^2 \times e^3$? Check any result by evaluating the old and the new expression with your calculator or computer.

7. Can you use the laws of exponents from algebra to simplify the expression: $e^2 + e^3$? Check any result by evaluating the old and the new expression with your calculator or computer.

3.5 One-Dimensional Diffusion

The top part of Figure 3-5 is reproduced as Figure 3-12. This is a one-dimensional diffusion situation because there is really only one dimension, which we shall call the x-dimension, in which the material can diffuse. (Of course, it could move in the negative x-direction or the

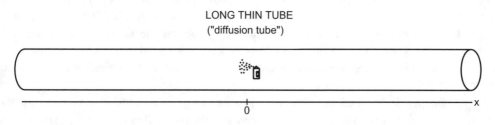

LONG THIN TUBE
("diffusion tube")

FIGURE 3-12
Basic one-dimensional diffusion situation

positive x-direction, but that is still a single linear dimension.) Here is the basic question about this situation. Suppose that we inject a total mass M of dispersible material into the tube, concentrated right at the point corresponding to $x = 0$. We will call the time at which we inject this material $t = 0$. What will the concentration of material *be at any point* in the tube *at any time* after the moment of initial injection?

We will assume throughout this discussion that the tube is "infinitely long," so that we do not have to take into account the effects on the concentrations that might result from its being blocked off at some point. Thus this would be a good approximation to cases where it is quite long by comparison with the distance scale of our problems, even if not actually infinite. Certainly for outdoor air pollution problems, there is indeed a large distance available for dispersion, and so this kind of assumption should apply to the cases we are most interested in.

You certainly know that as time goes on, the concentrated material at the center of the tube will gradually diffuse both to the left and to the right, so that the concentration at the center will gradually decrease and the concentration at points away from the center will gradually show some increase. As time goes on even further, and as the material diffuses out through a longer and longer length of the tube, the concentration at all points will gradually decrease to 0. This would all be a very reasonable and correct "qualitative" description of the process.

Here is a quantitative description of precisely what the concentration should be at any point in time and at any point along the tube:

$$C = \frac{M}{\sqrt{4\pi Dt}} e^{-\frac{x^2}{4Dt}}$$

This equation is known as the *one-dimensional diffusion equation.* Let us look at this equation and understand its individual parts. First, C, the left side of the equation, is a concentration. Since there is only one dimension for the material to diffuse in, the concentration will generally be expressed as units of mass per unit length, such as grams per centimeter or pounds per foot. You are probably much more familiar with thinking about the concentration of something in units of mass per unit volume, such as grams per cubic centimeter or pounds per cubic foot. So you want to be careful to adjust to these "linear" concentration units. If you say that C has the value 0.02 pounds per foot, you mean that in a one-foot length of the tube, there would be 0.02 pounds of the material of interest.

On the right side of the equation, you can see the exponential function that we have discussed above. But before getting to that function, let us look at the complex expression that has been placed in front of it and hence is multiplied by it. The number M is simply the mass of material that we instantaneously inject into the tube at the start of this hypothetical experiment. Thus its units would be grams or pounds or other units used to represent mass. The denominator contains the constants 4 and π, and two other quantities D and t. Obviously t refers to time, measured in whatever time units are appropriate for the particular problem you might encounter. Thus if you were interested in the concentration 10 minutes after the injection of the original material, you would substitute the value $t = 10$ or its equivalent in other units of time.

The quantity D is called the *diffusion constant* and gives an indication of how fast the diffusing material moves through the basic material in the tube, often called the *substrate*. For example, if the diffusing material is insect spray and the substrate is air, there would be a certain associated diffusion constant D that scientists would need to measure by laboratory experiments. If you used the same tube but filled it with a substrate such as water and used a diffusing material such as sugar, then the value of the diffusion constant D would generally be different and, in this case, would correspond to a slower rate of diffusion. So, in general, you would have to be given the diffusion constant in order to use the one-dimensional diffusion equation. One further observation about this equation is that the number 4 obviously could be taken out of the square root sign as its square root, 2, and the equation is sometimes written in this form. However, the expression is actually simpler to work with when all the quantities stay within the square root sign. With respect to the part of the equation involving the exponential function, note that the exponent portion simply requires the values of D and t, discussed above, as well as the value of x corresponding to the location along the tube at which you wish to calculate the concentration.

In principle, you should have no difficulty in applying the above diffusion equation. The only complication is that the calculations can become somewhat tedious. For this reason, if you have a computer available with either a spreadsheet program or some other kind of programming language with which you are familiar, it would turn out to be very useful in doing calculations involving this equation. Otherwise, every time you have a new value of t or x, you will need to redo any different calculations by hand, using your calculator for each step.

As an example of this equation, and to give you some typical values against which to test your own calculations, Figure 3-13 contains some representative values both in tabular and graphic form. From the first graph, which corresponds to a portion of the tube at a specific time t, you can see that the concentration of material has the general shape anticipated earlier. In fact, if you have ever heard of the "bell-shaped curve" often used in statistics, this is exactly what you have here. The diffused material spreads out so that at any point in time, its concentration follows the shape of this classic bell-shaped curve! The second graph in Figure 3-13 shows how the concentration at a fixed location will generally increase initially from zero, and then eventually peak and start to fade away over the long term.

Although at first the one-dimensional diffusion equation might look somewhat complicated, it is actually quite easy to use as long as you are given the required information. Some of the problems below ask you to investigate some properties of this equation further, and it is recommended that you try these problems before proceeding to later sections.

You might also be asking about where this equation comes from. It certainly is not obvious that the distribution of material in a one-dimension diffusion situation should follow this exact equation. It turns out that while the equation is relatively simple to use, it is much more complicated to see why it holds. For readers with a more advanced mathematical background, we will return to this question in Chapter 6; but the fact that you may not know how to derive an equation should be no obstacle to your going ahead and using it to calculate important information about environmental situations. (After all, you have probably used the equation $A = \pi r^2$ many times to find the area of a circle, but it is doubtful that you have ever seen a derivation of this common equation.)

Exercises for Section 3.5

The following problems all pertain to this situation: Consider the one-dimensional diffusion problem as discussed previously in terms of a long tube. Assume in this case that a solid is being released into a liquid that fills the tube, and that the solid is diffusing. Let $x = 0$

	1	10	20	30	40	50	75	100
-5	3.21E-27	0.002723	0.043821	0.101397	0.147826	0.180722	0.223832	0.238743
-4	1.89E-17	0.025834	0.134977	0.214657	0.259442	0.283429	0.302142	0.298984
-3	7.55E-10	0.148663	0.323794	0.384666	0.401832	0.402205	0.381545	0.356163
-2	0.000202	0.518884	0.604927	0.583498	0.549239	0.516442	0.450743	0.403586
-1	0.366125	1.098478	0.880163	0.749227	0.662509	0.600019	0.498148	0.435018
0	4.46031	1.410474	0.997356	0.814338	0.705237	0.630783	0.515032	0.446031
1	0.366125	1.098478	0.880163	0.749227	0.662509	0.600019	0.498148	0.435018
2	0.000202	0.518884	0.604927	0.583498	0.549239	0.516442	0.450743	0.403586
3	7.55E-10	0.148663	0.323794	0.384666	0.401832	0.402205	0.381545	0.356163
4	1.89E-17	0.025834	0.134977	0.214657	0.259442	0.283429	0.302142	0.298984
5	3.21E-27	0.002723	0.043821	0.101397	0.147826	0.180722	0.223832	0.238743

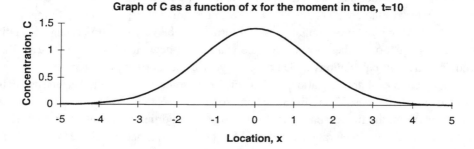

Graph of C as a function of x for the moment in time, t=10

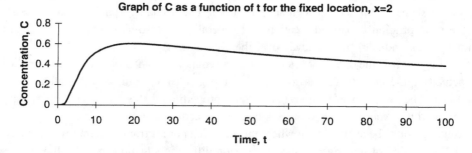

Graph of C as a function of t for the fixed location, x=2

FIGURE 3-13

Representative tabular and graphic values illustrating the use of the one-dimensional diffusion equation

correspond to the center of the tube and as usual use negative values to the left and positive values to the right. If a mass of 10 grams is released at the center of this tube, we are interested in knowing the concentration distribution along the tube as time increases. Assume that for the materials involved (the solid and the liquid), the diffusion constant $D = .05$ cm^2/sec.

1. Calculate the concentration at all of these locations: $x = -2, -1.5, -1.0, 1/4, +2$ at each of the following time values: $t = 1, 10, 100, 1000$. (Hint: it is strongly recommended that you use a spreadsheet program or other computer program to carry out these calculations easily. Furthermore, test a few of your calculations with a calculator to make sure you did not make a programming error.)

2. Graph these results, preferably by computer, in the form of curves showing concentration as a function of location, a different curve for each time value. Try to get all the graphs on one set of axes.

3. What are the correct units for the concentrations that you have calculated?

4. Explain in your own words why the graphs look reasonable (assuming that they do).

5. This problem makes reference to your work on the previous problems. For each of the following situations, you are asked how the original answer to the diffusion situation would be modified if certain input values were changed. You must provide three distinct lines of reasoning to support your answer: first, an explanation based on your intuitive understanding of the physical diffusion process; second, an explanation based on the actual diffusion equation; third, an explanation based on recalculating your spreadsheet or program under the modified condition(s).

 a) What should happen to all the entries if the mass M you start with is increased by a factor of two?

 b) What should happen to the entry for $x = 0$, $t = 100$, if D is decreased by a factor of two?

 c) What should happen to the entry for $x = 1$, $t = 100$, if D is increased by a factor of 10 (i.e., an order of magnitude) to a new value of 0.5? Will the same direction of change (i.e., increase or decrease) *always* be experienced at every x value (except perhaps $x = 0$) and at every t value (extending beyond those specific values in your table) whenever D is increased? Discuss why or why not.

6. The diffusion coefficient D shows up in two distinct places in the one-dimensional diffusion equation. The point of this question is for you to investigate what happens to the concentration value at a fixed location x and at a fixed time t as the value of the diffusion coefficient varies. Notice, for example, that as D gets small, its occurrence in the denominator of the first factor would tend to increase the size of the concentration. However, at the same time, when D gets small, it causes the exponent in the exponential term to become a very large negative number, thereby making that exponential factor get smaller. Thus these two effects tend to work against each other. Similarly, as D gets large, the opposite occurs. So the question is really how do these factors balance out. You may answer this question by using clear physical explanations and/or numerical experiments, or by using calculus (if you have studied this subject). Be precise in summarizing your conclusions.

3.6 Two-Dimensional Diffusion

In this section we consider the standard two-dimensional diffusion situation, for which purpose the top part of Figure 3-4 has been reproduced in Figure 3-14. Once again, we would like to be able to calculate the concentration of diffusing material at any point in time after its initial injection and at any point in the two-dimensional region through which it is diffusing. In order to characterize locations within this two-dimensional region, obviously we need both an x value and a y value, so we should expect both an x and a y value to be required as input values for our calculation.

Without further ado, here is the way the answer turns out:

$$C = \frac{M}{4\pi t \sqrt{D_1 D_2}} e^{\left(-\frac{x^2}{4D_1 t} - \frac{y^2}{4D_2 t}\right)}$$

This equation is called the *two-dimensional diffusion equation.* As expected, its required input values include the injected mass M, the time t at which you wish to calculate the concentration, and the x and the y values of the point at which you want the concentration. But in this case there are two diffusion constants, D_1 and D_2, which may be the same or may be different, but in any case which correspond to the rate at which the diffused material is expected to diffuse through the substrate in both the x-direction (D_1) and the y-direction (D_2).

Why should we allow for a different rate of diffusion in the two different directions? After all, if insect spray is sprayed into a room, one expects it to diffuse into the air at the same rate in all directions. To answer this, let us recall what constitutes a diffusion process. The diffusion process, as discussed earlier, is one in which the diffusing material moves through a substrate at a rate that is proportional to the concentration gradient or the change in concentration from one point to another. One of the physical processes cited earlier as leading to such diffusion was that of the kinetic energy of individual molecules, which tends to shake them up in a

CORNER VIEW OF ESSENTIALLY 2-D DIFFUSION EXPERIMENT

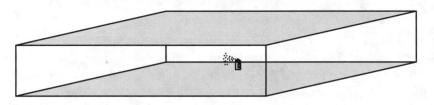

PLAN VIEW (I.E., TOP VIEW) OF SAME EXPERIMENT

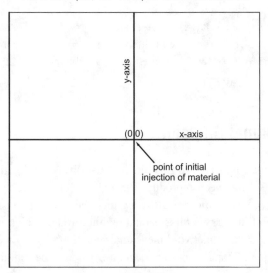

FIGURE 3-14

Basic two-dimensional diffusion situation

random way so they tend to mix with each other. This leads to the lowering of a concentration in the high concentration areas and a raising of the concentration in low concentration areas with a net transfer of material from one area to another at a rate proportional to the difference in concentration.

Molecular action is not the only process that can drive this kind of material transfer. For example, if the substrate itself is being agitated in a certain way, this could also accelerate mixing and shuffling of the diffusing material. If you look once again at Figure 3-14 and you imagine that the substrate is being agitated in the y-direction but not in the x-direction, then you could well imagine that material would be diffusing more rapidly in the y-direction than in the x-direction. That is because the shuffling of the molecules and the spreading out of the concentration in the y-direction would be driven by two processes: both molecular kinetic energy as well as the mixing of the substrate. Since this would lead to a higher rate of diffusion in the y-direction than in the x-direction, clearly the diffusion constants for those two directions would need to be different.

If you were successful in dealing with the problems involving one-dimensional diffusion, then it should not be much more difficult to deal with those involving two-dimensional diffusion. It is recommended that you try some of these problems before proceeding further in this chapter. After all, you want to be extremely comfortable with the one- and two-dimensional diffusion equations before proceeding to the very complex equation that you saw earlier for modeling an airborne plume (but a quick second look at this equation will begin to reveal many similarities with the equations that have just been discussed).

Exercises for Section 3.6

The first four exercises below all pertain to the following situation: Suppose that 800 kg of soluble material are spilled into the center of a large shallow lake and gradually diffuse out into the lake. Because of the wind and current patterns, the diffusion constants are different in the east/west and north/south direction. The east/west diffusion constant is .3 m²/min, and the north/south diffusion constant is .9 m²/min. Use a spreadsheet or similar aid for the calculations. Let the x-direction be east/west and the y be north/south, and assume the spill is at the origin. (Be sure to check a few answers with your calculator in case of programming errors, and be careful in using relative and absolute cell addresses if you use a spreadsheet program.)

1. Find the concentration at the point $x = 15$ meters, $y = 20$ meters, and time $t = 10$ minutes. Be sure to label your answer with the correct units.

2. Consider the point $x = 2$ meters, $y = 2$ meters. Calculate and graph the concentrations at this point for $t = 10, 20, 30, \ldots, 1000$ minutes.

3. Do the resulting values from Exercise 2 look like what you would expect from intuitive reasoning? Explain your answer.

4. Demonstrate by means of some concentration calculations of your own choosing and/or the corresponding graphs that in the situation under discussion, there really is a different rate of material distribution by diffusion in the x- and y-directions.

5. Use techniques from basic algebra, including one of the laws of exponents, to rewrite the two-dimensional diffusion equation as a product of three factors. The first factor should be the total mass M, and the second two factors should correspond to terms that involve diffusion individually in the x- and y-directions, respectively. Furthermore, the second two factors should also have an identical form, except for whether their input values correspond to the x- or the y-direction.

6.   Using the new version of the two-dimensional diffusion equation you developed in connection with the previous problem, determine for which values of x and y the second two factors have their largest possible values. What does this location correspond to physically, and why does this mathematical observation correspond to an intuitively obvious physical observation?

3.7 The Basic Plume Model

Consider the two distinct types of smokestack releases illustrated in Figure 3-15. The top part of the figure illustrates a type of release known as a "puff." This is essentially a momentary release of exhaust gases in the form of a discrete cloud that, at the effective height H (discussed earlier), moves horizontally with the prevailing wind. It should be obvious from looking at the sketch that the exhaust material in the puff is free to diffuse or disperse in three distinct directions: the two horizontal directions (i.e., the x-direction and the y-direction) and the vertical or z-direction. Thus the release of such a puff is a true three-dimensional diffusion situation (at the same time that it has the added complication that the whole expanding cloud is moving along with the wind).

But now consider the plume release shown in the bottom part of Figure 3-15. In this case there is a steady release of exhaust gases, and hence a continuous plume begins at the stack and continues for an indefinite distance in the downwind direction, where its concentration gradually decreases. At first glance it might appear as though the plume represents an even more complicated situation than the puff. However, to quite a high level of accuracy, this is not so, for reasons that will be described now.

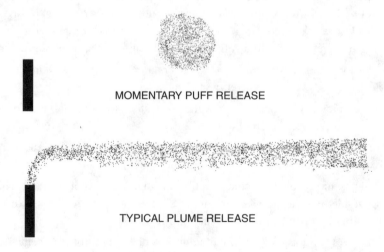

MOMENTARY PUFF RELEASE

TYPICAL PLUME RELEASE

FIGURE 3-15
The distinction between "puffs" and "plumes"

In the plume situation, there are really only two directions in which a significant amount of dispersion can take place. These include the vertical (z) direction and the horizontal direction that is *perpendicular* to the axis of the plume, generally denoted as the y-direction. (In Figure 3-15, the y-direction would be perpendicular to the plane of the figure.) There will be very little dispersion in the x-direction, along the axis of a plume, for the simple reason that there is only a very slight and gradual change in concentration as you move through the plume along this axis; and since a diffusion process is driven by a concentration gradient, this small concentration gradient implies a small amount of dispersion in this direction. To be sure, over long horizontal distances, certainly the concentration in the axial direction does decrease from a high level near the stack to a negligible level far from the source, but still this is a very gradual decrease and so the gradient, or the slope in the concentration function, in this x-direction will be very small. We will make use of this assumption in formulating our mathematical model, the result being that the model should look more like a two-dimensional diffusion situation than a three-dimensional one.

Here is the actual equation typically used in many air pollution calculations for determining the concentration of material in a plume of the type that has just been described:

$$C = \frac{Q}{2\pi\sigma_y\sigma_z u}\left[e^{-\frac{y^2}{2\sigma_y^2}}\right]\left[e^{-\frac{(z-H)^2}{2\sigma_z^2}} + e^{-\frac{(z+H)^2}{2\sigma_z^2}}\right].$$

This is called the *Gaussian plume model.* You can see some resemblance to the previous diffusion equations, but there are some additional variables as well as a sum of two exponential terms in the bracketed portion at the far right. Let us begin by describing what the various variables stand for:

Q The source term, representing the amount of mass of pollutants emanating from the stack per unit time. For example, this might be expressed in terms of grams per second, pounds per hour, or tons per day.

u The wind velocity in the prevailing direction, and specifically representing an average wind velocity over the kind of time periods you are interested in modeling. For example, if you are interested in calculating average one-hour concentrations at a given receptor location, you would also want to use a wind velocity value that represents an hourly average, rather than an instantaneous value.

y The horizontal coordinate measured in a direction perpendicular to the axis of the wind movement.

σ_y A "dispersion coefficient" corresponding to the amount of dispersion that should be expected in the y-direction under the given meteorological conditions. It is analogous to the diffusion constant discussed previously. This dispersion coefficient depends on the downwind distance x, so it is a function, not a constant. In fact, since the plume spreads out more as you move downwind, you would expect σ_y to increase with larger values of x. (Furthermore, just as the wind-speed values are based on a measurement time scale, the dispersion coefficients also have an implicit time scale, but this is a subject we need not discuss in detail at this point.)

z The elevation above the ground, measured at the location of the stack. Therefore, $z - H$, as occurs in the first exponential term involving z, represents the height above the effective top of the stack. In other words, since we want to use z in the usual sense as elevation above the ground, we have to use this modification, $z - H$, to convert our coordinate system to the effective top of the stack, which is on the plume centerline.

σ_z The dispersion coefficient in the vertical or z-direction, and again, this is analogous to the previous use of a diffusion constant. As with σ_y, this dispersion coefficient σ_z is also an increasing function of x.

H The effective stack height, meaning the original physical height of the stack, h, augmented by any additional portion intended to correspond to the extra distance it takes for the exiting exhaust gases to dissipate their thermal and momentum effects and thus begin to migrate horizontally with the prevailing wind.

You will doubtless need to refer to the above list of variables from time to time before you become completely comfortable with the plume equation. This is a very complicated equation, and when you first look at it, it can seem quite overwhelming. However, it turns out that the underlying ideas behind the equation are quite simple, and we will discuss them further below.

Using basic rules from algebra, the plume equation has been rewritten below in a form that is intended to sort out its various components:

$$
C = \underset{\substack{\text{source} \\ \text{strength}}}{\frac{Q}{u}} \times \underset{\substack{\text{diffusion effect} \\ \text{in } y \text{ direction}}}{\frac{1}{\sqrt{2\pi}\sigma_y} \left[e^{-\frac{y^2}{2\sigma_y^2}} \right]} \times \underset{\substack{\text{diffusion effect} \\ \text{in } z \text{ direction}}}{\frac{1}{\sqrt{2\pi}\sigma_z} \left[e^{-\frac{(z-H)^2}{2\sigma_z^2}} + e^{-\frac{(z+H)^2}{2\sigma_z^2}} \right]}
$$

In this form, the concentration of pollutant C is shown as a product of three quantities. The first of these quantities, Q/u, is a measure of the strength of the pollutant at the source itself. It takes into account two factors: the rate Q at which the pollutant is being injected into the atmosphere and the wind speed u, because, for example, if the wind speed is higher, then the amount of pollutant being released in a given amount of time is going to be diluted by a larger quantity of air blowing past the stack in that amount of time, and the hence the actual concentration at the initial point will be less. This is why the "source strength" must take into account both the source term Q and the wind speed u.

The second main factor in the expression for the concentration represents the diffusion effect in the y-direction. This looks quite similar to the one-dimensional diffusion equation or the first part of the two-dimensional diffusion equation. (The only really new component that has not been discussed in any detail yet is the diffusion coefficient σ_y.)

The last factor in the concentration equation involves components related to the elevation z. This would look almost the same as the y term except for the fact that there are two different exponential terms added together in this new case. Without going into complex mathematical details at this point, it will suffice to say that the second exponential term is an error correction factor intended to take into account the fact that vertical diffusion in the downward direction is blocked once the material reaches ground level, so material is not quite so free to disperse in the downward direction as it is in the upward direction. This has a relatively small effect on concentrations resulting from fairly tall stacks, but can be significant when the point of release of pollutants is lower. (For readers with further interest, this aspect will be discussed in greater detail in Chapter 6.)

There are three very natural questions that you might be asking yourself as you try to see if you understand the previous equation. These are discussed below.

1. Why doesn't x show up in this plume equation? After all, shouldn't the concentration of pollutant depend on the downwind distance x from the source of release?

It is correct that x does not show up explicitly anywhere in the above plume equations. However, x does actually show up in these equations in a hidden way, namely, through the dispersion coefficients σ_y and σ_z. All other factors being equal, for higher values of x (i.e., for farther distances from the original source), the appropriate dispersion coefficients σ_y and σ_z will be larger, corresponding to a higher level of dispersion by the time the material gets to the point corresponding to this x value. See below for further information.

2. How do you calculate the dispersion coefficients σ_y and σ_z?

Meteorologists have made extensive studies of the extent to which pollutants disperse in the atmosphere under a variety of meteorological conditions, characterized by both the atmospheric stability class as discussed earlier as well as the wind speed itself. (The wind speed does certainly affect the atmospheric stability class, but it also has other effects on the rate at which pollutants are dispersed.) These studies, some carried out theoretically using additional mathematical models and others carried out using experiments and measurements, have led to a set of well-accepted values of such dispersion coefficients as a function of both downwind distance, x, and atmospheric stability class. These are shown in Figures 3-16 and 3-17. For a given numerical calculation, you would need to know the downwind distance at which you wanted to calculate the concentration, as well as the atmospheric stability class. Using this downwind distance and going to the curve for the correct stability class (which resembles a straight line for certain ranges), you can then read over at the left side what the corresponding value of the dispersion coefficient is.

Figures 3-16 and 3-17 are specified in metric units, but you can convert to any units that might be appropriate for your own calculations. For example, if your distance unit is feet for a given problem, you would have to do two conversions. First, you would have to convert your x value to meters to apply these figures, and then you would have to convert the resulting σ_y and σ_z back to feet from the meter value you would estimate off the graphs. In applying these figures, you must also be careful about the logarithmic scales in which they are given, each small unlabeled subdivision representing successive factors of $2, 3, 4, \dots$ by which the base number is to be *multiplied*. For example, the subdivisions from 100 to 1,000 would stand for the values 200, 300, 400, etc. Thus, if you look carefully at the intervals, a point halfway from 100 to 1,000 on this scale would stand for some number between 300 and 400. (There is some further investigation of this in the exercises because logarithmic scales are so common in practice, and a numerical example will be introduced shortly.)

3. Where does the plume equation come from? How do you derive it?

The mathematics leading to the derivation of the plume equation is somewhat complicated, involving, for example, partial differential equations. Readers who are interested in pursuing this aspect will find further discussion of it in Chapter 6. However, the emphasis in this section

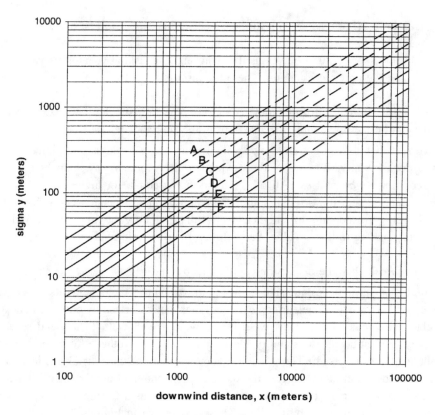

Adapted from D. B. Turner, *Workbook of Atmospheric Dispersion Estimates.*
Based on open-country conditions and ten-minute sampling times.
Dashed portions correspond to greater variability or uncertainty.

FIGURE 3-16

Horizontal dispersion coefficient σ_y as a function of downwind distance, for each stability class A through F

is on gaining an operational familiarity with the equation and understanding in a qualitative way the nature of its various components.

To review your understanding of the situation addressed by the above plume equation, it may be useful to consult Figure 3-18. This figure shows the spreading of the plume (represented conceptually in cross-section by an ellipse) as well as the gradual decrease in concentration as you move away from the plume center line, suggested by the bell-shaped curves in the horizontal and vertical directions on the figure. It also shows the distinction between the stack height h and the effective stack height H.

Now let us see what is involved in applying our basic plume equation to an applied problem. Consider the following example.

You live two miles due east of a coal-fired utility power plant that produces electricity for your city. The stack on the plant is 350 feet high, and the ground is level. On a given day, the sun is shining brightly and the wind is blowing from southwest to northeast at 10 mph. Measurements at the plant stack of the concentration of nitrogen oxides in the exhaust gas show that such pollutants are being released at the rate of

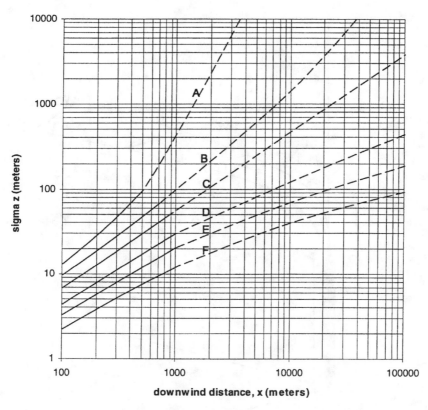

Adapted from D. B. Turner, *Workbook of Atmospheric Dispersion Estimates*.
Based on open-country conditions and ten-minute sampling times.
Dashed portions correspond to greater variability or uncertainty.

FIGURE 3-17
Vertical dispersion coefficient σ_z as a function of downwind distance x, for each stability class A through F

80 pounds per minute. What would you expect the concentration of nitrogen oxides to be at your residence?

Figure 3-19 shows the general layout described in this problem. Since the x-direction always corresponds to the direction in which the wind is blowing, and since the y-direction, the other horizontal direction, must always be perpendicular to the x-direction, you need to calculate the x and y coordinates using some simple geometry involving right triangles. In particular, in this case, the direction of the wind makes a $45°$ angle with a line from the power plant to your house, and this line from the power plant to your house is the hypotenuse of a right triangle, as shown in the figure. Using the fact that you know the hypotenuse (2 miles), and you also know the angles of this right triangle, you should be able to determine its other two sides. From these it should be simple to find the corresponding x and the y coordinates of the residence. See if you can verify the calculations represented by the figure.

With these preliminaries out of the way, let's make a list of each of the quantities you would need to have in order to apply the plume equation to this problem. We will work off the original first version of the plume equation because it is in a more consolidated form and

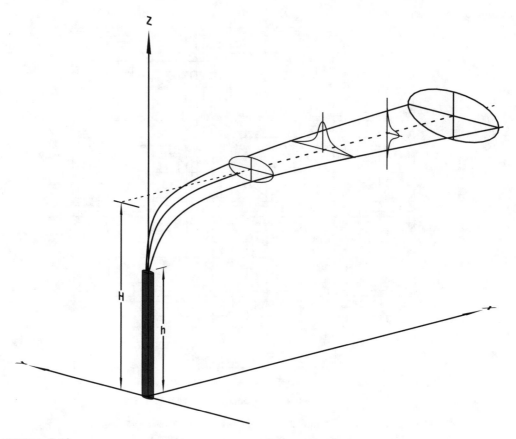

FIGURE 3-18
Three-dimensional representation of plume coordinate system and dispersion effects in horizontal and vertical directions

hence is easier to set up on a calculator or computer for numerical calculation. As a reminder, here is the equation again:

$$C = \frac{Q}{2\pi\sigma_y\sigma_z u} \left[e^{-\frac{y^2}{2\sigma_y^2}} \right] \left[e^{-\frac{(z-H)^2}{2\sigma_z^2}} + e^{-\frac{(z+H)^2}{2\sigma_z^2}} \right]$$

The determination of input values proceeds as follows:

Q You are given this value as 80 pounds per minute.

u The average wind speed is specified in the problem as 10 mph.

y The y-value has been calculated in Figure 3-19 using the geometric description of the problem.

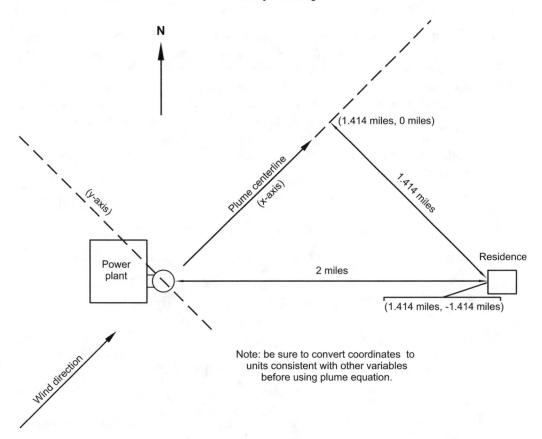

FIGURE 3-19

Geometric analysis of coordinates and distances corresponding to power plant problem described in text

σ_y You can figure this out by combining the downwind distance, or x-value, as calculated in Figure 3-19, with the atmospheric stability class corresponding to the description provided in the problem, by using Figure 3-16. In particular, from Table 3-1 we see that the stability class is B. The x-value converts to 7465.9 ft and then to 2275.6 meters. (You might ask why work in feet at all. Well, concentration in units of mass per cubic foot is quite a bit more meaningful than concentration in units of mass per cubic mile! Alternatively, you could work the problem in metric units.) For this x-value, Figure 3-16 indicates that σ_y is about 300 meters, which converts back to about 984 ft.

z Remember that the z-value is the elevation or the height above the ground at which you wish to calculate the concentration of the pollutant. Since your residence is (presumably) at ground level, the z-value that you would want to use to obtain the solution would be $z = 0$.

σ_z This is similar to the calculation of σ_y, except you would use the graphs on Figure 3-17. The resulting value of σ_z is approximately 230 meters or 755 ft.

H This is the last input quantity that you need in order to plug everything into the plume equation and calculate the concentration of interest. Unfortunately, all you are given in the problem is the value of h, the actual physical stack height. Since the actual calculation of H can be quite complex, *we will adopt the conservative approach of using the actual physical stack height h as the effective stack height H*. This is a conservative assumption because it effectively places the plume at a somewhat lower elevation, which would tend to overestimate the concentration of pollutants at receptors on the ground. (This issue is investigated further in problems 9 and 10.)

Putting the above information together, we obtain the concentration as follows[*]:

$$C = \frac{Q}{2\pi\sigma_y\sigma_z u}\left[e^{-\frac{y^2}{2\sigma_y^2}}\right]\left[e^{-\frac{(z-H)^2}{2\sigma_z^2}} + e^{-\frac{(z+H)^2}{2\sigma_z^2}}\right]$$

$$= \frac{80 \text{ lb/min}}{2\pi(984 \text{ ft})(755 \text{ ft})\left(\dfrac{10 \text{ miles}}{1 \text{ hr}} \times \dfrac{5280 \text{ ft}}{1 \text{ mile}} \times \dfrac{1 \text{ hr}}{60 \text{ min}}\right)} \times$$

$$\left[e^{-\frac{(-7465.9)^2}{2\times984^2}}\right]\left[e^{-\frac{(0-350)^2}{2\times755^2}} + e^{-\frac{(0+350)^2}{2\times755^2}}\right]$$

$$= 1.948 \times 10^{-8} \text{ lb/ft}^3 \times [3.158 \times 10^{-13}] \times [1.8]$$

$$= 1.1 \times 10^{-20} \text{ lb/ft}^3$$

Of course, now the question is: Is this concentration a lot or a little? The large negative exponent on the 10 at least suggests that it is very little, and in fact this is true. But to be sure, one needs to have some scale of comparison. Unfortunately, the standard units for such a scale, at least for nitrogen oxides, are not in the form of lb/ft^3. Rather, they are in the form of so-called "parts per million" or ppm.[†] For the time being, simply take as a given the fact that 1 ppm of NO_x corresponds to about 1.1×10^{-7} lb/ft^3, so that our above result, when converted to ppm units, is:

$$1.1 \times 10^{-20} \text{ lb/ft}^3 \times \frac{1 \text{ ppm}}{1.1 \times 10^{-7} \text{ lb/ft}^3} = 1 \times 10^{-13} \text{ ppm}$$

Concentration levels of concern for various pollutants often fall into a range within two or three orders of magnitude (i.e., factors of 10) of 1 ppm; and levels below 1 part per *billion* (10^{-3} ppm) are practically undetectable for most substances in standard analytical tests. Thus a value of 10^{-13} ppm is certainly of no concern. It is "off scale" by about 10 orders of magnitude.

---

[*] Note: no attempt will be made in these and other calculations to treat decimal places and significant digits systematically. The author believes that it is more important to give readers sufficient decimal values in the calculations to enable them to be sure that they are correctly following the calculations. In addition, calculations of this type almost always involve many assumptions and approximations, and so the answers are likely to be rounded or treated as rough estimates in the end.

[†] Units of ppm were introduced in Chapter 2 in the context of ground-water contamination; but, as mentioned there, the measurement in that case is by mass or weight, whereas for air dispersion, the measurement is by volume or numbers of molecules. The general unit conversion techniques used previously apply just as well, except that the conversion factors would have to relate to volume measurements. There is further discussion of this in the next chapter in Section 4.3.3.

That completes our example of how to apply the Gaussian plume equation to a numerical problem. Further examples will be given in the exercises. The exercises also invite you to use the plume equation to conduct a number of numerical experiments in order to answer questions about where the highest ground-level concentrations might occur and what kinds of meteorological conditions might be most undesirable in terms of such high ground-level concentrations.

Exercises for Section 3.7

Note: It would be very useful in carrying out the plume calculations in these exercises for you to program the Gaussian plume equation into a spreadsheet program, another computer program, or a programmable calculator.* This will eliminate the tedium of extensive hand calculation and will let you focus on using numerical experiments conveniently to answer some of the questions asked. If you are developing your own program and would like an analytical way to represent the curves given in Figures 3-16 and 3-17, see Exercise 11 below prior to undertaking the earlier exercises, or, if it is within your capability, use a curve-fitting method to set up your own approximation to the curves in Figures 3-16 and 3-17.

1.   Repeat the power plant example carried out in the text, using your spreadsheet or alternative program. (Hint: be especially careful to convert units when necessary.)

2.   For the power plant example considered in the text and in the previous problem, answer the following additional questions:
     a)   Would you say that the very low concentration value that was calculated is due primarily to dispersion considerations in the y-direction, dispersion considerations in the z-direction, or dispersion considerations in the x-direction?
     b)   Changing your focus now to concentrations right along the axis of the wind passing over the stack, determine the limits (in this direction) of the zone that experiences a ground-level concentration of at least 1 ppm. Explain whether the results seem reasonable.

3.   A coal-fired power plant in Ohio emits sulfur dioxide at the rate of 80 grams per second from a tall stack 120 meters above the ground. The average wind speed is 8 meters/sec, although the direction is variable. Consider the pollutant exposure of an individual living 1,000 meters from the plant. Assume that the individual's residence has an elevation of 35 meters above the ground level at the stack. Based on meteorological data for this site, you may use a stability class of C and assume that the wind blows in the direction of the residence 25% of the time.
     a)   During periods when the wind is blowing directly towards the individual's residence, calculate the ground-level concentration of this pollutant at the residence.
     b)   What are the correct units for the concentration you have calculated?
     c)   Given that the wind is blowing in this particular direction, does this location experience the maximum ground-level airborne concentration? Please explain your answer and illustrate it, if necessary, with additional calculations.

---

* In connection with use of this text, you may have received or have access to a special spreadsheet which already contains a program for the Gaussian plume model and that automatically incorporates the dispersion factors that you would otherwise have to calculate manually from Figures 3-16 and 3-17. If you elect to use this program, be sure to read the instructions on the top few lines.

d)   Suppose that when the wind is not blowing directly towards the residence, the pollutant concentration there decreases to a negligible level. What then would be the long-term average concentration at the residence, taking into account periods when the wind is blowing towards it and when it is not?

4.   A chemical plant emits ammonia into the air at the rate of 10 milligrams/sec from an open waste lagoon, which can be considered as a ground-level release point. The average wind speed at the site is 4 meters/sec. Assume that conditions at this site are generally quite overcast and that the wind over the lagoon blows directly at a school, located 2,000 meters away, 20% of the time (and that the rest of the time the wind has no impact on this receptor location).

a)   What stability class best describes the site conditions for this area? Please explain.

b)   Calculate the long-term average ground-level concentration of ammonia in the air experienced at the location of the school.

5.   A nuclear power plant emits radioactive krypton at the rate of 350 microcuries/sec from a tall stack 175 meters above the ground. The average wind speed is 10 meters/sec, and the wind generally blows from the west, although there is some variation in the direction along an arc ranging from southwest to northwest. To provide a simplified description, a meteorologist reviews the historical wind data from the site and summarizes it by saying that 60% of the time it comes from due west, 20% of the time it comes from 30° north of due west, and 20% of the time it comes from 30° south of due west. The meteorologist also suggests that a stability class of D would be most representative of this site's conditions.

a)   Calculate the long-term average ground-level air concentration of krypton experienced by an individual living 1,500 meters due east of the stack. The individual's residence is on a slight rise at an elevation of 20 meters above the ground level at the plant itself.

b)   In the calculation for part a, how would you describe the relative amount of radiation received at the residence from times when the wind is blowing from 30° north or south of due west, as compared to that when the wind is blowing directly from due west?

c)   Restricting your attention to the case when the wind is blowing directly from the west, find the downwind distance at which the concentration (at the same elevation as the residence) is largest. Explain what makes this location compare the way it does with the location of the residence.

6.   Analyze this situation and any assumptions made and decide whether the approach taken is logically valid. Give a *physically based intuitive explanation* for your answer, and then verify this with one or more *representative numerical calculations* using your spreadsheet or other program.

You are applying for a permit to build a wood-fired boiler to generate heat and electricity for an industrial complex in central Maine, so you want to do some calculations to see how high the off-site pollutant concentrations might be, in particular at points of maximum off-site concentration, which will control the regulatory acceptability of the plant. The two predominant atmospheric stability classes experienced at your site are C and D. You limited your calculations for airborne concentrations of fine particulate matter to a variety of weather conditions within the D stability class because you believed that this was a "conservative approach" and would yield the highest concentrations, since a higher stability class means that the plume stays together longer and transports material farther from the source in higher concentrations.

7. Analyze this situation and any assumptions made and decide whether the approach taken is logically valid. Give a *physically based intuitive explanation* for your answer, and then verify this with one or more *representative numerical calculations* using your spreadsheet or other program.

If a smokestack is emitting sulfur dioxide under a steady north wind of 6 mph, call the point of maximum ground-level concentration A and assume it is 1.3 miles downwind from the stack. Now suppose that the wind increases to 12 mph but that other atmospheric conditions remain the same. Since the same plume is just being moved along twice as fast, you assume that the maximum ground-level concentration will now be at a point B 2.6 miles downwind from the stack. (Two important cautions: First, be careful with your units. Second, be sure that any numerical example you give really satisfies every single aspect of the problem as described. For conditions or parameters not specified, you are free to choose your own values to illustrate your points as long as your values are reasonable.)

8. You are applying for a permit to build a hazardous waste incinerator in your community. You already own your site, and you are converting an old power plant on that site, so the height and location of the stack are already fixed. In particular, the height of the stack is 120 feet, and it is located in the center of your roughly circular piece of property, placing it 900 feet from the property fence-line. The surrounding topography is approximately level.

Your principal pollutants of concern are "polycyclic aromatic hydrocarbons" (PAHs), which can be produced from incomplete combustion of organics. (PAHs can cause cancer.) These are often a component of the smoke or soot that can be produced during periods when something goes wrong with the combustion system.

a) If your most common daytime weather condition at the site is strong sunlight and a wind of 12 mph from the west, where would the point of maximum ground-level concentration be?

b) What would be the *worst case* weather conditions for *off-site* ground-level concentration? You may assume that average wind speeds less than 4 mph are too rare to require consideration.

9. Based on the experience gained in the previous problems, write a concise summary of the relationship between atmospheric stability class, wind speed, stack height, point of maximum ground-level concentration, and actual concentration level at that point of maximum concentration.

10. One of the models that has historically been used to calculate the difference between physical stack height and effective stack height is called *Holland's equation,* which is given by

$$\Delta H = \frac{v_s d}{u} \left(1.5 + 2.68 \times 10^{-3} \times p \times \frac{T_s - T_a}{T_s} \times d \right)$$

where

ΔH = the amount by which the effective stack height
exceeds the physical stack height (meters)

v_s = stack gas exit velocity (meters per second)

d = inside stack diameter (meters)

u = wind speed (meters per second)

p = atmospheric pressure (millibars)

$$T_s = \text{stack gas temperature (degrees Kelvin)}$$
$$T_a = \text{air temperature (degrees Kelvin)}$$

and 2.68×10^{-3} is a constant having units of $1/(\text{millibars} \times \text{meters})$. (There is actually a much more widely used formula due to Briggs, but it would introduce some additional concepts that have little value for our specific purposes.)

You can see how complicated this calculation can be and that it introduces some additional physical parameters with which you are probably unfamiliar. Nevertheless, it will be useful to carry out one numerical example involving this equation, using realistic input values, to see how the effective stack height can vary under different conditions. In particular, a proposed source is to emit 72 grams per second of sulfur dioxide from a stack 30 meters high with an inside diameter of 1.5 meters. The effluent gases are emitted at a temperature of 250°F (394° Kelvin) with an exit velocity of 13 meters per second. The atmospheric pressure is 970 millibars, and the ambient air temperature is 68°F (293° Kelvin). Make up a table and graph to show the calculated difference between the physical stack height h and the effective stack height H for a range of wind speeds from 1 meter per second up to 20 meters per second.

11. This is based on the results you have obtained in the previous problem. Using your physical intuition, explain why the trend shown in the calculated results in the previous problem is very reasonable. Use this same reasoning to describe under what general kinds of conditions you would expect the use of physical stack height instead of effective stack height in the Gaussian plume equation to lead to perhaps overly conservative calculational results.

12. With further reference to the model in Exercise 10, consider the following three input parameters: v_s, d, and T_a. From looking at the equation, what would be the effect on ΔH of increases in each of these three parameters, considered individually? For each case, explain on physical grounds why the direction of increase is what you would expect.

13. The use of the curves shown in Figures 3-16 and 3-17 would be much easier if you had an analytical representation for them (i.e., actual equations for the curves.) A number of different approximations, as well as refinements in the original curves for certain kinds of applications, have been developed over the years. One simple approximation is based on generating curves of the type found in Figures 3-16 and 3-17 in the form $\sigma = ax^b$, using different pairs of constants a and b to generate the six curves in each of the two figures. Generate figures comparable to Figures 3-16 and 3-17 by means of this approach, using the following table of values for the constants:

Stability class	Horizontal dispersion, σ_y		Vertical dispersion, σ_z	
	a	b	a	b
A	.5169	.8689	.006948	1.638
B	.3585	.8748	.08336	1.0517
C	.2190	.8949	.11296	.9102
D	.1323	.8998	.2686	.6597
E	.0969	.9026	.3230	.5725
F	.0667	.8992	.2760	.5281

Use logarithmic axes, as in the original figures, so as to be able to make a comparison. (Hint: such axes are generally a standard option on spreadsheet chart or graph menus.)

14. It was mentioned in the text that there should be a reasonable correspondence between the sampling times to which the (empirically determined) dispersion factors (σ_y and σ_z) correspond, and the length of the time-averaging periods for which the calculated concentrations will be interpreted to apply. For example, if the dispersion factors are based on 10-minute measurements, then if you use them in the plume equation to calculate a "long-term average concentration C," what C really corresponds to is a long-term average of 10-minute averages. (This is a subtle point, and you may need to think about it before addressing the following question.) Suppose you had two sets of curves of σ_y and σ_z values, one like Figures 3-16 and 3-17, based on 10-minute averages, and another new one based on one-day averages. How would the new set of curves compare with the original set? If you have studied statistics, can you relate this to any concepts from that field?

3.8 General Comments and Guide to Further Information

Through your reading of this chapter, you have encountered some of the fundamental methods used to model air pollution issues today. However, when one sets out to use models to resolve an issue of planning, permitting, or design, one needs to know a great deal about the limitations of the models, ways to handle special situations, and whether the results from the model calculations appear reasonable based on past experience. After all, major investments in plants, additional expenditures for pollution control, decisions where to locate, etc., are often tied into the results of the air modeling predictions. It seems today that no matter what kind of facility a company may want to build, there will be opposition among some quarters, often representing real, valid concerns but also often suggestive of the NIMBY* principle. Thus the arguments over whether to go ahead must often be resolved in the framework of contentious litigation or licensing procedures. If the models are too conservative, the plant will appear worse than the actual case, and hence less desirable. If the models are too complex or use new methods that are not yet part of well established practice, then they are more subject to challenge, and the entire process becomes very burdensome and costly for all participants. The "bottom line" is this: mathematical modeling to support real world decision-making is both an art and a science.

Perhaps nothing would be more valuable in providing a "real-world" perspective on some of these issues than a visit to an operating facility that deals with air pollution and other environmental issues on a regular basis. Personnel at such facilities often have extensive experience with siting, regulatory, public interaction, financial, and technical aspects pertaining both to their facility and to other facilities at which they may have worked. A power plant would be a prime target for a tour and meeting, but many other industrial facilities face these same issues as well. To orient you to such a plant prior to any tour, Figure 3-20 provides a schematic diagram typical of many power plants—namely, those that operate on the so-called "steam cycle."

The process is very simple and easy to understand. We begin with a source of heat (shown on the lower left of the figure). This could be from the burning of a wide variety of fuels—such as coal, oil, gas, wood chips, trash, and others—or from another source such as nuclear fission or concentrated solar energy. In any case, the heat is used to boil the water in a "pressure

---

* NIMBY = Not In My Back Yard

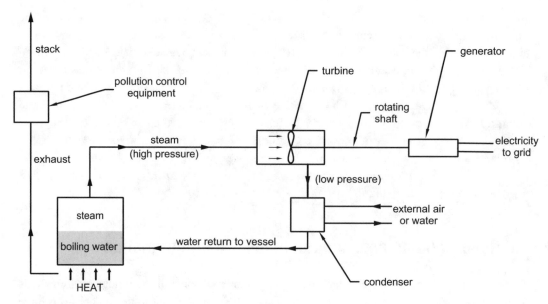

FIGURE 3-20
Schematic diagram of typical steam-cycle power plant

vessel" (like a pressure cooker on a stove), and then the high pressure steam that is produced flows through pipes to a turbine, which is just a very fancy version of a big fan. The only difference is that with a fan, you put in electrical power, which turns the fan and blows the air. In a turbine, the process is reversed. You "blow" at it with high pressure steam, and this turns the turbine blades. The turbine is attached to a long shaft, and so this causes the shaft to turn, which is connected to the electrical generator. When you turn the generator (just like the way the engine and V-belt turn the alternator in an automobile), it produces electricity, which is then fed out through transmission lines to the power grid.

In order to keep the turbine going, you have to get rid of the steam that has passed through it. One way to do this would be to let it go off into the air. But the problem is that it takes extremely purified water (much cleaner than drinking water) to run the turbine without damaging the delicate blades, and a large volume is passing through all the time. All that purified water would be wasted by letting the exhaust go off into the air, and you would have to keep preparing clean water at the other end—a far too prohibitive process. So you clearly want to save the water in the steam. However, if you just send it back as steam to the pressure vessel, the pressures would all balance out in the system, and there would no longer be any pressure imbalance left to drive the steam through the turbine. The solution to this is the condenser, where the steam passes through a radiator or cooling coil arrangement and is thus cooled by air or water pumped through from the external environment. The result is that it "condenses" back to water. When it changes from a gas back to a liquid, naturally its volume shrinks tremendously. This greatly reduces the pressure in this particular part of the system. It essentially produces a vacuum or suction effect on the downstream side of the turbine because the pressure there is now so much less than in the high pressure steam lines feeding into the other side. This large pressure differential is what causes a large volume of steam to keep flowing through the turbine, thus providing a strong force for turning the shaft and producing electricity. The water from the condenser is then pumped back into the pressure vessel and reused.

Naturally, the primary air pollution issue is usually associated with the original generation of heat by burning. The burning process produces exhaust gases, which will vary in pollutant content depending on the fuel used and on the temperatures of combustion. In any case, pollution monitoring and control equipment is usually installed on this exhaust stream, after which the hot gases (and perhaps some particulates) are released through the stack. Both the burning process itself as well as the pollution control equipment may also produce solid waste (e.g., ash and fine dust) that needs to be disposed of in an environmentally acceptable fashion. (These materials can cause ground-water problems if disposed of improperly.) If you visit such a plant, you may see a modern engineered landfill on the same site for disposal of these solids.

To return to the issue of models, many assumptions have been built into the models in this section, some stated and some more implicit. For example, with respect to the dispersion coefficients in the Gaussian plume model, these will certainly have some dependence on terrain (a factor investigated to a limited extent in the exercises). A more subtle limitation is the fact that air pollution measurements are usually average measurements over the period of time in which you collect or analyze the sample. Thus, when we try to predict the concentration C at a certain point, we are really more interested in the average over a period of time than in an instantaneous value. After all, a high level of pollution for only one second may not bother us too much, but a high average level over a day might. In fact, predictions of pollution levels from plants and other sources (e.g., cars) are often calculated separately for different lengths of time. If you were to do these calculations and compare, say, the 10-minute averages with the quarterly averages, you would of course find a higher degree of variation in the 10-minute averages, because over longer periods of time things tend to even out. Working backwards, then, we would really like to be sure that we know what our dispersion coefficients actually represent. Are they idealized instantaneous values? No, they are based on field experiments and represent discrete sampling periods. But then, did we really use them correctly in our calculations in this chapter? What do the calculations that we made really correspond to physically?

We could also ask many more penetrating questions about these models, and the person who uses them must be sure to understand all the underlying assumptions. If you would like to pursue these aspects, the following two references would provide excellent readable introductions to a number of associated issues:

Turner, D. B., *Workbook of Atmospheric Dispersion Estimates,* US Public Health Service, Publication 999-AP-26, 1967, reissued by U. S. Environmental Protection Agency.

Hanna, S. R., Briggs, G. A., Hosker, R. R., Jr., *Handbook on Atmospheric Diffusion,* Technical Information Center, U. S. Department of Energy, Publication DOE/TIC-11223, 1982.

For more advanced treatment of the mathematical derivations and for a discussion of many related issues, one may wish to see Pasquill's important treatise:

Pasquill, F., *Atmospheric Diffusion: The Dispersion of Windborne Material from Industrial and other Sources* (several editions, beginning in 1961).

There are several other excellent texts on the modeling of air pollution, such as:

Lyons, T. J., and Scott, W. D., *Principles of Air Pollution Meteorology,* CRC Press, 1990.

It should also be pointed out that the calculation of air pollutant concentrations is not really the end of the line as far as modeling is concerned. Often it is important to estimate what the health impacts might be on the population. Sometimes this comes up as a very contentious issue even if the plant is predicted to meet all applicable air pollution standards. For example, it may be that the standards were set before some more recent health studies on a certain pollutant, or there may be no standard for some specific pollutant of concern. The first of the final two modeling steps in such a case would include estimating the *exposure* of an individual or an entire population segment to the pollutant, which might involve modeling how members of the population move in and out of the polluted area during times when the plant would be operating and how much pollutant would actually be taken into their bodies. And this latter might not involve only inhalation! There could be important secondary routes, such as deposition of pollutants on agricultural crops and subsequent ingestion of agricultural products. Following the estimation of exposure, then the problem is to use modeling to estimate adverse *health effects,* generally on a statistical basis, using medical or other scientific studies, usually available only for other distinct population groups or, worse still, only for animals. How to extrapolate from one group to another is itself a very challenging issue. To pursue these latter aspects in any degree of detail, texts on epidemiology usually have very interesting case studies and general methods of approach. An alternative approach for an interesting introduction would be to talk to, or invite to give a lecture, a member of your state department of public health, which is generally the lead agency in resolving many of these issues.

Another topic that has not been discussed much in this chapter is the nature of the pollutants themselves. For modeling purposes, we have treated them as though they are all gases whose molecules behave essentially the same as the various molecules in the air. While this is reasonable for a first introduction to the simplest models, the reader should realize that not all air pollutants are gases, and those that are gases may be "heavy gases" or "light gases" and may have an important buoyancy force acting on them when they are released in the air. An example of a non-gaseous pollutant would be the black soot particles that form the smoke released from poorly operating diesel vehicles or from fireplace chimneys. Fortunately from the modeling standpoint, the size of the particles in many such "aerosols" is sufficiently small that they diffuse through the air similarly to gases. Some pollutants actually change their chemical form as they move through the atmosphere. This has been one of the difficult issues in understanding the ozone problem. In such cases, complex chemical transformation models may need to be superimposed on the transport models. Many introductory environmental science texts give a nice introduction to these aspects without getting technical about the modeling. (There is also an introduction to some heavy gas modeling in Chapter 7 of this text.)

Then there is the issue of pollution control. What can be done to reduce the level of pollution from plants, vehicles, or other sources? For plants, there are a number of devices for pulling a large fraction of the pollutants out of the exhaust stream. How much an operator should invest in this depends on how high the pollution levels are that are predicted. It would be naive to suppose that costs should not be taken into account in these decisions, however, because running a profitable plant (and hence being able to continue to employ people and provide services) can be much more of a challenge than is often realized by those who may not ever have had the experience. But suppose you do pull out a large quantity of pollutants—what is to be done with it next? If you bury it in a landfill, you may contaminate someone's drinking water. If you transport it to some distant destination, the required trucks will have to generate a certain amount of new pollution in this process. These kinds of questions about tradeoffs are one of the "hottest" topics in the environmental field, and they are being incorporated more and more in the development of regulations. Mathematical representations of costs, risks, benefits, and uncertainties are key to such work. For an introduction, a text on environmental economics could be of considerable interest. Alternatively, either on-line or via your library, read about the official record leading to the development of any specific new environmental regulation, and you will surely find analyses of this type. You may also be surprised at the wide range of groups who have contributed to this record.

Although it also goes beyond air pollution, naturally one would like to do all that is reasonably possible to avoid creating pollution in the first place. Thus there are many efforts underway to reduce the use of materials or fuels that lead to pollution. How does the government provide incentives for this? Regulations? Tax advantages, and if so, how and how much? Publicity opportunities? Free technical support? Grants? Even in evaluating its strategies, the government needs to use quantitative methods to assess the efficacy and the possible impacts of such strategies.

4

Hazardous Materials Management

4.1 Background

Hazardous materials are all around us. We use them for fueling our vehicles, cooking and heating, maintaining our lawns and gardens, and keeping pests out of the house. However, we may pay little attention to their associated risks—fire, explosion, poisoning, asphyxiation, and others—and we may too quickly forget the lessons of the past, when many lives have been lost through mishandling and accidents.

The combination of rapid technological advances during the twentieth century, leading to some disasters of epic proportion and many more smaller ones, and the increased environmental awareness and activism that took shape in the latter quarter of the century have caused us as a society to develop a much more elaborate framework of regulations and technical tools to enable us to manage hazardous materials safely. The reader may be surprised at the degree to which mathematical analysis permeates this development, not only at the level of research laboratories and universities, but even right in the local communities in connection with their planning and decision-making.

But first, a brief look at history: In 1979, the Three Mile Island nuclear power plant, just outside Harrisburg, Pennsylvania, experienced an unexpected sequence of events eventually leading to a partial meltdown of its radioactive core, the kind of event that nuclear experts had always said was virtually impossible. The nation huddled around TV screens as the accident unfolded, watching utility spokesmen tell us that everything was under control, when it wasn't, watching politicians and regulators debate the need to evacuate the area, and watching experts who were most uncertain about what should be done to bring things under control. Fortunately, there was no real physical health impact from this event on the surrounding population, but it taught many lessons, among them:

- Carefully designed but complex technological systems can fail in unexpected ways.
- Planning for a response to low probability emergencies had not been give sufficient priority.
- We cannot have blind faith in the safety assurances of credentialed experts; even they make mistakes.

Move ahead to December 12, 1984. The evening news carried a late-breaking report of an evolving chemical plant emergency in Bhopal, India, and the next day we learned the gruesome details: a toxic gas cloud released silently in the night from a pesticide plant had killed or severely harmed thousands living in its vicinity as it floated through their neighborhood. The aftermath of this event was huge, not only for the community itself, where the long-term death toll eventually exceeded 3,000, but also for the government, for the multinational corporation, Union Carbide, that owned the plant, and for the entire chemical industry, which almost overnight came under the most extreme scrutiny it had ever experienced. The story of Bhopal became every chemical company CEO's worst nightmare, and company leaders became much more interested in investigating the potential risks at their own plants.

As time went on, it seemed as though there would be no stopping the sequence of technological disasters. In April of 1986, the Chernobyl nuclear plant in the Ukraine was rocked by two explosions and a fire, and the world learned of the seriousness of this event only after radioactive fallout was detected in Scandinavia. In November of the same year, a fire in a chemical warehouse in Basel, Switzerland, led to the release of over a thousand tons of dangerous chemicals into the Rhine River, which wreaked havoc on that river all the way to its mouth in the Netherlands. In 1988, a newly reconstructed million-gallon diesel oil tank, surrounded by a supposedly protective berm wall, failed catastrophically in Pennsylvania, allowing the oil to splash right over the berm and into the Susquehanna River, along which it moved down to and along the Ohio River, forcing the closing of water intakes and upsetting the local ecology over a long distance. The year 1989 brought the Exxon Valdez tanker spill off the coast of Alaska—232,000 barrels of crude oil released within a few miles of the shore. And the list goes on.

A more complete review of the record would surely surprise the reader in terms of the number and extent of failures of systems for handling hazardous materials. (This is pursued in the problems at the end of this section.) Every year there are many evacuations of people as a result of spills, fires, explosions, or gaseous releases of such materials. Sometimes these occur at fixed facilities (e.g., warehouses, manufacturing plants, chemical terminals, and also hockey rinks, swimming pools, and other less likely facilities); at other times they occur during transportation (e.g., truck rollovers, train derailments, airline crashes, ship groundings, pipeline ruptures). It seems that no matter how hard society tries to avoid accidents, the unexpected continues to happen.

The kinds of events referred to above are often called "episodic events," referring to their occurrence in more or less discrete episodes. Combined with the further discovery of long-term environmental problems during the same period (see Chapter 2 for some examples), there began to develop a much heightened national effort to improve the environmental performance of our companies. This derived from several forces:

- Public pressure. People had become better educated, more skeptical, and had perhaps gotten some hint of the power of public pressure from their experience in the Vietnam war era.
- Financial concerns by company management. The financial aftermath of the TMI and Bhopal accidents for the owners of those plants was extreme. This involved not only liability and clean-up costs, but also, very importantly, precipitous drops in stock value, shareholder confidence, and access to capital.
- Regulatory developments. Regulatory agencies at both the federal and state level, surely feeling somewhat empowered by the passage of the Superfund Program (see Chapter 2) and the Resource Conservation and Recovery Act (RCRA), moved to expand their attention from hazardous wastes to all hazardous materials.

The industry efforts were the fastest to take form, with a wide range of large companies, often multinationals, developing internal management and control systems to better guard against environmental surprises. A number of industry organizations, such as the Chemical Manufacturers Association, sponsored unified efforts that significantly expanded the range of companies where meaningful reforms could realistically be expected (since small and medium-size companies usually could not afford the development work themselves). Examples of tools that were developed include:

- More extensive written policies and procedures governing company operations that could impact the environment, usually made essentially uniform for all plants around the world. An important area for this involved the approval process for plant modifications, to be sure that all the environmental and safety implications were thoroughly considered.

- Formal oversight processes, such as process safety reviews and environmental audits, often carried out at least in part by outside personnel with a reporting line directly to top management.

- Quantitative methodologies, such as probabilistic safety assessment and risk analysis, for addressing quantitatively both the probability and the consequences of undesirable occurrences.

- The identification of so-called "critical operating parameters" which, when out of the acceptable range, would automatically trigger corrective action or shutdown.

- Extensive emergency planning, including both on-site and off-site notification and response, so that an emergency could be better managed if it did in fact occur.

It is worth noting at this point that mathematical modeling was key to many of these developments in an usually wide variety of ways. Examples will be seen later.

At the same time that industry was trying to get its house in better order, the regulatory agencies were developing long-term strategies to assure a higher level of environmental and personnel protection. (Examples of such agencies at the federal level would be the Environmental Protection Agency, the Department of Transportation, the Occupational Safety and Health Administration, and the Nuclear Regulatory Commission. At the state level, they might include the state departments of environmental protection or health.) Examples of developments during this period include:

- Wide availability of so-called "material safety data sheets" or MSDSs, listing the properties, hazards, and protective measures appropriate for a wide range of chemicals in everyday commerce.

- The formation of local and statewide emergency planning committees to collect data on the use of hazardous materials in states and communities and to plan for potential emergencies involving such materials.

- Strict reportability requirements for companies or other organizations on their inventory of hazardous chemicals and on incidents involving releases to the environment.

- Mandated use of techniques such as environmental audits and quantitative risk analysis in certain situations.

- Modified routing, labeling, and packaging requirements for the transport of hazardous materials.

Aside from the regulatory side of governmental operations, most agencies also expanded their programs of technical assistance to communities and industry. Examples of these contributions include:

- The "orange book," a first responder's guide to dealing with a hazardous materials emergency. This includes chemical data, evacuation decision guidance, and other basic information. You would find it in common use among firefighters, police, and other emergency responders. (Mathematical modeling was key to its development.)
- The development of risk analysis computer programs to enable a broad range of users to prioritize and plan for risks. (This chapter will include the use of such a program.)
- Programs to help businesses evaluate cost-beneficial opportunities for reducing their use of hazardous chemicals.

The examples given above and throughout this section are only isolated examples. Many others could have been chosen and many other agencies and organizations could have been cited. However, these should serve to provide a reasonable representation of the kinds of activities that have been energetically pursued over the last ten to fifteen years. With this background, we will discuss the scientific and engineering concepts, and then the mathematical modeling aspects, in subsequent sections.

Exercises for Section 4.1

The problems below ask you to find information about various classes of episodic events. You should be able to find such information fairly easily using your library's resources or via on-line searches. Newspaper searches, either computerized or via paper indexes, can be invaluable when you do not otherwise locate a database containing what you are looking for.

1. Aside from the hazardous materials incidents discussed in the text, identify one major incident in each of the categories below and provide a concise description of the incident, size of the affected population, materials involved, cause, and other relevant factors:
  a) rail transportation
  b) highway transportation
  c) oil well or pipeline
  d) fixed facility (e.g., factory, chemical plant, refinery)
  e) marine spill

2. For either your home state or the state in which your institution is located, find five examples of hazardous materials incidents that have occurred in the last two years and provide brief summary information on each.

4.2 Hazardous Materials Handling Practices and Potential Accidents

Hazardous materials may be encountered at fixed locations where they are stored or used, or in transportation systems as they are moved from one location to another. You have no doubt often seen some examples of the latter, such as tank trucks on the highway and tank cars as part of freight trains on the railroad. Other common examples of transportation situations include:
- closed trucks or vans carrying drums, bottles, or other containers of chemicals;
- pipelines used to move chemicals and fuels either locally or even across the country;
- ships and barges used to move chemicals into ports, along coasts and rivers, or across the sea; and
- airplanes, which sometimes carry hazardous cargoes.

You may be less familiar with the storage of chemicals at fixed facilities, where they may be in "containerized" forms, such as racks of 55-gallon drums, or in "bulk" tanks. Several bulk tank configurations are shown in Figure 4-1. Note that all such tanks need to have pipe connections for moving material in and out, and they also generally have some kind of vent or pressure release device on the top to keep the pressure from building up to excessive levels under normal or emergency conditions. Two of the three tanks in the figure are shown with an accompanying dike surrounding them to contain the material, at least temporarily, in the event of some failure of the tank itself.

It seems that there is hardly any end to the list of things that can and have gone wrong during the transportation, storage, or use of hazardous chemicals. Table 4-1 lists some typical accident scenarios.

While there are many forms that the resulting consequences can take, the following capture most of the dominant hazards:

- *Pool fire.* A pool of liquid chemical on the ground or floating on top of a surface water body ignites. The hazard is from the thermal radiation (heat) given off by the fire.
- *Vapor cloud fire.* The spilled chemical evaporates, forming a flammable cloud of vapor that leaves the site and blows into some surrounding neighborhood. If it reaches an ignition point, such as someone cooking or smoking a cigarette, or even a hot catalytic converter on a vehicle, then the resulting cloud may ignite, possibly causing a flash fire all the way back to the source.

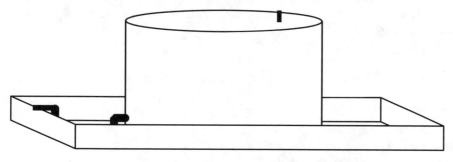

Large vertical storage tank, common in tank farms and terminals

Narrow vertical tank, often found
at user location or where space is limited

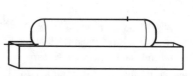

Horizontal tank, rounded ends often
indicating that it is pressurized, as for a
material with a low boiling point

FIGURE 4-1

Typical bulk chemical storage tank configurations

- *Vapor cloud explosion.*   This is similar to a vapor cloud fire, but under certain conditions the cloud will actually explode. Such conditions are usually associated with some kind of confinement and are difficult to achieve outdoors.

- *Flame jet.*   Material released through a broken pipe or tank perforation can ignite and have the effect of a long blowtorch anchored to the release point. This could impinge on adjacent tanks with hazardous chemicals in them as well.

- *BLEVE.*   This is a Boiling Liquid Expanding Vapor Explosion. When a pressurized tank of flammable chemical is immersed in a fire for a long period, it can eventually release its contents catastrophically.

- *Toxic vapor cloud.*   This is similar to the flammable vapor cloud scenario, except that the vapor cloud itself can cause illness or death through its toxic or poisonous effect on the body.

- *Suffocation.*   Chemical fumes that are not themselves toxic or that do not ignite may nevertheless block out sufficient air and oxygen so that people within the cloud die of asphyxiation.

The bulk of our work will be with vapor cloud fire and toxicity risk, as this is a highly mobile source of risk and one that can easily be encountered. However, we will encounter some of the other hazard types as well in connection with some of our accident scenarios.

We will not explore in this book the subject of hazards from radioactive materials, both because this is a very specialized subject with its own set of scientific background principles, and also because it is much more rare for a radioactive materials accident to present a risk to members of the public.

TABLE 4-1
Examples of hazmat accident initiators

Hose ruptures during filling of storage tank.
Inattentive operator overfills tanker truck; material pours out the top.
Backhoe ruptures underground pipeline.
Incorrect valving in tank farm directs material to full tank, which overflows out the vent.
Bird's nest in pressure relief valve prevents release of pressure when tank is filled; tank ruptures from overpressure.
Improperly set brake causes vehicle to roll into tank fill pipes; tank drains through broken fitting.
Careless welding of pipe fittings ignites residue from past leakage; starts major fire around tank.
Earthquake causes narrow vertical tank to lean and fail.
Lightning strikes storage tank.
Static electricity generated during material transfer causes spark, ignition, and major fire.
Valving error causes incompatible materials to be mixed, leading to runaway reaction.
Heavy winds and waves cause barge to break loose while transferring chemical to shore.
Sabotage is caused by disgruntled employee or terrorist group.
Tank truck overturns on icy road.
Tank truck gets stuck at railroad crossing; hit by train.
Tank car derailment; tank is punctured by coupler on adjacent railcar.
Tank car derailment; tank car is surrounded by flames from burning material released from another car.

4.3 Physical Principles and Background

4.3.1 Basic Material from Physics and Chemistry

4.3.1.1 Atoms and Molecules. The purpose of this section is to provide a small amount of basic background material from physics and chemistry that you need in order to understand the nature of the chemical hazards we will be modeling.

You surely know that matter is made up of atoms and molecules. In particular, atoms are the smallest components of the basic "elements," such as hydrogen, carbon, iron, and others. There are only slightly over 100 such elements in the universe! Atoms themselves are made up of a nucleus consisting of protons and neutrons (and other smaller components we shall ignore here), surrounded by orbiting electrons. In a normal atom, the number of electrons is the same as the number of protons in the nucleus, and this latter number is called the *atomic number*. The atomic number determines the chemical properties of the element. See Table 4-2 for a list of some common elements, along with their atomic numbers. For example, you can tell from this table that a carbon atom always has 6 protons, since 6 is its atomic number.

The *atomic weight* is also an important characteristic of an element. It is essentially the total number of protons plus the total number of neutrons in the nucleus of the atom. But this would make you think that it should be a whole number, just like the atomic number, and yet the atomic weight values in the table all show a fractional component! How can we explain this paradox?

Consider the example of carbon. A typical carbon atom has 6 protons (so the atomic number is 6, as in Table 4-2) and also 6 neutrons. This would make its atomic weight 12. The value in the table is 12.011, which is close to 12 but not identical to it. This is because some of the carbon atoms in the universe have a different number of neutrons. For example, some have 7 neutrons and some have 8 neutrons, and their atomic weights would then be 13 and 14

TABLE 4-2

Selected elements, showing their atomic numbers and atomic weights

Symbol	Name	Atomic Number	Atomic Weight	Symbol	Name	Atomic Number	Atomic Weight
H	Hydrogen	1	1.008	Ca	Calcium	20	40.08
C	Carbon	6	12.011	Cr	Chromium	24	51.996
N	Nitrogen	7	14.007	Mn	Manganese	25	54.94
O	Oxygen	8	15.9994	Fe	Iron	26	55.85
Na	Sodium	11	22.99	Co	Cobalt	27	58.93
Mg	Magnesium	12	24.3	Ni	Nickel	28	58.69
Al	Aluminum	13	26.98	Cu	Copper	29	63.55
Si	Silicon	14	28.086	Zn	Zinc	30	65.39
P	Phosphorus	15	30.97	As	Arsenic	33	74.92
S	Sulfur	16	32.07	Au	Gold	79	196.97
Cl	Chlorine	17	35.5	Hg	Mercury	80	200.59
Ar	Argon	18	39.95	Pb	Lead	82	207.2
K	Potassium	19	39.1	U	Uranium	92	238.03

respectively. In fact, you may have heard of carbon-14 from radioactive dating methods. Thus different carbon atoms, while they always have 6 protons, may vary in total atomic weight. These different variations on carbon are called *isotopes* of carbon, and are usually written C^{12}, C^{13}, and C^{14}. Scientists have estimated the general fraction of each occurring naturally on the earth, and the atomic weight in the table is just the "weighted average" of their different values. Since, C^{12} is by far the most common form of carbon, it is not surprising then that the atomic weight for carbon should be close to 12. Remember that the chemical properties are always the same, although certain physical properties (such as whether they are radioactive) may vary from one isotope to another.

Although a few substances are made up of just plain atoms, it is much more common for them to be made up of *molecules,* which are collections of atoms bound together in particular combinations and structures. For example, you are no doubt familiar with water and its chemical formula H_2O. This means that water is made up of molecules, each of which consists of 2 hydrogen atoms and 1 oxygen atom. The natural gas that you may use for cooking or heating is primarily methane, whose formula is CH_4, so you know it is made up of molecules with one carbon atom and 4 hydrogen atoms. Even many of the individual elements, such as the oxygen in the air we breathe, are organized into molecules rather than individual atoms. Thus the oxygen in the air has the form O_2, whereas the oxygen in ozone has the form O_3. So now it's easy to see how it can be possible that with only about a hundred basic elements, we can still have the very wide variety of chemical substances we encounter in everyday life. It mostly depends on the myriad ways in which they can be combined into molecules.

One property that will be important for us is *molecular weight.* Quite logically, it is just the sum of the atomic weights of the individual atoms making up the molecule. Thus the molecular weight of water is: $2H + 1O = 2(1.008) + 1(15.9994) = 18.0154$. The higher the molecular weight, the more we might refer to the molecule as a "heavy" molecule, and this might impact how the material behaves. For example, ordinary hydrogen gas in the form H_2 is very light, with a molecular weight of about 2, and in fact the gas itself is very light and wants to rise in the air. This has made it popular in the past for balloons and blimps, except that it has another important property of being highly flammable and explosive!

To return to the example of methane (CH_4) mentioned above, note that it is composed of carbon and hydrogen. In fact, it is one the simplest hydrocarbons, this latter term referring to the class of chemicals made up of carbon and hydrogen, often with some additional components. Other typical examples would include acetylene (C_2H_2), trichloroethylene (C_2HCl_3), propane (C_3H_8), butane (C_4H_{10}), and ethanol (C_2H_5OH). Many of the hazardous materials in common use are in fact hydrocarbons, as they make good fuels, solvents, and other useful materials.

Exercises for Section 4.3.1.1

For some of the exercises below, you will need to consult some reference material of your own choosing, such as in your library or via computer. Sometimes even an ordinary dictionary or encyclopedia can be the most convenient source for data on elements or chemical compounds, or for conversion factors. The material needed to help answer these questions should not be difficult to find, and this initial practice in finding it will help guide you to sources that will be of further use later in this chapter.

1. Find the molecular weights of benzene, sulfuric acid, and ammonia. (Reread the instructions at the top if you're puzzled on how to start.)

2. The atomic weight of carbon (and many other elements) has often been revised in the scientific literature. Doesn't this seem strange for such a common element? What do you think is the dominant reason why such values are revised from time to time?

3. Research the famous "Hindenberg disaster" and relate it to some of the material in this section.

4.3.1.2 Physical Properties of Matter. You are familiar with the three common *states* of matter: solid, liquid, and gas. We shall be mostly concerned with liquids and gases, and especially with the transition from liquids to gases as spilled hazardous liquids evaporate or otherwise react to form gases. Not only can these gases be toxic or flammable, but they can move silently with the wind away from the location of an accident to an area with an unsuspecting and unprepared population.

You are also familiar with the basic concept of *density,* referring loosely to how much a given volume of material weighs. For example, the density of water is 62.4 pounds per cubic foot (lbs/ft^3), and the density of solid rock is around 200 lbs/ft^3. Even air has a density, around 0.004 lbs/ft^3 at sea level and decreasing as you move higher into the atmosphere. Density can, of course, be specified in different sets of units. For example, in the metric system we would find that the density of water is 1 gram per cubic centimeter (g/cc). (This is the value at certain "standard conditions" and it could go up or down slightly depending primarily on temperature.)

Density is sometimes spoken of in terms of an equivalent property: *specific gravity.* The specific gravity of a substance is the ratio of its density to the density of water. Since it is a ratio of like quantities, it has no units. For example, if we wanted to find the specific gravity of a certain kind of rock that weighs 200 lbs/ft^3, we might calculate it as follows:

$$\text{spec.grav.(rock)} = \frac{\text{density of rock}}{\text{density of water}} = \frac{200 \text{ lbs/ft}^3}{62.4 \text{ lbs/ft}^3} = 2.5$$

where the identical units in numerator and denominator have canceled out. We could also conclude from this that the density of this kind of rock in metric units would be 2.5 g/cc, since the density of water is 1 g/cc and rock is 2.5 times as dense. Because density and specific gravity values are numerically equal in this system of metric units, it is not uncommon in the industrial literature to find the terms "density" and "specific gravity" used interchangeably and not necessarily in precise conformance with the above definitions.

You probably also know that most substances, upon being heated, tend to expand. Thus the same amount of mass will be taking up more volume, and the density will decrease. That is why density values are usually specified together with the temperature value they correspond to unless only relatively rough numbers are needed for one's purposes. This phenomenon has been encountered in Chapter 3, for example, with the rising of air as it is heated by being in contact with the warm ground. Are you familiar with a rather famous example of a substance that does not necessarily expand when heated?

Now let us focus on the process of liquid evaporation. We begin with a container of some chemical liquid sitting on a table. The liquid could be water, antifreeze, rubbing alcohol, or anything else. You would expect it to evaporate eventually, and you would probably expect the alcohol to evaporate the fastest among these examples. We need to understand what controls the rate of evaporation because if there were a large spill of hazardous chemical, such as gasoline from an overturned tank truck, we probably would want to try to keep it from evaporating too fast into a concentrated chemical cloud that could carry the danger offsite.

First, we have two simple principles that should be in keeping with your physical intuition and should help lead to further insights about the underlying physical processes:

1. The *rate of evaporation is proportional to the surface area* (all other factors being equal). You can see this illustrated in Figure 4-2, where you can imagine that the two containers hold the same volume of liquid although one is tall and thin and the other is short and wide. The container on the right provides a much larger surface area, and thus the liquid will evaporate faster. Similarly, if you took a glass of water and poured it out all over the floor, it would "dry up" (i.e., evaporate completely) faster than if you left it in the glass just sitting on a table. The underlying physical principle is roughly this: evaporation is a process whereby molecules within a short distance of the surface that have sufficient kinetic energy to break through the surface (where there is a generally cohesive effect called *surface tension*) do so and escape individually into the space above the liquid, thus becoming part of the gaseous or vapor component of the chemical. So if you double the surface area, for example, you double the number of molecules in this zone, and hence you would expect them to escape at twice the total rate.

2. The *rate of evaporation increases as the temperature of the liquid increases.* This is illustrated in Figure 4-3. Temperature is generally defined as a measure of the average

SLOWER EVAPORATION

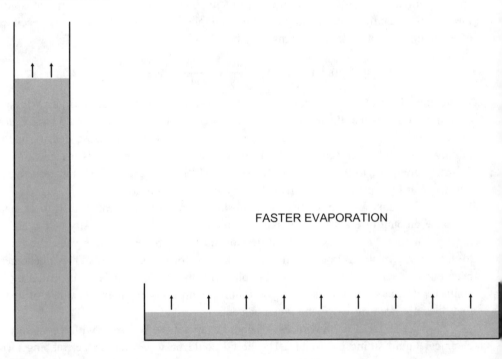

FASTER EVAPORATION

FIGURE 4-2
Effect of surface area on evaporation rate

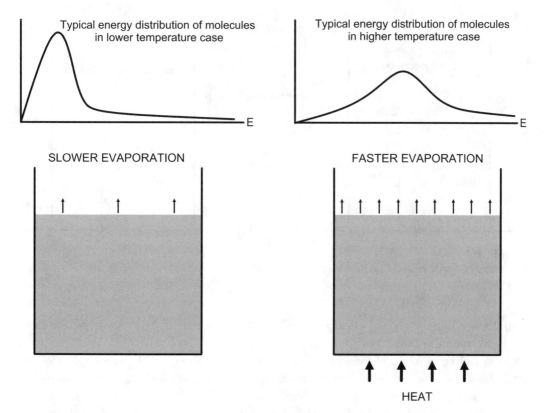

FIGURE 4-3
Effect of heat on evaporation process

kinetic energy of the molecules. (Of course, some have higher amounts and some lower, as shown in the graphs on this same figure.) When you apply heat, you are applying further energy to the material and hence increasing the movement of the molecules, so that their average energy becomes higher. With a higher average energy, you would expect that the proportion of those with the extra amount needed to break through the surface would also go up, and hence the evaporation rate would increase.

Now we shall conduct a simple experiment to investigate the key topic of *vapor pressure*. See Figure 4-4. The left side of this figure shows a beaker of chemical just having been placed under a larger closed glass cover. Initially, all the chemical is in the beaker, and the other space inside the cover is just filled with air or with some other gas that is assumed to be "inert" with respect to our chemical, meaning that it doesn't react or interact with it at all. But now the chemical in the beaker begins to evaporate, and thus some of its molecules join those of the other gas in the vapor space inside the cover. None of the original gas molecules have any place to go, so there is an increase in the total number of molecules in the same space. The net effect of this is that the pressure under the cover increases. This process eventually slows down and reaches an equilibrium point, however, when there are so many chemical molecules in the vapor space that the chemical has no further propensity to evaporate.

This is just like having a high "relative humidity" on a summer day: there is so much water vapor in the air that the sweat produced by the body no longer wants to evaporate, and

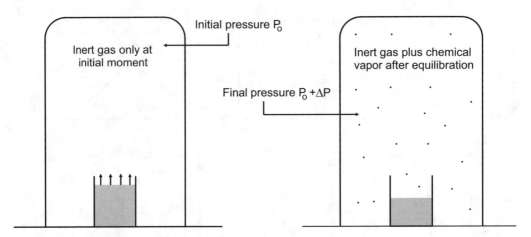

FIGURE 4-4
Illustration of vapor pressure of an evaporating material

with that we lose the cooling effect produced by such evaporation. To be more precise about the equilibration process, what is really happening is that at all times and conditions there are both molecules of the chemical coming out of the liquid and into the vapor space (due to sufficiently high energies in the layer near the surface, as discussed earlier), and molecules in the gas space colliding back with and rejoining the liquid. Early in the experiment, when the gas concentration of chemical is low, the first group far outnumbers the second, so there is *net* evaporation. But as the gas concentration increases, the rates eventually balance out, and that's when we would say we are at equilibrium.

When this evaporation process reaches equilibrium, since there are more total gas molecules than at the outset squeezed into essentially the same space, the total pressure will be higher. The amount of increase, denoted on the figure by ΔP, is called the *vapor pressure* of the chemical at whatever temperature the system is at. Naturally, if we were to increase the temperature, that would tend to favor more evaporation, and thus we would expect the vapor pressure to increase as well. Thus we would say that vapor pressure is generally an increasing function of temperature.

What if the original experiment depicted in Figure 4-4 were begun with a higher or lower initial pressure P_0 of inert gas? (For example, after sealing off the dome over the beaker, we could pump gas out with a vacuum pump or pump additional gas in.) What would happen to the value of ΔP? The answer is that *as long as* the other gas is truly inert and undergoes no interactions with either the liquid or gaseous form of the chemical, then the increase in pressure ΔP due to the chemical molecules entering the gaseous state would be the same. Now this idealization of total inertness is never quite true in practice, and for this reason the precise conditions under which vapor pressure is measured can change the results somewhat. Therefore, vapor pressure values are usually reported in connection with certain well-defined test conditions, but *except for the temperature dependence,* the variation in values is not generally significant enough to affect in any important way the general modeling calculations to be carried out in this chapter. Incidentally, the value ΔP would also be called the *partial pressure* of the chemical component in the vapor space. In general, under idealized conditions, the sum of the partial pressures of each distinct gaseous component will equal the total pressure of gas in the space. (This is called Dalton's law of partial pressures.)

The last basic scientific concept we must address is that of the boiling process. Consider Figure 4-5, which illustrates the gradual heating of a beaker of a liquid chemical to higher and higher temperatures, just like heating a pot of water on a stove. Remember that the ordinary atmospheric pressure of the air is always pushing down on the liquid, as is illustrated in the figure by an imaginary piston. If we raise the temperature of the liquid so high that its vapor pressure increases beyond atmospheric pressure, then you might imagine that the confining force of atmospheric pressure no longer can hold it in its normal stagnant-looking form, and vapor bubbles of the chemical can expand rapidly causing the effect that we call *boiling*. At this point, the chemical can enter the vapor form from throughout the liquid, not just at the surface, since the bubbles essentially create their own vapor space wherever they develop.

Much larger quantities of chemical can move into the vapor state rapidly under boiling conditions, and the primary limiting factor is the amount of heat that is being supplied. This is because it takes a certain amount of heat energy to change a fixed amount of a given chemical from liquid to gaseous form, whether it be by evaporation or boiling. For water, for example, it takes 540 calories of heat energy to convert one gram of water from liquid to gas under normal conditions. This is called the *heat of vaporization*. Once a liquid reaches its boiling point, all the heat energy being applied to it is used up in converting more and more of it to the gaseous form, so the temperature stays essentially constant right at the boiling point, rather than continuing to rise higher.

To illustrate the above ideas, just think about the process of boiling an egg. Suppose you try to do this in New York, which is at sea level and thus has a certain associated atmospheric pressure. Say it takes 4 minutes to boil the egg the way you want it. Remember that the boiling point of water under these conditions is 212°F (or 100°C). Now try the same thing

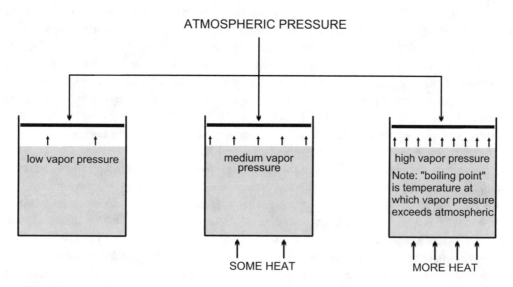

FIGURE 4-5
Illustration of the boiling process

in Denver, which is at an elevation of about a mile. There the atmospheric pressure is less, since there is one mile less atmosphere on top of you, and so you have to heat the water only to a temperature somewhat lower than 212°F before reaching the point where the vapor pressure equals atmospheric, and thus boiling begins. But now you would be cooking the egg for 4 minutes at a lower temperature, and thus it would not be so well cooked as you wanted. Equivalently, you would have to boil it for a longer time to achieve the same result. On the other hand, you may be familiar with a "pressure cooker" (referring to a cooking implement and not to your school or work conditions!). This is a strong pot with a top that screws on tightly, and with a weight over a small hole that serves as a "pressure control valve." When you start to cook, or essentially heat water, in this pot, the pressure builds up inside higher than atmospheric pressure. Since the pressure is higher, it takes a higher temperature to bring water to a boil under these conditions. Therefore, when you reach the boiling point, you are at quite a high temperature and thus the food cooks much faster.

We are accustomed to thinking of a boiling point as some kind of high temperature achieved by the application of heat. But many important (and dangerous) chemicals have such high vapor pressures that even at normal everyday temperatures their vapor pressures are already higher than atmospheric. In other words, they want to boil at room temperature! You are familiar with some of these, although perhaps you have not considered them from this point of view. For example, the propane gas that is used for gas grills, soldering, cooking, and heating in many places comes in small metal cylinders or "bottles" that are just like the pressure cooker described above. If you were to pour the propane out on the ground, you would see it boil just like water poured on a hot griddle. But inside the container, the pressure is at or slightly above the vapor pressure, and this keeps the liquid propane from boiling. Just think then what would happen if a tank truck of propane were to break open in an accident. The liquid would pour on the ground, boil very rapidly, form a highly flammable gas cloud, and float off into the surrounding neighborhood. Analyzing important risks like this is where we are heading in this chapter.

All the scientific concepts discussed above were for "pure substances," referring to materials consisting of a single chemical. All the molecules in containers of such substances are the same, even if the molecules themselves are made up of a combination of various elements. For example, water is a pure substance, as all of its molecules have the same composition, consisting of two hydrogen atoms and one oxygen atom. On the other hand, a mixture of water, say, and acetone is not a pure substance. There are two kinds of molecules present even though they may be thoroughly intermixed with each other. The everyday term "mixture" is the correct chemical description of this latter situation. Many important hazardous materials are actually mixtures, common examples being gasoline and diesel fuel.

How do the chemical properties of mixtures relate to those of their basic components? For example, what about boiling points, vapor pressures, etc.? The answer is that mixtures may not really have the same unique physical properties as pure substances have. For example, instead of a boiling point, mixtures may have a range of temperatures through which they boil, with the more volatile components tending to boil off first. While the properties of the individual components may give some initial indication of the range of properties of the mixture, there can be certain interactions between the components that can modify these properties significantly. The important point here is to be aware of the complications involved with mixtures; we will return to them again in a later section.

Exercises for Section 4.3.1.2

1. Consider the following five common elements: gold, silver, iron, lead, and uranium. Find their densities or specific gravities and place the elements in order from the lightest (i.e., the least dense) to the heaviest.

2. Sometimes the line between different states of matter is not so clear. For example, some people think that ordinary window glass is a solid, while others think it is a liquid. (Yes, it says "liquid." After all, if you look at very old windows, you might even see horizontal lines in them where they appear to have "slumped" over the years.) Investigate this issue with a science text or teacher, and summarize your conclusions.

3. What is the atmospheric pressure in Denver compared to that in New York? How long would you have to boil an egg in Denver to reach the same result as with a given boiling time in New York? (Hint: you will have to be a clever investigator, chef, or scientist to answer this latter question. It is a well-known issue.)

4. If you had two identical stoves and two identical pots of water on them in both Denver and New York, which one would come to a boil faster? Explain your reasoning.

5. Return to the potential accident discussed in the text, where liquid propane is poured out on the ground and boils off rapidly. Even though it boils at ambient temperature, it still takes a fixed amount of heat (the heat of vaporization) to convert each gram to gaseous form. Thus it would take a huge amount of heat for a whole tank-truck load to boil off. Where would this heat come from? Explain how you would envision the scenario.

6. Aside from propane, find another relatively common chemical that is used in liquid form but that boils at ambient temperature.

7. Can you identify the common substance alluded to in the text that does not necessarily expand when heated (or contract when cooled)? What is the significance of this behavior?

4.3.2 Characterization of Flammable Vapor Hazards

We are going to begin our discussion of chemical hazards by focusing on their gaseous forms, because in general these forms are the most mobile and have the ability to float into unexpected and undesirable places.

Imagine the following experiment. (Just **imagine** it; *don't try* it!) Roll up a few pieces of newspaper to make a torch. Light one end. Now open the gas cap on your car and shove the lighted torch all the way into the tank. Brace yourself. What do you think would happen?

Figure 4-6 shows a schematic of this situation. Gasoline is a highly flammable material, and as mentioned earlier it is actually a mixture of a number of different hydrocarbons. It makes a good fuel because it ignites so easily in your engine, and when it burns it releases a large amount of energy. But when you plunge the torch into the tank, the torch should go out! Yes, that's right, it should go out.

What's going on here? As you know, for a material to burn, there needs to be oxygen present, for burning is simply a chemical reaction under which oxygen combines chemically with some fuel to give off a substantial amount of heat. This heat is usually enough to keep the burning process going, so you don't have to keep lighting the fuel material over and over. Now there is certainly oxygen in the vapor space of the fuel tank, as there is always going to

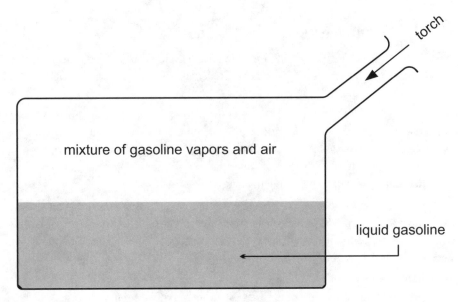

FIGURE 4-6
Plunging a torch into a gasoline tank

be some air mixed with the fuel vapors, but the key fact is that there is *not enough oxygen* there for this burning process to keep going, so the torch goes out.

People frequently ask about the moment when the torch is first plunged into the tank entrance. Might there be more oxygen there? A good question. That's why you shouldn't try this experiment! By the way, if you have any experience with engines that have carburetors, you may know that it is important to adjust the carburetor so there is neither too much nor too little air mixed with the gasoline as it goes into the cylinders. This is exactly the same issue that we are discussing here.

To cast this in a more scientific framework, consider the situation shown in Figure 4-7, which is similar to an earlier figure. If we start with air under the dome and then let evaporation of the chemical take place until we reach equilibrium, the result will be that a certain *percent* of the molecules in the vapor space will be those of the chemical. (When we discussed the situation earlier, we focused on the pressure, which is of course closely related.) If this percent is *too high*, then the vapor will not be able to burn, as in the case of the gasoline tank. We call the maximum percent of chemical vapor at which burning can take place the *upper flammable limit*. Similarly, if the percent in the vapor space is too small, naturally burning would also not take place because in this case not enough heat would be released to keep the process going. The percent value below which burning will not take place is called the *lower flammable limit*. These two limits for the concentration of the chemical in the vapor form, expressed in percents, are called the *flammable limits*. You can well imagine that in determining the risk of fire from chemical vapors, we will want to know whether the concentration is within the flammable limits. These flammable limits vary from one chemical to another, and some representative values are shown in Table 4-3.

Thinking back to the gasoline tank experiment, you can well imagine that in the name of caution and common sense we should never assume we are safe because there is so much

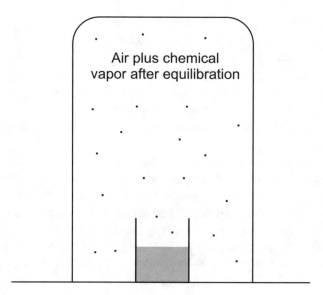

FIGURE 4-7
Equilibrium point in the evaporation process

flammable chemical in the air that it is above the upper flammable limit (UFL)! Thus it is really the lower flammable limit (LFL) that attracts our greatest interest. For example, we could make certain assumptions about safety if we had good reason to believe that the flammable concentration was below the LFL.

One important way to see if the LFL might be present in a potential emergency situation is to look at the temperature. In particular, once again consider Figure 4-7. The percent of chemical vapor in the vapor space in this experiment depends on temperature. As discussed earlier, as you raise the temperature of a chemical, this increases its propensity to evaporate, thereby putting more molecules in the vapor space. Therefore, considering the experiment in the figure, if we could keep the temperature low enough, we could keep the amount of chemical in the vapor space below the LFL, and we would no longer have a flammable vapor hazard. Thus there must be some temperature for any given flammable chemical that forms the dividing

TABLE 4-3
Selected examples of upper and lower flammable limits

Chemical	Lower Flammable Limit (LFL), Percent	Upper Flammable Limit (UFL), Percent
Acetone	2	13
Butane	1.8	8.4
Gasoline	1.4	7.5
Methanol	7.3	36
Vinyl acetate monomer	2.6	13.4

line between the vapor being below the LFL and the vapor reaching or exceeding the LFL. This critical temperature is called the *flash point*. Naturally it will vary from one chemical to another. Table 4-4 shows the flash points for the chemicals listed earlier in Table 4-3. Since gasoline is a mixture, the value given may vary with the formulation.

Thus in handling flammable chemicals, all other factors being equal, you would rather be handling them at temperatures below their flash points because in this situation, it is physically impossible to achieve a flammable concentration of vapor, even in a closed space with lots of time allowed for evaporation.

The flash point of a flammable chemical is often used as a valuable indicator of its level of hazard. For example, all other factors being equal, a material with a *higher* flash point would generally be thought of as *safer* because it would have to be at a higher temperature to put enough vapor out to be flammable, and thus it probably does not produce flammable vapor as easily or as quickly as a lower flash point material. In fact, quite often the term "flammable" is reserved for chemicals with a flash point below some threshold such as 73°F or 100°F, with the term "combustible" being applied to materials that will burn but that have a higher flash point. However, in this text, we will continue to use the term "flammable" generically for a burnable material with any flash point.

Materials that quite readily evaporate are called *volatile*. This is a qualitative term in the sense that we might use it to say that one substance is "more volatile" than another, or that a material is, say, "very volatile." You might reasonably expect, although of course there could be exceptions, that a material with a *lower flash point* would tend to be *more volatile,* and would also tend to have a higher vapor pressure at any fixed temperature. Think about this relationship for a moment to be sure you see the underlying rationale.

You should also know that in some data sources, the upper and lower flammable limits may be called the upper and lower "explosive" limits, denoted UEL and LEL instead of UFL and LFL. These terms mean the same thing; you should not assume that the explosive limits refer to explosion instead of fire. This usage occurs for historical reasons.

Where should you turn to find the values of various physical and chemical parameters such as LFL, UFL, flash point, density, boiling point, etc.? There are certainly some very good handbooks available with extensive data summaries. However, the best source is probably a *material safety data sheet,* otherwise known as an *MSDS*. These will become vital tools for us in this chapter as we perform modeling calculations for hazardous materials release scenarios.

TABLE 4-4
Selected examples of flash points

Chemical	Flash point (°F)
Acetone	0
Butane	-101
Gasoline	-45
Methanol	51
Vinyl acetate monomer	18

They contain basic physical and chemical information about the materials, as well as safety and health considerations that apply to their use.

Where do you go to see such MSDSs? Almost any organization that uses, buys, sells, or otherwise handles a chemical or potentially hazardous material will have on file an MSDS for that material. The law is rather wide ranging in this respect, with the intention of protecting workers, customers, or others who may become exposed to such chemicals. Thus you may find relatively easy access to such information even in stores that sell such materials. In addition, MSDSs are readily available by using standard search tools on computer networks. You are asked to gain some direct experience with these sources in the exercises that follow, and you should find it quite easy and interesting to do so.

Exercises for Section 4.3.2

1. [Institutional availability of MSDSs.] Identify two chemicals that you believe are in use in some quantity in your institution, at least one of which should not be primarily used in a laboratory setting. Find out if and where your institution has the MSDSs for these chemicals, review the MSDSs, and summarize in one paragraph the nature of the hazards identified there. (Hint: chemicals are used extensively in our homes as well as our institutions, so the identification of such use should not be difficult. MSDSs may be kept in various locations, such as the departments where the materials are used, the organization's safety or security department, the purchasing department, or elsewhere.)

2. [Library availability of MSDS-type information.] Determine the best sources within your library or, if not readily available there, elsewhere within your institution, of chemical properties information and associated safety information. (Hint: if it would be helpful for you to have specific target chemicals in mind, consider Exercise 6, below, at the same time.) Summarize the availability of sources that provide the values of the parameters discussed in this section for multiple chemicals.

3. [Vendor availability of MSDSs.] Identify two chemicals that you can buy in your local community and ask the vendors to let you see their MSDSs. Summarize how this process goes and how hard you or they had to look to find these MSDSs.

4. [On-line availability of MSDSs.] By using standard on-line computer search tools, determine the availability of one or more sets of MSDSs that you can access by computer.

5. [US regulations regarding MSDSs.] Outline the regulations issued by the Occupational Safety and Health Administration (OSHA) with respect to MSDSs, providing further detail on the subject of who is required to possess MSDSs for chemicals in their possession. (Hint: many libraries have the Code of Federal Regulations, a multivolume series containing the regulations issued by each federal agency. Alternatively, such regulations can be found and searched on-line by computer.)

6. Using any of the sources of MSDSs or chemical information that you pursued in connection with the above exercises, or any other source of information, fill in as many of the blanks as you can for the chemical properties shown in Table 4-5. For any values that do not seem to be available, indicate why it is reasonable that they are not provided on the MSDSs.

TABLE 4-5
Selected chemicals and their properties to be determined in Exercise 6

Chemical	Molecular Weight	Specific Gravity	Boiling Point	Flash Point	LFL (%)	UFL (%)	Vapor Pressure at about 68°F	Vapor Pressure at alternative temp.
Acrylonitrile								
Anhydrous ammonia (liquid)								
Benzene								
Cyclohexane								
Hexane (normal)								
Kerosene								
Nitric Acid (fuming)								
Turpentine								

7. Within a given family of chemicals, it is not uncommon to find a general relationship between molecular weight, volatility, and boiling point. For example, the boiling point might a good indicator of volatility because it reflects the fact that more volatile substances, in that they vaporize more easily, are likely to reach a vapor pressure equal to atmospheric at a lower temperature. On the other hand, heavier molecules might correspond to lower volatility because they would require more energy before they can become freed up and leave the liquid state. Explore the validity of this idea by considering the following family of chemicals: methane, ethane, propane, butane, pentane, and hexane. This family, called *alkanes,* all consist of chains of carbon atoms (ranging from 1 to 6 in the above list) surrounded by the maximum complement of hydrogen atoms that can attach to them under the principles of chemical bonding. (Note: You should be able to find the chemical data you need by the methods practiced above. If you encounter variations on these chemicals in your database, use the values for the "normal" or "(n)" form of the chemical, which has the simplest single-chain molecular structure in each case.)

8. When the concept of LFL was introduced in the text, it was referred to as a percent "of the molecules." For example, an LFL of 5% would mean that at least 5% of the molecules in the vapor space would need to be those of the flammable material in order to support combustion. However, the MSDSs you have examined probably refer to the LFL as a "percent by volume," a distinction that may or may not have caught your attention. These two concepts are actually essentially the same because of a physical principle applying to "ideal gases" that says that an equal number of gaseous molecules take up the same space under the same conditions, regardless of their mass or chemical composition. Given this principle, consider a vapor space that contains 2% acetone molecules and 3% methane molecules, so that they are in a volume ratio of 2:3. What is the ratio *by mass* or *by weight* of these two components in the vapor space?

4.3.3 Characterization of Toxicity Hazards

The previous section focused on flammability hazards, and the data presented there suggested that flammable vapors occur when a few percent of the vapor space by volume is occupied by the flammable material. Toxic hazards, to be discussed in this section, sometimes result from much lower concentrations, and so the whole framework for measuring toxic hazard levels is somewhat different from that encountered above.

To begin, keep in mind that there can be *acute toxic hazards,* such as when a poisonous gas causes severe injury or death almost immediately. For example, a strongly acidic gas might damage the respiratory tract so much that a person could no longer breathe effectively. Or a gas like carbon monoxide might attach to the hemoglobin in the blood, making it impossible for the body to transfer oxygen to the vital organs. There can also be *chronic toxic hazards,* such as when longer-term exposure to a gaseous chemical might cause gradual organ deterioration or some specific disease, such as cancer, even though a short-term exposure may have no identifiable effects.

Thus there is no single measure of concentration that represents "the" hazardous level. Rather, the levels of exposure that may constitute a hazard are determined by a combination of both concentration and exposure time. In fact, you may have already noticed while perusing MSDSs in connection with the previous exercises that under a topic such as "health hazards" a number of exposure levels may be discussed. For example, very often you will find *TLVs,* standing for *threshold limit values,* which represent levels of acceptable exposures under normal working conditions, and which may be further subdivided into daily time weighted averages as well as limits for short-term exposures, generally 15 minutes. TLV values are issued as recommendations by the American Conference of Government Industrial Hygienists (ACGIH).

While such TLV values are much in keeping with the use of MSDSs to protect workers in the workplace, they do not provide much useful information for determining acute risk levels associated with episodic events. For example, if such exposures levels are permitted day in and day out in the workplace, one would expect that much higher exposures could probably be tolerated on a one-time basis by members of the public or the emergency response agencies in connection with a hazardous material spill. Various organizations have estimated such levels for some toxic chemicals, although often they are not listed on MSDSs.

One set of somewhat more relevant measures are the *IDLH* values, referring to concentrations that are *immediately dangerous to life or health.* These values have been published by NIOSH, the National Institute for Occupational Safety and Health. Although these values still are not precisely oriented toward emergency planning and management, as they have been developed to indicate when occupational workers should use respirators to protect against airborne toxics, they are both relatively widely available, and in the right general category, so they represent a "compromise" set of values that might be used at least for some practice or rough emergency planning calculations.

There are many other measures of threshold or dangerous concentration levels, some even specifically oriented to emergency planning. However, they are not always widely available, and both the values and the organizations issuing them tend still to be in some state of flux. Thus we shall not review them here. This is because our primary interest is in using mathematical modeling to determine the zone within which such dangerous levels might be encountered. We will certainly want to turn to MSDSs or other sources so as to use reasonable values in our calculations, but the fine points of choosing just the right value for a real case are too intricate to treat here. The best guidance for our work is to review the MSDS and other readily available information and choose a toxicity level of concern based on a description of the various values that may be given there. Obviously, if an IDLH value is given or if some concentration level is referred to as an "emergency action level" or something similar, then it would be a reasonable value to use in our practice calculations.

In any case, you will find that such toxic levels are generally expressed in different units from the levels discussed in the previous section on flammability risks. Recall, for example, the use of "volume percents" to characterize the LFLs and UFLs, which we said was roughly equivalent to "percent of molecules in the vapor space." Thus an LFL of 5% would correspond to having 50,000 molecules of the hazardous material for every million molecules in the vapor space, since $50,000/1,000,000 = 0.05 = 5\%$.

The short calculation above leads to a new and useful unit: *ppm*, or *parts per million*. In particular, the 5% concentration referred to above could be restated as "50,000 ppm," meaning that there are 50,000 molecules of the chemical for every total of one million molecules in the vapor space. For chemicals that have a significant toxic risk, you will generally find that the actual toxic concentrations (e.g., IDLH values) will be in the tens or hundreds of ppms (sometimes even lower), and hence much lower than the values customarily encountered when dealing with flammability risks. (Note, this ppm unit is different from the ppm unit customarily used with water contamination, where it is based on weight percentage, as discussed in Chapter 2.)

On some occasions, toxic concentration levels in the air are expressed in units of mg/m^3 (milligrams per cubic meter) or perhaps some other units that involve mass per unit volume. For example, suppose a level of concern for ethylene glycol (automotive antifreeze) in the air was published as 110 mg/m^3. How does this relate to ppm units or volume percent units, as discussed in the previous paragraphs? The answer involves some very simple but fundamental principles from chemistry, namely:

a) *Under fixed conditions of pressure and temperature, a given volume of gas will contain the same number of molecules of any gaseous substance whether the molecules are small, light molecules or larger, heavier ones.* If this sounds questionable, you should know that it was also soundly rejected by the scientific community when first proposed by Avogadro in 1811, and that it did not reach wide acceptance until about fifty years later. The basis for "Avogadro's principle" is largely that the molecules in a gas are sufficiently far apart that their individual size is not an important factor in how much volume the gas takes up at given conditions of temperature and pressure.

b) *If M is the molecular weight of a gaseous substance, then M grams of that substance should take up the same space and contain the same number of molecules as N grams of a substance whose molecular weight is N.* You can see this better by looking at an example: If the molecular weight of nitrogen gas (N_2) is about 28 and the molecular weight of hydrogen gas (H_2) is about 2, then nitrogen molecules are about 14 times as heavy as hydrogen molecules (since $28/2 = 14$). Therefore 28 grams of nitrogen gas should take up the same space as 2 grams of hydrogen gas, since they should involve the same number of molecules. (Make sure you understand this before moving on. It is the hardest step in the logic.)

c) *The actual number of molecules contained in M grams of a substance with molecular weight M is about* 6.022×10^{23}. This is called *Avogadro's number.* This quantity of any substance is called one *mole* of the substance. (By the previous step, you know that this number is the same no matter what substance you have or what M is.) In other words, *one mole is M grams of a material, and the number of molecules in it is given by Avogadro's number.*

d) *One mole of a gas under "standard temperature and pressure conditions" occupies a volume of 22.4 liters.* You might wonder where this number 22.4 popped out of. Well, this value can be determined both by experiment and by theoretical calculations, but you have to go and take a chemistry course if you want to know more about it!

e) *If you have x grams of a gas and if its molecular weight is M, then you would have* x/M *moles of the gas and the number of molecules would be* $(x/M) \times 6.022 \times 10^{23}$. Look at this in terms of a specific example. If you have 56 grams of nitrogen gas, then you know you have $56/28 = 2$ moles, or $2 \times 6.022 \times 10^{23}$ molecules.

It can be quite hard to absorb all the above steps until you really see how these principles are used, so here is an example based on the problem that led to all this. Work through it very slowly and make sure you understand the logic.

Suppose an emergency level of concern for ethylene glycol is published as 110 mg/m^3. What does this correspond to in ppm units? To answer this, we will need the molecular weight of ethylene glycol, whose formula is $C_2H_6O_2$. Therefore, using the atomic weights in Table 4-1, we have:

$$M = 2(12.011) + 6(1.001) + 2(15.9994) = 62.03$$

Now we can proceed in just the same way as in many of our previous unit conversion problems:

$$\frac{110 \text{ mg glycol}}{\text{m}^3 \text{ air}} \times \frac{(1 \text{ m})^3}{(100 \text{ cm})^3} \times \frac{1000 \text{ cm}^3}{1 \text{ liter}} \times \frac{22.4 \text{ liters air}}{1 \text{ mole air}} \times \frac{1 \text{ mole air}}{6.022 \times 10^{23} \text{ molecules air}}$$

$$\times \frac{10^6 \text{ molecules}}{1 \text{ million molecules}} \times \frac{1 \text{ g}}{1000 \text{ mg}} \times \frac{1 \text{ mole glycol}}{62.03 \text{ g glycol}} \times \frac{6.022 \times 10^{23} \text{ molecules glycol}}{1 \text{ mole glycol}}$$

$$= \frac{39.7 \text{ molecules glycol}}{\text{million molecules air}}$$

$$= 39.7 \text{ ppm}$$

So we can conclude now that the level of 110 mg/m^3 specified for ethylene glycol could also have been listed as 39.7 ppm.

A few comments about the above calculation may be helpful as you try to get it organized in your mind. The first group of factors came from our strategy of converting the denominator (m^3 air) into millions of atoms of air, since that's really the denominator in ppm units. Then we went on in the next group of factors to deal with the numerator (1 mg glycol) and convert that to molecules of glycol, which is the numerator we want for ppm units. Some people wonder whether, in the first denominator, we should have m^3 of the air/glycol mixture, rather than just air. This distinction between pure air and air/chemical mix is usually not made because, since we are only going to be using the ppm units when the chemical concentration is quite small, they have essentially the same properties. Therefore, people often write "air" as above and mean "mix," and you can review the above calculation with this substitution and you will see that the answer is exactly the same. We did not really use any special properties of air at all, only Avogadro's principle, which applies as well to the mix. To be precise, we wrote "air" in keeping with standard usage, but technically we meant "mix."

Note also how useful our standard unit conversion scheme is for a problem like this, and by writing out all the factors before doing any of the arithmetic, you can see how many of them cancel, including ones like Avogadro's number, making the math much simpler. This layout also makes it easy to check your work and retrace your logic.

Exercises for Section 4.3.3

1. Search the literature, either on-line or with printed sources, to determine reasonable concentration levels that you might use when trying to assess the risk from airborne concentrations of each of the following: ammonia, carbon monoxide, chlorine, methanol, nitrous oxide, and phosgene. (Hint: don't be surprised if you find a wide variation in the amount of information on MSDSs for the same chemical, but from different manufacturers or different sources. Also, if you are accustomed to on-line information searching, do not completely forget other kinds of sources that may exist at your institution in case you run into difficulty.)

2. Suppose two different organizations publish emergency action concentration levels for a given chemical. Organization A publishes a value of 10 ppm based on their analysis of the research data, and Organization B publishes a value of 50 ppm based on their interpretation of the same data.

a) Which organization would you say has the more *conservative* standard? Explain.

b) If a toxic vapor cloud of this material is forming around a spill and you calculate the extent of the hazard zone of the cloud based on the two different published concentration values, which value will lead you to identify a larger hazard area? (Note: the identification of hazard levels is somewhat judgmental, generally involving qualitative or semi-quantitative extrapolation from data that may be based on animal studies or very different conditions or concentrations. This is why there can be different interpretations.)

3. The first two parts of this problem involve specific conversions to and from ppm units. The last part asks you to derive a general formula that could be used for all such problems. You may wish to do the specific parts first to improve your mastery of the logical process, or you may wish to do the third part first and then use that result for the first two parts. (Note: the molecular weight values for these materials should have been determined earlier in Exercise 6 of Section 4.3.2.)

a) Convert a 75-mg/m^3 concentration value for ammonia to ppm units.

b) Convert a 200-ppm benzene concentration to mg/m^3.

c) Consider a chemical whose molecular weight is M. Find a simple equation that can be used to convert between a value A, representing its concentration in air in terms of mg/m^3, and B, the equivalent concentration in ppm units.

4. In Chapter 3, just at the end of Section 3.7, the ppm unit was briefly introduced in connection with nitrogen oxide concentrations. At that time, you were simply given the conversion relationship that 1 ppm of NO_x was equivalent to 1.1×10^{-7} lb/ft^3. Based on your experience in the current section, can you reconstruct the basis for that conversion factor?

5. We have encountered several units for specifying the concentration of a chemical vapor in the air: ppm, lb/ft^3, mg/m^3, and volume percent. Suppose that you have actually measured all of these in the field at a temperature of $68°F$ and atmospheric pressure of 14.7 psi. Now suppose that night arrives and the temperature of the same air mass drops, so that it becomes more dense, while the atmospheric pressure remains the same. If you were now to repeat your measurements of the concentration expressed in each of the four sets of units listed, which ones would still have the same value, which would change, and in which direction would these latter change?

4.3.4 Other Forms of Hazards

Other forms of hazards have also been referred to earlier, such as pool fires, vapor explosions, BLEVEs, and others. Since these are often included in the computerized models used for emergency planning and risk analysis, you may encounter them as you apply such models later in this section. The principal risk from pool fires would, of course, be the intense heat output, and the hazard zone around such fires is generally calculated by applying standard thermal radiation criteria associated with the burning of skin or the ignition of combustible materials. These criteria may be built into the model itself, and so you may find that the hazard zones are calculated automatically without any additional input data being requested from you as the user. The other scenarios are somewhat more specialized, and you may wish to read the documentation for the model you are using before trying to carry out any meaningful calculations with submodels for such phenomena.

4.4 Typical Quantitative Issues

Let us begin by reviewing a variety of situations in which some quantitative analysis of a hazardous material release scenario might be needed.

Situation 1. You are the mayor of a small city, and you have just received a phone call from the fire chief, telling you that there has been a derailment on a freight line that runs through the city. It appears that some chemicals might be leaking from the overturned railcars. A few minutes later your press secretary runs in and tells you that the phone is ringing off the hook with media inquiries about the "fire and chemical spill" at the rail accident, and about whether you are going to order an evacuation. At this point the fire chief calls back and tells you that the situation is getting more serious, and that it may be necessary to evacuate the surrounding area if the situation is not brought under control quickly. He recommends that you come down to the site and meet right away with him and the police chief to plan your strategy.

Many thoughts are going through your mind as you are driven to the site in a police cruiser. There had been a big controversy during the last election, when your opponent argued against subsidizing new retail development in that part of town, so near to the rail line and a number of industrial plants. But you had prevailed and had gone ahead and built a pedestrian mall and other public facilities there. One of the old factories had even been totally renovated into a facility for assisted living. Some opponents had focused on the safety issue, since everyone in the city knew that many freight trains with dangerous cargoes moved through the city. But you had argued that they were being ingenuous by pushing the safety issue; there had never been a significant chemical spill or other major problem associated with the rail or industrial operations in that part of town. Was fate now turning the tables and showing that they were right after all?

You arrive at the site and are quickly hustled into a meeting in a makeshift operations center in a local building about two blocks away. The deputy chief reports that there are at least three leaking cars, and the material is starting to form a large pool as it runs down the bank by the tracks. Because of the angle of the overturned cars and the dense smoke from a separate fire involving two boxcars, they can read the identifying placard on only one of tankcars. The number corresponds to vinyl acetate. The chief has already ordered the evacuation of the buildings on the block facing the incident, but he is thinking about moving the evacuation distance to half a mile. The police chief says that his force is ready to implement the evacuation, but it is very complicated because of the elderly-housing project and the dense development. Looting is a concern if the buildings are all abandoned too. A radio call comes in to say that the other cars have been identified by a fireman on the roof of a building on the other side of the incident. The materials are acetone and nitric acid.

The state hazmat unit has been called, but their response time is expected to be at least another hour because they need to be called in from diverse units that may be some distance away and they have to go through an elaborate mobilization procedure. A contract laboratory team arrives with further monitoring equipment, so the chief tells them what chemicals are believed to be leaking, and they put on breathing apparatus as necessary and head out to take measurements of concentrations in the vicinity. But their mobilization takes about half an hour, so in the meantime you decide to see what information your team can contribute to a quick analysis of the situation. Table 4-6 lists a number of the items you might like to know and how the information might be obtained. Ideally, you would like to put the information together to answer questions along the following lines:

- Taking into account the spill rates and the size of the pool, what vapor concentrations should you expect in different directions and at different distances from the accident site.
- How do these concentrations compare with the flammable and toxic limits?
- Is there significant potential for a flammable or toxic vapor cloud to extend some distance from the location of the accident?
- Are there any other important hazards, such as toxic combustion products if the pool or vapor cloud ignites, reaction products from interaction between the chemicals, BLEVE formation, etc.?

These are largely quantitative issues, and if you knew the answers, it might help you decide about evacuation or otherwise handle the emergency more effectively.

TABLE 4-6

Selected information needs during evaluation of an emergency

What you would like to know	Source of information
Flammability properties of chemicals	MSDS, in fire department file, or call the chemical industry emergency 800 number.
Toxicity properties of chemicals (e.g., various hazard levels)	Same as above.
Other special hazards of chemicals	Same as above.
Wind direction	On scene personnel slightly removed from incident. Call local weatherman.
Wind speed	Same as above.
Expected changes in weather, if any	Same as above.
Availability of foam to reduce effective area of pool	Fire department personnel. Need to check compatibility with chemicals.
Location of ignition sources	Estimate significant sources downwind. Try to eliminate.
Potential for chemical interactions	Chemical industry emergency number or local chemist on call.
Other basic chemical data	Same as above.

Therefore, what if you had at your fingertips a computer and a menu-driven software package that could generate these answers almost instantly? It would be very helpful.

Understand that no one is going to manage such an emergency simply by looking at the output of a computer program. What the program might do, though, is give you fast access to some very useful quantitative estimates. For example, if a toxic vapor cloud could extend only a few feet from the pool under the conditions you have, then certainly this would argue against evacuating a large area. After all, evacuations have their downside, not just along the lines listed earlier, but in terms of human risk from accidents or emotional stress, especially to more fragile groups such as you might have in the assisted living facility. As a compromise, you might decide to "shelter in place" by arranging for people to close their doors and windows, and by shutting off air-handling equipment in larger buildings. On the other hand, if the flammability risk extends some distance from the accident, it might be better to get people out right away, before a larger fire might make this very difficult.

Is it realistic to expect that some quantitative calculations along the above lines might be carried out during an emergency? Yes, but only if there is some specialist in the modeling participating in the emergency response. Larger fire departments, as well as hazmat units, might well have such specialists, as well as having the computers and programs available right on the emergency vehicles.

But certainly, even if such resources are available, it would be far better if much of the analysis and planning had been carried out well in advance of the incident, where there are not the same pressures of time and where a wide range of technical quantitative analyses might be carried and reviewed with all parties. This idea leads to the next hypothetical situation.

Situation 2.   Imagine that you live near a large manufacturing plant, and you observe that tank trucks enter and leave the facility with some frequency. You also observe a number of chemical storage tanks near some of the buildings on the site. From time to time there are peculiar smells in the air, and when you call up the company to express concern, they readily admit that there was some kind of a spill or leak, but they assure you that it is under control and that there has been no danger to your health.

But then you see on the news from time to time stories of chemical leaks in other places, and you keep wondering whether your situation is any different. Might the same kind of major incident occur at this plant, and might you be at some risk from it? This is a very reasonable question for you to ask, although you should hope that someone has already asked it before and has taken the answer into account in building or operating the plant.

Since you're not sure whom to call and you don't really want to come across as a "trouble maker," you sit on the matter for a while until one day you're at a little league game, and you find that you're sitting next to one of your town's selectmen. You mention just in passing about what a bad accident you heard had happened in some far-off place, and you ask whether your town is really on top of the issue of possible similar incidents here. The other person doesn't know much about it, but says that he will ask about it at the next meeting.

At the next selectmen's meeting, no one knows much about the town's preparedness for hazardous materials emergencies, so the selectmen decide to invite the fire chief to the next month's meeting to address the subject. An announcement in the paper attracts a number of citizens, and, in fact, when the chief comes, he is accompanied by a member of a regional emergency planning committee and a member of the state hazmat team from a neighboring town's fire department. They explain that they conduct a number of hazardous materials training exercises each year, based on the kinds of materials that are used in the various towns or are likely to be shipped through on the highways or rail lines. Often, in conjunction with industrial representatives, they identify certain plausible accident scenarios, and they often use computer models to estimate the range of hazards to expect (e.g., how far a toxic vapor cloud might extend). Based on this, they may conduct an actual field practice session, or they may run a "table top" exercise which involves talking their way through how they should respond to such an incident.

The range of participants in these exercises is quite interesting, often including for example: fire department, police department, state hazmat team, state police, EPA, FEMA, state environmental agency, state transportation department (since many accidents are on state highways), civil defense organization, town public works department, local hospital emergency room representative, local company representatives, private contractors (who would be called for actual clean-up), local university representatives (who help with the modeling), school department representative (because of special evacuation problems), and others. This wide variety of participants in emergency planning or in an actual emergency underscores the need for careful advance planning, clear lines of communication and authority, and good overall organization and management of the response to an incident.

A key observation from this "Situation 2" is that modeling can (and does) play a key role in the local planning and training processes when, for example, you are trying to analyze the hazards of a hypothetical incident and you need to know how extensive these hazards might be. This would also be true within an individual company that handles significant quantities of such chemicals. They certainly have a keen interest in the risks from their operations, and many of them do indeed conduct risk assessment studies that include as one part the development and analysis of potential spill or release scenarios. (In fact, in some cases this is required by law.) Some also conduct public meetings for nearby residents and town officials to let them know of the efforts they are making to protect public health and safety, and to hear if there are concerns or questions from the community.

Situation 3.   This is still quite a different kind of situation. Imagine now that you work for a government agency and you have been asked to develop a priority ranking of hazardous chemicals that are shipped "in bulk," meaning in tank truck or tank car or barge tank loads. Your agency is planning to review the regulations for the transport of such materials, and it wants to start with the most dangerous ones first. How might you proceed?

You might think of using some parameter such as flash point, which we discussed previously as a rough measure of flammability risk, but that would not give you a way of comparing flammable materials with those that primarily pose a risk of toxicity. You might develop a point system for ranking each chemical with respect to each kind of risk, and then add up the total points for each chemical. These approaches have been used, and their simplicity is a real advantage. But if you were to have mathematical models available for assessing the evolution of an accident scenario, you might also define a "reference spill scenario" and accompanying weather conditions intended to be somewhat average conditions, and then compare the "hazard distances" calculated by the model for the various kinds of risks, especially toxic and flammable vapor risks, but also perhaps including pool fires and other risks. You could then rank the chemicals, or put them into categories, by the hazard distance calculated for each one. This approach would integrate a number of the chemical properties that contribute to the hazard, and would give an overall quantitative measure that would be reasonably reflective of hazard level.

This last approach to Situation 3 has been used widely, not only in the context of that specific situation, but also by private companies trying to prioritize their risks so as to decide how to best invest their resources in safety improvements at what might be hundreds of plants around the world. Even the so-called "orange book," the first responder's guide mentioned earlier, was developed by putting chemicals into categories according to their hazard distances and other properties, and then including recommended evacuation distances and other protective measures based in part on the results of model calculations.

The main conclusion from this discussion of quantitative issues arising from hazardous materials emergencies is that there are several distinct levels at which the models can be useful, ranging from detailed site-specific response to long-range planning, training, and prioritization of risks. And even if you are using a model that is somewhat simplified and thus may have fairly large error bands or uncertainties associated with the calculated results, it may have important uses for so-called "comparative risk assessment" such as in Situation 3, even if you would not want to rely heavily on the results during an actual emergency situation such as in Situation 1.

Exercises for Section 4.4

1. For either your hometown or for the town or city in which your institution is located, find out if there is a "local emergency planning committee" or equivalent organization responsible for preparedness for hazardous materials emergencies. Summarize their scope of activities. Find out to what extent they use mathematical or computer modeling in their work.

2. Contact a local fire department to inquire about the degree to which they have equipment and procedures for dealing with hazardous materials emergencies. Summarize their observations on the potential for such incidents in the local community and what they consider to be the largest sources of risk. (Note: an individual or class visit to a fire department, especially a "hazmat" unit, can be very interesting and would touch on many of the topics to be discussed later in this chapter. Fire departments usually welcome such interactions with members of the community.)

4.5 Structure and Use of Hazmat Computer Modeling Packages

This section is intended as an introduction to publicly or commercially available computer software packages for analyzing release scenarios for hazardous materials. In order to repeat the calculations reported in the subsequent sections or to do the exercises there, it will be necessary for you to have such a package available.* While the many available modeling packages differ in a number of details, the overview in this section should be readily adaptable to most or all of them. Figure 4-8 shows the general structure of a typical hazmat modeling computer package.

Let us discuss each of the main components of such a model. First, such models are usually oriented around the analysis of a real or hypothetical accident scenario, and you have to be very specific in completely defining all aspects of the scenario that would be relevant to the nature and severity of the consequences. There may, of course, be some variation in input requirements depending both on the modeling approaches chosen by the developers and also on the level of detail intended to be incorporated in the model. (For example, one model might ask for detailed information on some family of parameters, while another more general model might have some "average" values for those parameters already built in.) While the background discussion in the earlier sections of this chapter should cause you already to be reasonably aware of the kinds of scenario characteristics that are relevant, a brief review of the list given in the top box of the figure will let you anticipate some of the areas that often cause difficulties or questions:

Chemical involved. This may sound straightforward, but it isn't always so easy for users who are not chemists, the reason being that many chemicals can exist in a number of forms and concentrations, and these are highly relevant to their properties and their hazards. For example,

---

* See the comments in Section 4.7.

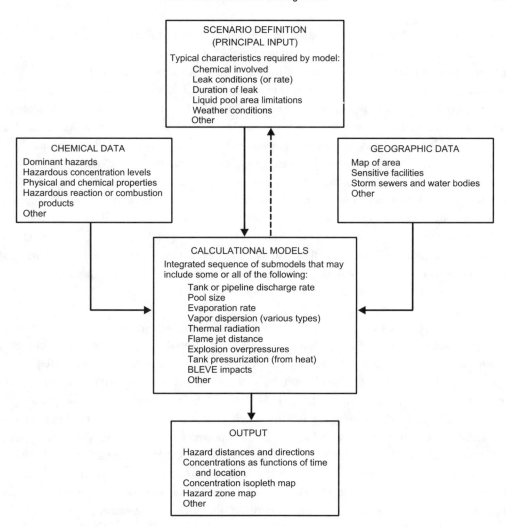

FIGURE 4-8
General structure of hazmat modeling packages

if you were analyzing the hazards from a leak of ammonia, you really would need to know whether the leak was of liquid ammonia or gaseous ammonia, and if of the former, whether the chemical was anhydrous ammonia (pure, liquefied ammonia) or ammonia solution (as is used in doing laundry and cleaning). If the incident involved nitric acid, say, then the concentration would be highly relevant because you could be dealing with fuming nitric acid, which gives off very aggressive acid vapors, or dilute nitric acid, which is much less dangerous when spilled. The United Nations ID numbers for hazardous materials are generally shown on their packaging and on the placards of vehicles carrying them in bulk quantities, and these are the most commonly accepted method for identifying them unambiguously for emergency planning purposes. (Some ID numbers refer to groups of similar chemicals.) Naturally, the quantity of chemical available for discharge would also be needed.

Leak conditions.   A basic scenario may involve a rupture of a tank or a break in a hose, pipe, or fitting. Obviously the size of the opening is highly relevant to how fast the material leaks out, and most models will want some estimate of this. After all, if a large tank of volatile chemical leaks through a 1" diameter instrumentation port, it is well possible that the material will evaporate so quickly after exiting that no large pool will form. Thus the amount of material in the air in the vicinity at any given moment might be relatively small. On the other hand, if a massive tank failure leads to the instantaneous formation of a pool the full size of the diked area, then because of the large surface area, much material would be evaporating at the same time, leading to a much more concentrated and dangerous vapor cloud. For an irregular type hole in a tank, such as from a tank-truck traffic accident, you may have to estimate the size and characteristics of the opening to most reasonably suit the requirements of the model. You may also need to specify the size and shape of tank that is leaking and whether it is a pressurized tank or pressurized pipeline, as all these factors would obviously affect how fast the material would be pushed out (by gravity or internal pressure) through the opening. Alternatively, some simplified models might just ask you to specify an overall discharge rate.

Duration of leak.   Models may contain various options for treating the duration of the leak for calculational purposes. For example, the duration of the leak from a large tank discharging through a broken pipe fitting could be calculated along the lines suggested above, based on the discharge rate (which may go down with time as the weight of material in the tank diminishes) and the amount of material in the tank at the start of the scenario. On the other hand, you might want to make the assumption that by a certain amount of time after the beginning of the scenario, the emergency responders would be expected to stop the leak. (They carry tools and many kinds of clever patching equipment.) Thus you might impose a cap on the discharge time, letting the model compare that with the time it would take to release all the material. Then again, you might be more interested in calculating some kind of "worst case" scenario, thus perhaps preferring to tell the model to consider the release to be essentially instantaneous and to include the full contents of the tank.

Liquid pool area limitations.   First, keep in mind that not all leaks form liquid pools. For example, the original material may be gaseous to start with, or it may be a liquid with such a low boiling point that it vaporizes immediately upon discharge. It may also burn immediately upon discharge, such as a flame jet at the rupture in a pipeline. However, in the situation where a liquid would be expected to form a pool on the ground, the size of that pool may be limited by topographic constraints (e.g., it may collect in a ditch) or by a dike or berm constructed for containment purposes. When such constraints do not exist, the model would still need to assume some pool size in order to calculate the evaporation rate, and to do this it might look for some equilibrium size at which the evaporation rate would equal the discharge rate, or it might use a time-dependent pool size based on some spreading model. Spills onto water can be treated similarly, but there are additional physical and chemical considerations one would need to account for to estimate the size of a floating pool or a "slick," meaning a mass of chemical that is immiscible with and lighter than water, so that it spreads out on the surface. In any case, the definition of a scenario for input to a model must consider such pool area issues.

Weather conditions.   There are a number of ways in which the weather can affect the evolution of a hazmat scenario. Perhaps the most obvious is that the temperature will affect the evaporation rate from a pool because, as has been discussed earlier, if you raise the temperature of a liquid, you increase its propensity to evaporate. Temperature may also affect the pressure in the vapor space above a liquid in a closed tank, and this may influence the rate at which it discharges from a release point. (Of course, if the tank is heated by a burning pool fire, that will have a much larger effect.) Wind also obviously will affect the distribution of toxic or flammable vapors, although the exact effect of wind speed may not be clear. For example, if the wind blows harder, it will carry the vapors farther faster, but in the process it may cause more dilution that could have a counterbalancing effect on the hazard distance. In fact, in the previous chapter on air pollution, you have already seen how a key meteorological input to the concentration values is the atmospheric stability class, which combines the effects of wind and sunlight into a parameter value that roughly captures the degree of dispersion of the vapor in the downwind direction. (Recall Table 3-1 there.) These same considerations apply to the present situation, except that for some chemicals, their vapors may be sufficiently dense that the Gaussian plume equation (discussed in Chapter 3) does not apply.

Other scenario characteristics.   Given the wide range of both general purpose and specific hazmat models that are in use, there can certainly be a number of additional input requirements. Some models might need more information on the leaking container or the container exposed to a fire, an example of the latter being the size and setting of any pressure relief valves. Other models might also be configured to calculate probabilistic aspects or overall risk values; these might need further information to assign the scenario to the right probability class. Many other examples could be listed, but they should all have a fairly clear logical connection to the final objective of the model calculations.

Note that the top box in Figure 4-8 is connected to the central model box by two arrows. The solid arrow indicates that the scenario characteristics are provided as input to the model calculations, as you would expect. But the dashed return arrow indicates that the provision of this information is often an interactive process such that the model asks the user for some part of the input and then gives some feedback about the assumption that might cause the user to refine the original input. For example, a user might specify a scenario under which a million-gallon tank is leaking out through a broken pipe for whatever duration it takes to empty the tank. But the model might reply with a message to the effect that this would take a certain number of hours, during which it might be reasonable to assume that some corrective action has been taken. Thus the user may decide to input a maximum time on the leak duration, which could substantially (and realistically) reduce the estimates of the hazard distance and duration.

Aside from the definition of the scenario, there are other inputs that the model needs in order to calculate the hazards. For example, the left box in the figure points out that a variety of chemical information is needed in order for the model to calculate how the material would discharge, evaporate, disperse, or otherwise cause one or more types of hazards. Some models require the user to provide this information as part of the input process, often in an interactive mode as the scenario definition is provided. (That is, the model would ask only for chemical information relevant to the modeling of the type of scenario being defined.) Other models have a built-in database covering a very wide range of chemicals for which calculations are likely to be carried out. These are more convenient because the model developers will have been sure to provide the specific parameters needed for the calculations. When there is no internal database, the user will need to turn to MSDSs or other reference sources for the needed information. As discussed earlier in the text, often the emergency toxic hazard thresholds will be the most difficult to find. Another problematic area can be whether the released chemical has important reaction or combustion products that also need to be modeled, and some discussion with chemists familiar with the material may be necessary in order to investigate such aspects.

The last category of general input information is shown in the box on the right side of Figure 4-8. If a site-specific real or hypothetical scenario is being modeled, then it will generally be relevant to see how the hazard zones overlay on the surrounding geographic area. This requires that a good map of the site be provided so that the user can plot the final results on it, or, alternatively, so that the model package itself can do the plotting, providing direct graphical output in the form of a site area map with hazard zones indicated. The availability of atlases of digitized maps makes it relatively straightforward for models to carry this information as an internal database, which some do. Note that the possibility that spilled material may leak into storm sewers or flow into surface water bodies is of constant concern to emergency responders, and this kind of information should help the modelers make realistic assumptions about pool formation and location.

The central box in Figure 4-8 represents the heart of the modeling package. It is important to realize that for the various phenomena under consideration, there is usually not just one single way to model them. In fact, every model is really just an approximation to reality, and various approximations may have different positive and negative aspects. For example, one may be expected to be more accurate but have extensive data requirements that might be difficult to meet for many important chemicals. Or another may be a good "compromise" for a wide range of conditions, but be quite rough for conditions near the extremes. One may be based on empirical experiments for certain materials, then extrapolated by judgment to other classes of materials, while another may be based on more fundamental scientific principles and mechanisms, although it may be less "validated" by comparison with experimental results. Still another may be designed for general ranking or planning problems (as for example in "Situation 3"), where simple but robust modes of approximation may be appropriate, while another may be planned to assist in facility design or on-site response, where details are much more important. Since the users of such models also may range from sophisticated modelers or engineers to regulatory or planning personnel with much less technical background in modeling, this factor would also imply the need for different levels of models and user interfaces.

Given this relatively wide range, it is not practical here to investigate the structure of the individual submodels for the various phenomena under consideration. For the modeling package that you intend to experiment with in connection with this chapter, the model documentation should provide a summary of the approaches taken in each case. Furthermore, Chapter 7 does develop some examples of such approaches for readers with sufficient technical background to study these aspects. But the key point of this chapter is that you do not have to be an expert in all the details of the modeling approaches in order to use a model package effectively. What you do need is a good basic understanding of the fundamental principles governing the processes being modeled, similar to the material covered earlier in this chapter. This will let you surmount many of the problem areas typically encountered in applying modeling packages.

Let us summarize here some suggestions that can help you avoid obtaining inaccurate and misleading results from hazmat modeling packages, or otherwise help you make the most of your modeling package:

Units.   Be sure that you input your data in the units requested by the model. For example, there is a big difference between a distance of 10 feet and 10 miles, or between a wind speed of 5 mph and 5 m/s. Make sure you understand any abbreviations for units involved in the input data.

Chemical data.   Make sure you use the data for the right chemical, and ask advice if you are confused by some of the distinctions you find on MSDSs. If you're dealing with a 5,000-gallon storage tank of anhydrous ammonia liquid, for example, and you find a density of 0.7714 g/l, not even the units you are used to, you might just do a mental check by saying that a liter is about the same size as a quart container of milk, but this liter of material would weigh less than a single gram! This doesn't sound like any normal liquid. Or you might say that one liter is a thousand cubic centimeters, which would weigh 1000 grams if it were water, but this material weighs less than a thousandth of that! Since you would hardly expect another liquid to be a thousand times lighter than water, you might read more thoroughly and see that you have the data for ammonia gas, not what you wanted. If you were to use this value in your calculations, the reduced source mass that would be calculated and used by the model would cause the hazard to be grossly underestimated. Thus it is important always to be thinking about the values you are using, to be sure they are what you really want for your situation.

Feasible scenarios.   Be sure that you think in careful physical terms about any scenario you are analyzing or defining. For example, for a leaking tank scenario, the internal pressure is relevant to the discharge rate, since a high internal pressure will tend to push the liquid out faster through the leak. The model will generally ask you about the type and dimensions of the leaking tank and whether it is pressurized or essentially at atmospheric pressure. If you are dealing with a chemical that boils below ambient temperature, such as propane or ammonia, and if, without thinking carefully, you say that the tank is at ambient temperature and is unpressurized, then you will have defined a physically meaningless scenario, since under atmospheric pressure these kinds of materials would boil off rapidly and could not be stored. Some models will identify the problem and call it to your attention, asking you to change some aspect of the input, but others might generate meaningless results. A more subtle problem in defining hypothetical scenarios for planning and practice exercises is to assume tank conditions or quantities that may not be encountered in practice. For example, most tank trucks actually are divided into a number of separate compartments, and so the entire volume of the tank truck might considerably exceed the amount that could leak through a broken hose or fitting, which would drain only a single compartment. Furthermore, there may be check valves or other engineered safety features that would reduce the maximum amount of leakage. Naturally, you would not be aware of many of these aspects from your own background, but in real applications it would be vital to talk to specialists in the field to identify these finer points.

Care with illogical input requirements.   Sometimes a model may ask you for an input parameter that you do not believe should be relevant to your specific scenario. For example, the model might ask you for the physical size of a tank leak, even though you are also telling the model to assume the instantaneous discharge of the entire contents. This may be perfectly reasonable, because some "general purpose" models work their way through a standard set of inputs before they start to look at the details of the scenario you want to analyze. In fact, this may then make it easier for the model to prompt you later to analyze some variations on your scenario that may need the additional input values. However, you should leave nothing to chance, because it is also possible that through some oversight on your part, you did not input the scenario that you really had in mind. (This latter error is really not hard to make, and most computer modelers have made many of them when learning to use a new package or programming language. The key is to know how to detect such errors.) If you think a certain input value should not affect your scenario, try inputting some different values, perhaps even some widely divergent ones, and see if you still get the same answer from the model. If you do, that would tend to confirm that the value was not really used. Along similar lines, if you are asked for a value that you are confident will not be needed for the actual calculations, don't waste your time trying to find it in a database; just put in a spurious value to keep the program happy. It shouldn't affect the final answer.

Adapting the model.   On some occasions it may be necessary to adapt your modeling package to situations slightly different from those for which it was designed. For example, the geometry of a spill may be different from the simple circular or rectangular options provided by a model, so a user may need to do some area calculations, perhaps dependent on topography near the spill, and then use something roughly equivalent within the model's framework. Or it may happen that the biggest risk is from a toxic combustion product, and thus it might be necessary to define an imaginary release scenario of that product to simulate the combustion scenario for the original material. Or there may be several relevant temperatures, such as chemical storage temperature, ambient air temperature, and ground temperature, but only one or two of these might be allowed as model inputs. For these and other similar situations, a good understanding of the physical principles governing release rates and material movement will usually allow one to make reasonable model adaptations to simulate the given conditions. In circumstances where it might be quite difficult to capture the real conditions with good precision, it may still be possible to make "conservative" assumptions about the release so as to obtain model results that should provide a *bound* or *limit* on the actual values, which is often sufficient. (If it is not sufficient, one will have to investigate the phenomenon further so as to fit the model more closely, or work with a more detailed model tailored to the specific phenomenon or situation.)

This brings us to the final box at the bottom of Figure 4-8, namely, the model output. Remember that the impacts of a hazmat scenario will vary both with location and with time. Furthermore, it may be necessary to consider the impacts of alternative response scenarios or alternative weather conditions, especially when the modeling is being carried as part of a planning program. In any case, the final quantitative output generally has the form of well-defined impacted locations at various points in time, or maximum impacted zones, considering all times. For many purposes, the most desirable way of presenting these results is in the form of site maps on which concentration zones or concentration isopleths (lines of constant concentration) are drawn or colored in, with each such map labeled as to the model assumptions and conditions under which it has been constructed. Some models produce such maps automatically, whereas others compute tabular data which the user then has to incorporate into the map framework.

However the model results are presented, one of the most important things is to give them the "reasonableness test," meaning to scrutinize them carefully and make sure that the model really did what was intended. One small typographic error in an input value, or some other error in applying chemical data or in converting units, could yield grossly incorrect and misleading results. Therefore one generally wants to test the results of a model run in several ways:

- *Against physical intuition.*   The user should simply look at the results, think about them, and see if there are any aspects that seem peculiar or unexpected. These bear careful investigation to make sure they are understood and convincing.
- *Against "back-of-the-envelope" calculations.* A good way to check the reasonableness of the model is to see if there are some simple hand calculations you might carry out to estimate whether the model results are in the right range. For example, while the model may treat the leak from a tank as a time-varying process depending on the gravitational force of the material left in the tank at any moment, a user might check results by selecting some average discharge rate, and then see if the results are at least comparable.
- *Against additional model calculations.*   It may be possible to compare the current model calculations with some that have been carried out previously, or with additional test runs for the current situation. For example, you may feel that you are dealing with a more dangerous chemical than for some earlier calculations, and yet you are getting a smaller hazard distance. Even though the release scenarios are slightly different, it would be good to look at the intermediate model results (provided by most modeling packages) to understand why this difference has appeared. Another common test method is to perform additional "sensitivity" runs to see how the results vary as you vary certain parameters or assumptions. First, the trends in these results should appear reasonable. Second, if the model results are highly sensitive to certain specific inputs, then you should give careful thought to making sure that you are using the most reasonable values for such inputs.

The comments provided throughout this section will take on greater meaning as the reader actually delves into real model calculations. Therefore, if problems occur, it may be useful to return to the above material to look for clues or advice on how to proceed.

4.6 The Analysis of Typical Scenarios

The purpose of this section is to apply the ideas of the previous sections to specific concrete scenarios, using a computer modeling package. There are two sample scenario analyses, in increasing order of complexity, that should give the reader an introduction to the modeling process in concrete terms, and also provide some "benchmark" problems to use with one's own modeling package in order to gain familiarity with it and compare results. You may even wish to try out your own model and compare results as you read this section. Naturally, different modeling approaches would be expected to yield at least slightly different results.* To make it

---

* The numerical calculations in the example problems have been obtained with the ARCHIE program, which is discussed further at the end of this chapter.

easier to trace calculations and results, more precision is provided in the numerical values than one would want to place on their real world significance.

Example 1. A chemical tank truck is used to transport acetone from a port city on the Mediterranean Sea to an inland industrial plant where it is used in the manufacture of paint. When the truck arrives at the paint factory, the driver connects a 3-inch hose from the truck to a valve on the plant's storage tank, and then he turns on his pump and goes inside for a cup of coffee during the transfer, which usually takes about half an hour. Unfortunately, near the end of the transfer, the truck is sideswiped by the boom on a backhoe working in the area, and it starts to roll. The hose breaks off at its fitting to the tank's piping, and a stuck check valve (which is supposed to prevent flow in the reverse direction) now allows the plant's storage tank to start to empty out through the broken hose. (The truck tank was almost empty, so the discharge from that half of the hose is only of minor concern.) Fortunately, this particular tank and its pipe connection points are in a large diked area 80 feet by 120 feet. The tank itself is a vertical cylinder 18 feet in diameter and 26 feet high. It is 75% full. Evaluate the hazard distances associated with a pool fire, a vapor cloud fire, and with chemical toxicity. Assume that the temperature is 85°F in full sun and that there is a slight breeze blowing at about 5 mph from the west.

Obviously one of the first steps in analyzing this situation would be to obtain information on acetone, which may be included in an internal database associated with the modeling package or need to be extracted from an MSDS or other source of chemical data. Table 4-7 summarizes

TABLE 4-7
Typical chemical data for acetone required for modeling Example 1

Parameter	Value	Comments
Normal boiling point	133°F	Gives the model some information on vapor pressure and hence on evaporation rate.
Molecular weight	58.08	Likely to be incorporated in model's vaporization rate calculations since substances with heavier molecules tend to vaporize more slowly (other factors being equal).
Liquid specific gravity	0.79	Should affect total mass present as well as leak rate through opening, as the latter is driven by the weight of the chemical still in the tank.
Vapor pressure at some given temperature	180 mmHg at 20°C	By using this value with the boiling point, the vapor pressure at ambient temperature can be roughly estimated. Some sources may give an additional data point, which some models can use.
Flash point	0°F	Used by some models to provide further information about the vapor pressure and temperature relationship.
Lower flammable limit (LFL)	2%	The model will generally give the option of computing the hazard distance based on some fraction of this limit, such as $\frac{1}{2}$ LFL.
Toxic vapor threshold	12000 ppm	This value is the IDLH value. Half this value will be used as the calculational threshold of greatest interest.

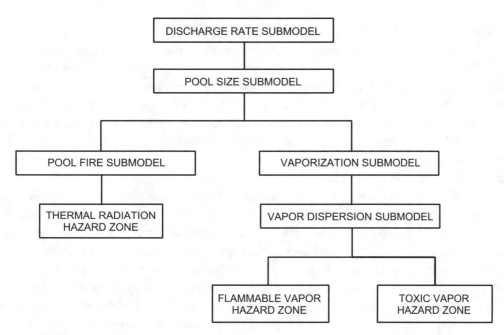

FIGURE 4-9

Typical submodel structure used in analyzing Example 1

some of the key information that a typical modeling package might require in order to complete the analysis. Depending on the nature and sophistication of the particular submodels in a given package, more, less, or different data might be requested by the model package during the course of the analysis.

A typical calculational sequence for this example is shown in Figure 4-9. The actual computational results using one such modeling package are discussed below.

The discharge rate submodel first calculates the total mass of acetone in the tank, based on the size of the tank, the percent of capacity at which it stands at the beginning of the leak, and the specific gravity or density of acetone. Using the gradual decrease in the height of liquid in the tank as it empties to reflect the force on the liquid at the leak point, the submodel uses certain fluid mechanics principles to estimate the discharge rate as a function of time. In particular, the total amount of acetone present at the outset is found to be about 244,600 pounds, and if allowed to go unstopped it would take about 153 minutes for the tank to empty. The average discharge rate over this period would be 1,594 pounds per minute.

Next, the pool size submodel would estimate the size of the pool being formed by the discharging liquid. Because this leak takes place within a diked area, and because the rate is sufficiently large that liquid would be expected to spread over the area fairly soon after the initiation of the leak, the submodel simply uses the total area within the dike, minus the area occupied by the tank itself, as the pool area. This value is 9,346 ft^2. (There would probably be other tanks within such a large diked area, but not subtracting their areas from the total diked area is just slightly conservative. If there were no dike, then the model might offer options for evaluating the expected size of the pool.)

With respect to the various hazards identified for this situation, first consider the possibility that the pool ignites. Then the entire surface would be burning and this would create a very large flame. The heat, or *thermal radiation,* given off by this flame would be intense and could kill or burn people in the vicinity. It could also damage property (and perhaps even cause other tanks to catch fire). In this case, the flame is calculated to be over 100 feet high. Based on generally accepted criteria that relate thermal radiation to personal injury, a fatality zone of about 190 feet from the fire is calculated, with an injury zone extending out to 275 feet. Thus people working in the vicinity to bring the situation under control would be at severe risk if ignition of the pool were to occur.

If the pool does not ignite, but rather the material continues to evaporate, then the acetone vapors will begin to move in a downwind direction. The vaporization submodel estimates the rate at which the material evaporates, taking into account its estimated vapor pressure at the ambient temperature, the windspeed, and perhaps other factors that may vary from one model package to another. In this case, the given pool size should lead to an evaporation rate of about 500 pounds per minute, or about one third of the discharge rate from the broken connection. This amount provides a source term for a vapor dispersion submodel, which may be based on Gaussian dispersion (as discussed in Chapter 3) or on some other mathematical framework. Thus, one would expect the atmospheric stability class to be relevant and for it or for the basic data leading to it to be requested by the model as input. We will use stability class A for this problem since the conditions are so close to the A end of the A-B range identified in the stability class chart.

The output of the vapor dispersion calculations would generally take the form of concentrations of acetone gas at various locations and times. These could then be compared to certain threshold values to determine the areas that are of greatest risk. In particular, for the flammable vapor hazard, the results of the vapor dispersion calculations would be compared to the LFL, and any zone where the predicted concentration is over the LFL would certainly be considered at risk. If an ignition source were to exist in that zone, the vapor would ignite and the flames would flash all the way back to the pool as well. In this case, the downwind extent of this zone is calculated to be about 55 feet from the center of the tank, which is not a great deal. It certainly does not seem to have the potential to leave the immediate plant site. To be conservative, it is actually more common to calculate the distance to a concentration equal to one half the LFL, since there are considerable uncertainties in the models and there is also the possibility that the vapors may be more concentrated in one zone than another. In this case, the distance to $\frac{1}{2}$ LFL is still only about 79 feet directly downwind, with even lesser distances as you move off the direct line of the wind.

For calculating the toxic vapor hazard distance, the model compares the vapor concentrations against the specified toxic thresholds provided by the user or taken from the internal database. Using the IDLH value shown in Table 4-7, the downwind hazard distance is calculated to be 72 feet, and using the more conservative value of $\frac{1}{2}$ IDLH, the hazard distance turns out to be about 102 feet.

In summary, Example 1 represents a situation where the risks are essentially confined to the plant site itself. The reader will be asked in some later exercises to return to this situation and see what the effect would be of changes in the accident conditions or the weather.

Example 2. [Note: this is an actual scenario used for an emergency training exercise, except that the place names have been changed. See Figure 4-10 for a map of the area.

At 3:15 P.M. on a summer Thursday afternoon, a loaded tank truck with an 8,000-gallon capacity is involved in a sideswipe accident on the southbound side of Interstate 97 just south of the Reservoir Road intersection. It overturns as it strikes the right-hand guardrail and comes to rest between the guardrail and the reservoir at a point that is 0.4 miles south of the Reservoir Road overpass. Liquid begins to flow out through a ragged crack in the underside of the tank that is irregular in shape but roughly one-half inch wide and eight inches long. The driver is rendered unconscious and trapped in the vehicle. Four other cars are involved in the collision sequence, effectively blocking the southbound roadway. The accident is called in by cellular

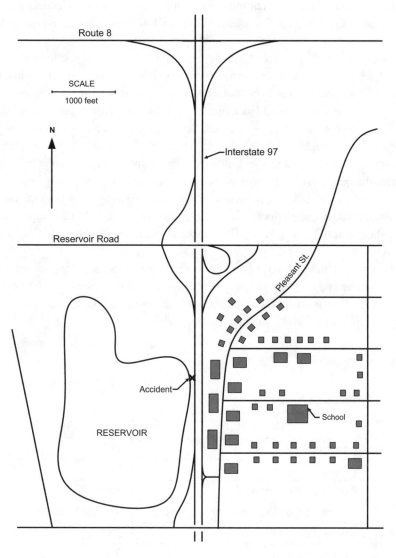

FIGURE 4-10

Location map for Example 2

phone to the state police at 3:16, and the caller reports its location as "a little bit south of Route 8." The caller also refers to the overturned tanker as a "gasoline tanker." Other calls by 3:20 by both cellular phone (again to the state police emergency number) and from the office buildings on Pleasant Street (to town police) improve the description of the location. No placard information is conveyed by the callers. The temperature at the time of the accident is 82 degrees Fahrenheit, and the air is quite still, with a very slight breeze blowing from the west at about 5 mph. There is a heavy overcast, and the humidity is high. Rain and possibly thunderstorms are predicted for the late afternoon or early evening. The local fire department arrives at 3:30 and is able to observe a placard number of 1093 with binoculars from south of the site, and they determine that this is acrylonitrile. They then plan a cautious approach.

We will of course focus on what information might be obtained by modeling this situation, but after this analysis, it will also be useful to discuss briefly some other non-modeling aspects of the exercise.

Table 4-8 summarizes the same kind of chemical information for acrylonitrile that was used in Example 1 for acetone. An initial inspection of the table suggests that acrylonitrile is in some respects comparable to acetone. It is slightly less volatile, as seen by the higher boiling point and lower vapor pressure values. It has a higher LFL and higher flash point. However, the toxic vapor threshold is about 4% that for acetone. So, in general, *under equivalent conditions,* the flammable vapor risk might be slightly lower, but the toxic risk might be considerably higher, depending on how the lower threshold balances out against the lower volatility. Moreover, in this scenario, the conditions differ in a number of important respects, such as:

- There is no dike or other equivalent containment system to limit the size of the pool of spilled material. A large effective area for evaporation increases the evaporation rate.
- The spilled material will actually flow into an adjacent reservoir and initially spread as a floating slick.
- Members of the public may be in close proximity to the accident scene, both in their vehicles on the highway and in the nearby residential and commercial buildings. Of urgent importance is the need to keep ignition sources away, including hot catalytic converters likely to be found on many vehicles.

TABLE 4-8
Key acrylonitrile data

Parameter	Value
Normal boiling point	172°F
Molecular weight	53.06
Liquid specific gravity	0.81
Vapor pressure at some given temperature	83 mmHg at 20°C
Flash point	30°F
Lower flammable limit (LFL)	3%
Toxic vapor threshold	500 ppm (IDLH)

- If the vapors or the liquid pool ignite, there is the additional hazard of toxic gaseous combustion products, including hydrogen cyanide (HCN). We will limit our analysis to this one additional toxic material, using it as a rough surrogate for the entire collection of such byproducts.
- The material may have carcinogenic (cancer-causing) properties not reflected in the published toxic vapor limit.

The actual model application to this scenario proceeds along lines similar to those for Example 1, except that it is more complex and requires multiple model runs to simulate the hydrogen cyanide release in the event of ignition. In the dialog between user and model, certain assumptions and approximations need to be made. The results presented in Table 4-9 are based on the following additional assumptions:

- The atmospheric stability class is D because of the overcast conditions.
- The tank is a horizontal cylinder 7 feet in diameter and 40 feet long.
- The rupture in the tank is equivalent to a circular leak with the same area.
- If it remains unignited, the liquid will form a floating slick on the reservoir that increases in size until the evaporation rate equals the discharge rate. Its shape is assumed to be circular.
- If the spill ignites, the burning rate is assumed to be equal to the discharge rate. (This would be the case, for example, if ignition took place as part of the accident or soon thereafter.)
- If the spill ignites, 10% of the acrylonitrile is assumed to be converted into HCN, and it is treated as a release at a height above the ground equal to the flame height. (This 10% "yield factor" is highly variable and difficult to predict; it is only a rough estimate based on discussions with chemists at one of the companies that manufactures acrylonitrile.)
- No corrective action is taken during the incident to stop the discharge or otherwise limit the spread of the material.

Figure 4-11 shows a graphic representation of the numerical results in the form of hazard zones superimposed on the original site map, using a typical 60° sector to identify the impacted downwind areas. We may make certain observations about these results:

TABLE 4-9
Results of model calculations for Example 2

Output variable	Value
Average discharge rate from tank	569 lb/min
Duration of discharge	95 minutes
Slick diameter	204 feet
Vapor cloud fire hazard distance (i.e., distance to $\frac{1}{2}$ LFL)	230 feet
Toxic vapor hazard distance (i.e., distance to $\frac{1}{2}$ toxic vapor threshold)	2,288 feet
Fire hazard zone (if material burns as it is released)	89 feet
Toxic hazard zone for hydrogen cyanide (assuming 10% maximum yield)	1,977 feet

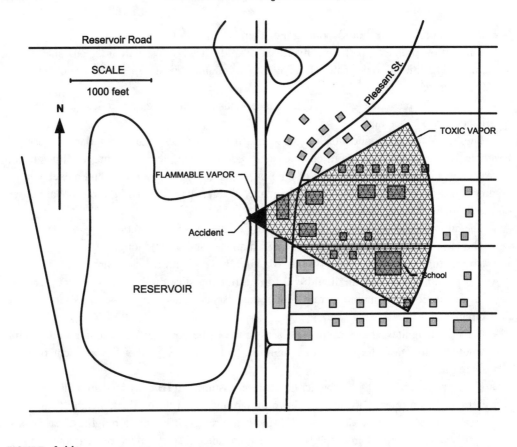

FIGURE 4-11
Toxic and flammable vapor hazard zones for Example 2

- This incident reaches its maximum impact in less than an hour, which is generally a shorter time than that required by many specialized hazmat teams to arrive on site and get mobilized to enter the area. Early efforts would probably be focused on removing the driver and other persons directly involved in the accident.
- The toxic hazard distances for acrylonitrile and its combustion byproduct HCN control the maximum impacted zones, which can be considerable. Given the rapid evolution of this incident, sheltering the population indoors with windows closed and air exchange equipment turned off would be more practical than a mass evacuation.
- The vapor fire hazard zone does not reach any buildings and barely makes it to the northbound side of the divided highway.
- The most urgent operational concern would be to evacuate the drivers on the southbound side who are blocked by the accident. Since this is the beginning of rush hour, the road will be busy and a large traffic jam will be forming. Drivers would need to be told to leave their vehicles and walk back to the Reservoir Road area.
- If ignition were not to take place until the pool approached its maximum size, then, in addition to toxic or flammable vapor cloud effects, the pool fire itself would have flames about 155 feet high and a fatality and injury zone in all directions of about 488 feet due to the heat. This helps one to appreciate the danger of this kind of situation.

This is quite a complex scenario, even in the simplified form presented here, and further discussion of the combustion byproduct modeling may be useful as an indication of a principle discussed in Section 4.5, namely, the need sometimes to adapt or "force" a model to fit an unusual situation. In fact, several distinct calculations and model runs needed to be made to simulate conditions similar to those involved in the combustion gas scenario. These were:

a)   An initial acrylonitrile run assuming ignition after the full pool formed, to get maximum spill size as well as flame characteristics for this case.

b)   A second acrylonitrile run with ignition assumed at time of release, to get equilibrium pool size and flame characteristics when burning rate equals spill rate.

c)   Hand calculation of the HCN production rate as a function of the corresponding acrylonitrile burn rate, taking into account the yield factor of 0.1 and the stoichiometric relationship between acrylonitrile and HCN. In particular, every pound of acrylonitrile that is converted to HCN converts to 0.509 pounds of the latter. This ratio of .509 is simply the ratio of the molecular weight of HCN (27) to that of acrylonitrile (53) because in burning, one molecule of HCN comes from one molecule of acrylonitrile, as can be seen from looking at their molecular formulas (C_3H_3N and HCN), each of which contains a single nitrogen. (The molecular formulas are also generally given on MSDSs.)

d)   A toxic vapor dispersion model run for HCN, treating it as an elevated source of gas being released from a height equal to the flame height, $48'$, in part b, above, and at a rate derived from the burning rate in that run. (In fact, in that run, the burning rate was assumed equal to the average spill rate of about 569.2 lb/min.) Therefore this would correspond (at 10% yield) to an HCN production rate of about 29 lb/min. A somewhat higher source height could probably be justified, based on thermal rise due to the heat, but lacking further data on this, we took the conservative approach of using the flame height itself.

While the logic leading to each of the individual steps above can probably be well understood, the original process of fitting a model to a new situation along the above lines is partly a science and partly an art. While the reader would not be expected to tackle a complex situation like this one without considerable advice, it is good to recognize here that the variety of real-world problems requires the constant development of new models and the adaptation of old ones, which makes this an interesting field. (In fact, a number of simplifications were made in the foregoing analysis, and one could already imagine model enhancements to deal with some of them, such as the effective height of release of combustion products from a flame.)

Naturally the modeling results depend both on the particular scenario conditions and on the modeling assumptions, and these aspects will be investigated further in the exercises. There is also the possibility of additional hazards with this particular chemical, such as the rapid exothermic (heat producing) polymerization of the material still in the tank. (This would generally be mentioned on MSDSs.) This could be caused by exposure of the tank truck to flames, and it could take place with the force of a major explosion, hurtling debris a large distance and with great force. Modeling the potential for this would be too technical for our current level of treatment. Another problem is the spill of a carcinogen into a drinking water reservoir. The fate of the material in the water would need to be evaluated and remedial plans evaluated, perhaps also with the aid of modeling. This is a risk that could be dealt with on a somewhat slower time scale as long as the reservoir pumping station were shut down during or shortly after this incident.

Model calculations such as those described above probably have their greatest current utility during planning or training activities. For example, as was mentioned earlier, Example 2 has been used for training sessions involving emergency responders, giving them an opportunity to see how incidents of this type might develop so that they might be alert to the dimensions of the risks both to themselves and to the public. Various response activities, such as evacuation, sheltering in place, putting foam on the pool to decrease its effective surface area, stopping the leak, and moving people away from the scene, can all be evaluated, discussed, and practiced in the framework of such a scenario. From a different point of view, such calculations might be useful in estimating the worst case public impact from a proposed plant or plant expansion, and this might let the designers evaluate the need for additional risk reduction measures or the municipality decide whether to approve the project. In a real incident, it would be more difficult to apply the models in an appropriate time frame, although there are some cases where this might be done to evaluate certain risks, especially if they represent contingencies associated with a more slowly evolving situation. (Some railroad derailments lead to hazmat fires that burn for days, with uncertain effects on unruptured tank cars; similarly, some plant incidents last for long periods before they are finally brought under control.)

Exercises for Section 4.6

These exercises require you to use a modeling package for hazmat emergency situations, as well as to have access to data for the chemicals under discussion. Some models contain internal databases that will meet your needs, whereas others require you to obtain the desired information from reference sources such as MSDSs. Such sources have been discussed further in previous sections. See Section 4.7 for further discussion of available modeling packages. Be very careful with your units and unit conversions in applying data to the models. In addition, be sure to check the reasonableness of your assumptions and model results at every step of the calculations. In some problems, you will find that you need to make additional assumptions beyond the information stated in the problem. State your assumptions clearly in these cases, along with a brief rationale about why you believe them to be realistic.

1. Estimate how long it would take to empty a vertical cylindrical tank 40 feet in diameter and 50 feet high. The tank contains #2 diesel fuel and is half full at the outset. It is being discharged from a broken 4-inch pipe at the bottom that was accidentally hit by a payloader working in the area. You can assume the outside temperature is around 68°F.

2. A westbound gasoline tank truck overturns on Interstate I-44, and its cargo of 7,000 gallons spills out almost immediately and collects in a pool in a depression just off the right shoulder of the road. This pool of gasoline is roughly circular and has a diameter of 20 feet, and it is located off the side of the westbound lanes, about 25 feet at its nearest point from the westbound roadway and 150 feet from the eastbound lanes. There is a slight wind blowing from the north at 4 mph, and it is very sunny. You are the local fire chief, and you arrive on the scene within 10 minutes of the accident. Your first decision has to do with whether to close down the *eastbound* side of the road for fear of ignition of the vapors from the evaporating pool by a passing car. It's a hot day (80°F), and the smell of gasoline is very strong. Is this a significant hazard? (Hint: gasoline is a mixture of various chemicals, each with its own physical and chemical parameters.

TABLE 4-10
Recommended gasoline modeling parameters

Molecular weight	90.9
Boiling point	114.9°F
Specific gravity	0.64 at 68°F
Vapor pressure at 0°F	82 mmHg
Vapor pressure at 68°F	343 mmHg
Vapor pressure at 100°F	595 mmHg
Lower flammable limit (LFL)	1.4%
Upper flammable limit	7.6%

Table 4-10 provides certain composite modeling parameters for this common material that have been found by practitioners to give good equivalent hazard analysis results.)

3. You have recently moved to Connecticut to take over the management of a municipal sports center which includes an ice rink. Your staff tells you that there has long been a problem of refrigeration breakdowns, so you decide to request funds from the town to upgrade the aging refrigeration system, which uses ammonia as its refrigerant. During the public budget review process, several residents who had been unaware that the plant had a large ammonia inventory express concern about this system, not only because of their own proximity but because there is a new nursing home and rehabilitation center located just about a mile away. (At this point, you may be beginning to wish that you had never even suggested any changes to the system.)

Therefore the town decides to do a more thorough review of options for the plant, including closing the rink and contracting ice time from a newer rink in the next town, upgrading the ammonia system according to your recommendations, or replacing the system with a freon-based system. A consultant is hired to help evaluate the risks of the second option, and in a scoping meeting with the fire chief it is decided to define a "worst case" spill scenario as the short time frame (say, one minute) release of the full contents of one of three liquid ammonia receiving tanks on the high pressure part of the cooling circuit, and the further assumption that the material boils off instantly as it is released. Each such tank holds 80 gallons. Furthermore, even though this part of the system is located in a closed room on the back of the building, it has a large garage door that would probably be open during the kinds of maintenance and repair activities that might lead to such an accidental release, so the release should be assumed to be directly to the outside air.

Could the toxic impact of such a release impact a location as far away as the nursing home, in which case you would probably also be responsible for developing emergency procedures for that facility?

Note: this is a typical framework for such a question, but you will find that not all the input data for the model have been provided. In this situation, you must make reasonable assumptions, generally trying to err on the conservative side if necessary. In addition, you need to be sure to find a way to have the model simulate this one-minute release and vaporization scenario.

4. A 5,000-gallon tank truck of concentrated nitric acid is pulled over into a roadside rest area by a state trooper who is responding to a cellular phone call from a concerned motorist who noticed that it was dripping liquid along the highway. Upon inspection, it appears that a faulty valve or cracked fitting is the problem, and the material leak rate seems to be increasing and looks like it's now up to about 5 gallons per minute. What is the nature of the hazard and the size of the hazard zone? Assume moderate sunshine with scattered clouds, a temperature of 75°F, and a slight breeze of 4 mph.

5. With respect to Example 1 in the text (not Exercise 1), provide your best guess about the answers to the following questions, and provide a supporting argument for each. You are not being asked to rerun the model for these situations. In particular, what would you roughly estimate to be the type of effect on the hazard zones for toxic and flammable vapors of each of the following individual changes from the conditions of the original example?
   a) Increasing the broken hose diameter to 4 inches.
   b) Having the tank 95% full at the outset of the leak.
   c) Reducing the temperature to 50°F.
   d) Eliminating the dike and assuming the ground is level.
   e) Basing the hazard zones on 10% of the LFL and toxic thresholds.
   f) Changing to overcast conditions.
   g) Decreasing the windspeed to 2 mph.
   h) Increasing the windspeed to 10 mph.

6. For each of the situations in the previous exercise, carry through the required model calculations and compare your answers to your prediction. Provide a physically plausible explanation for any differences from your expectations as stated in the previous exercise.

7. With respect to Example 2 in the text, provide your best guess about the answers to the following questions, and provide a supporting argument for each. You are not being asked to rerun the model for these situations. In particular, what would be the nature of the estimated effect on the toxic hazard zones for hydrogen cyanide only (a combustion product of acrylonitrile) of each of the following individual changes from the conditions of the original example?
   a) Increasing the yield factor of HCN to 50% from the combustion process.
   b) Assuming that ignition takes place only after the pool reaches maximum size.
   c) Changing the weather conditions to moderate sunshine.

8. For each of the situations in the previous exercise, carry through the required model calculations, and compare your answers to your prediction. Provide a physically plausible explanation for any differences from your predictions. (Hint: this is a challenging exercise; you should be sure to be able to reproduce the results in Example 2 before considering these modifications. You may have to "force" the model to fit the situation you want to apply it to.)

9. Under what weather conditions would the toxic hazard zone from acrylonitrile vapors in Example 2 extend the farthest distance from the scene? What about the case when the accident could also occur at night? (Hint: these two questions may require some experimentation on your part; keep your results well organized and present them in tabular form.) Give a plausible physical explanation for your conclusion.

4.7 General Comments and Guide to Further Information

The reader might be surprised at the extent of commerce in hazardous materials, entailing well over a million annual bulk shipments of flammable and toxic cargoes on highways and rail lines through virtually every community in this country, as well as handling and storage at manufacturing locations, industrial plants, distribution/storage facilities, and retail locations (e.g., the local gasoline service station). Early in this chapter, you were asked to review available accident and evacuation data for incidents involving such materials, and the frequency might have been surprising. This explains why there have been such widespread and concerted efforts to develop models to aid in the evaluation of the safety of such activities. But who are the actual developers and users of these models?

First let us treat the developers. These are often consulting companies, who undertake model development either to enhance their own consulting services (e.g., risk assessment, regulatory support, facility design) or in direct fulfillment of contracts with clients who want to "buy" such models for their own use. Developers may also consist of universities working under research contracts from organizations who want to sponsor such work so as to benefit from the results. This has been a strong business area for many years, and understandably it has also grown quite competitive. Thus it is not surprising that there has been quite a proliferation of models and modeling packages, some in direct competition with each other.

How does one modeling package compete against another? You might think that accuracy would be the overwhelming factor, but in fact there are a whole host of additional considerations, such as:

History of acceptance by regulatory bodies.   For model applications that involve some kind of regulatory review or that may confront legal challenges, it can be better to use a model that has an established history of acceptance, rather than one that is the most recent or most complex and probably representative of the "state-of-the-art."

Simplicity and ease of communication.   People don't like to base major decisions on things they don't feel they have some understanding of, so it can be better to use a simpler model whose logic is reasonably clear than a more complex one that is inscrutable.

User friendliness.   For models being developed for use by others, ease of use is a prime consideration. Models differ widely in this respect. You may have gotten some sense of this by applying your own modeling package throughout this chapter.

Cost.   The cost of hazmat modeling software ranges from public domain packages at less than twenty dollars to complex proprietary packages that may sell for over a hundred thousand dollars. The cost of hiring a research organization to develop even part of a package could easily run to several hundred thousand dollars.

Thus the best modeling package for a specific organization or project would generally be chosen taking all these factors into consideration.

Now let us discuss the users of the models and the sponsors for model development, which are not always the same groups. Some typical examples of users would include:

Chemical companies.   Model calculations might be used in the design of parts of a plant (e.g., are the chemical storage tanks far enough away from residences outside the fence line?) or even in the choice of a site (e.g., is the site large enough and the typical meteorological conditions such that off-site impact from a plant accident is extremely unlikely?). They might also be used to obtain regulatory approval for construction and operation. In some cases they are used to investigate the interactive effects of several different plants in the same area.

Regulatory agencies.   These agencies may wish to do some of their own modeling before deciding on applications for new plants, plant modifications, or new transportation operations. In a different vein, they may use modeling to identify general regulatory issues worthy of their future attention (e.g., whether there should be specific routing guidelines for certain kinds of shipments).

Other government agencies.   Some agencies need to have models available for emergency planning and response. For example, the Coast Guard has sponsored the development of many models to help them deal with marine spills. The EPA provides modeling support in response to local hazmat emergencies. On the other hand, agencies like the Department of Energy and the Air Force use models regularly to evaluate the safety of their own hazmat activities.

State and local emergency planning groups.   There are probably over 10,000 copies of hazmat modeling packages in the hands of "Local Emergency Planning Committees" and coordinating state agencies. Since that averages to 200 per state, you can imagine that many communities have received such materials. Naturally, the degree of actual use varies widely, depending on computer capabilities, availability of personnel, training, and other factors. However, it would not be unlikely for you to find local fire department personnel in your community already familiar with some of the same modeling concepts and techniques that were treated in the first half of this chapter. In some larger cities, computers loaded with modeling packages are actually carried on fire trucks for use on-site.

There are other users as well, but the list above is fairly representative. It should be noted that quite often the end users hire consultants to help them with the model applications.

The last group to be considered are the sponsors for model development, that is, the people who pay for the development of these models. Aside from some of the users themselves, model development, because of its substantial costs, is often sponsored by industry organizations or company consortia. Furthermore, in the case of local emergency planning tools, these have generally been sponsored by Federal agencies, such as the EPA, FEMA (Federal Emergency Management Agency), NOAA (National Oceanic and Atmospheric Administration), and DOT (Department of Transportation).

Where can you get access to modeling packages of the type referred to in this chapter? A good source for local access to currently used packages might be the Local Emergency Planning Committee or equivalent agency serving your community or a nearby large city. They almost certainly have such software and may even encourage the development of a collaboration with a local academic institution (although you may need to obtain your own licensed copy). You can track this organization down by contacting the local fire chief. Another good source of support is your regional EPA office. The two models in most common use among such organizations at the time of writing of this book are ARCHIE (Automated Resource for Chemical Hazard Incident Evaluation) and CAMEO (Computer-Aided Management of Emergency Operations), the latter coming with companion programs, ALOHA (Areal Locations of Hazardous Atmospheres) and MARPLOT (Mapping Application for Response, Planning, and Local Operational Tasks), for performing vapor dispersion calculations and automatically overlaying the resulting hazard zones out on maps of your local area. Either of these programs would fit well with the kinds of calculations called for in this chapter. You may also wish to consult publications or on-line information emanating from EPA's Risk Management Program (RMP) or successor programs, where recommended models and their characteristics are discussed.

For further information about some of the topics treated in this section, you may want to consult the following references:

Center for Chemical Process Safety, *Guidelines for Chemical Process Quantitative Risk Analysis,* American Institute of Chemical Engineers, New York, 1989

This is a general summary of the state-of-the-art practice in quantitative risk assessment in the chemical industry, published out by an industry organization to help facilitate the use of these tools by its members.

Federal Emergency Management Agency, US Department of Transportation, and US Environmental Protection Agency, *Handbook of Chemical Hazard Analysis Procedures,* US Government Printing Office: 1990-725-600/20672.

This is the ARCHIE manual, but it is actually also an excellent self-contained background document on chemical emergencies. It can be read and understood independent of your having access to this computer program. This is written at a slightly more elementary level than the previous reference.

The references listed at the end of the previous chapter, on air modeling, also contain material relevant to the vapor dispersion models discussed in this chapter.

In addition to reviewing written materials, most people find it quite interesting to obtain some field experiences in connection with this material. Here are four suggestions:

- Arrange a visit to your local fire department or regional hazmat unit. They have extensive equipment and other resources for dealing with hazmat emergencies and are usually very interested in educating the local community about these activities. They may even be able to arrange for you to visit an actual field practice session. It is very informative to get the perspective of the first-line responders.
- Arrange a visit to a local industrial plant, tank farm, fuel distribution company, or port facility for chemicals. You would be able to see the precautionary steps taken in these facilities to avoid accidents or to deal with them if they occur. Watch the loading procedure for tank trucks. Inspect a dike. See what foam equipment looks like. Set up such a visit by contacting either the Plant Manager or the Safety or Environmental Manager (or some similar titles). (The fire department regularly inspects most such plants and may be able to suggest a plant and contact person.)
- Attend a meeting of your Local Emergency Planning Committee, or the regional equivalent, or meet with a representative. See what the current issues are in your area.
- Obtain a copy of the little "orange book" (US Department of Transportation, Emergency Response Guidebook) and carry it with you when you travel by highway. When you see a tank truck drive by, note the placard number and look up the chemical. Over time, you will note quite a range.

Part 2

*Further Development
of
Modeling Concepts*

5

Additional Topics in Ground Water

The purpose of this chapter is to provide a brief introduction to a number of additional aspects of ground-water modeling, ranging from elementary to advanced levels. At the beginning of each individual section, some discussion will be given in a footnote as to the level of mathematical background required for that section. This and the other chapters in Part 2 are organized somewhat differently from previous chapters in that the exercises are interspersed throughout the sections so that they pertain to one or two closely related ideas at a time. While this creates the burden that you really have to take time out from reading to work the problems along the way, it has the advantage that you will be able to master the individual concepts in small pieces one at a time, and your overall understanding should be significantly enhanced. In general terms, all but two of the sections in this chapter use basic calculus (derivatives but not integrals), and some later sections use concepts from multivariable calculus and make connections to other more advanced subjects for readers who may have this additional background. For readers with no calculus background, Section 5.3 (on retardation factors) and Section 5.4 (on constructing head contour diagrams) should still be accessible.

5.1 The Continuous Form of Darcy's Law for One-Dimensional Flow*

Recall the basic form of Darcy's law that was used throughout Chapter 2:

$$q = Ki,$$

where the hydraulic gradient i was defined by the equation

$$i = \frac{\Delta h}{L}.$$

We also made heavy use of the interstitial velocity equation:

$$v = \frac{Ki}{\eta},$$

---

* This section uses the basic concept of the derivative as the limiting value of the rate of change of one quantity with respect to another.

which also involved the hydraulic gradient i. You probably gained considerable experience in Chapter 2 in applying these equations to a variety of problems in which volumetric flow rates, ground-water velocities, travel times, and related quantities were needed. In this chapter, we will develop a more general interpretation of the hydraulic gradient simply as a derivative of the "head function," and this will let us solve a wider range of important ground-water problems.

Since we will be taking the derivative of the so-called head function, we need to go back and look a little more carefully at just what this concept of hydraulic head means. We originally *defined* the head to be the *height of the water in an aquifer,* measured with respect to some arbitrary datum level. How did we talk about *measuring* this level of water in an aquifer? We said that you just drill a well down to the aquifer and measure down to the level of water you see in the well. That was all perfectly reasonable for the so-called "water table" aquifers treated in Chapter 2 because the water level in a well in such aquifers is always at the same height as the top of the aquifer. The trouble is that some aquifers are much more complicated. In fact, for some aquifers, if you drill a well down into them, the water will shoot right out the top of the well casing! (This is similar to an oil well 'gusher'.)

For example, look at the two situations depicted in Figure 5-1. On the top, you see the kind of system that is often used to pressurize the water in municipal water systems. The water is pumped from a reservoir or well up to a tank, often on the top of a hill or tower. The extra elevation provides the kind of pressure needed in the pipeline that supplies the needs of the community. If the horizontal section of pipe had a lot of big leaks in it, then of course the pressure in the pipe would be less because it would be being dissipated, but if the pipe is in good condition, then the pressure over at the lower right end would be considerable.

An analogous natural phenomenon is shown in the lower part of the figure. Here the results of long-term geologic deformation, sedimentation, and erosion can be seen to have created a pipe-like situation. A continuous aquifer formation has risen over millions of years on the left side with the geologic uplift that formed the mountainous zone. Meanwhile, the much more impermeable sediments above it (perhaps clays and silts or their corresponding rock forms) eroded away in the more uplifted areas, allowing water from precipitation to recharge the aquifer there, but leaving them to serve as a kind of insulator or seal over the aquifer in lower sections to hold in the pressure resulting from the water entering that aquifer up in the mountains. Therefore, if you put a well into the aquifer as shown at the lower right, the water level there would almost certainly be higher than the physical top of the aquifer, and it might even be enough to flow out the top of the well casing. Whatever the height is that the water would rise to in that well (assuming we made the casing as high as necessary), that is what we will call the hydraulic head in the aquifer at the location of the well. Incidentally, if water would actually flow out the top all by itself, the well would technically be called an *artesian well,* although in some parts of the country this term is used to refer to any deep, drilled well, as opposed to a shallow, "dug" well.

Given these possibilities, we must now extend our original definition of *hydraulic head* at any point in an aquifer by saying that it is the height to which water will rise in a well that penetrates the aquifer at that point. This covers the new case, as well as the original case of a water table aquifer, where the water in the well will simply rise to the same level as the top of the aquifer.

There is one more issue we need to think about before moving on. In using a well, as above, to measure the head in an aquifer, do you suppose it makes any difference how deeply

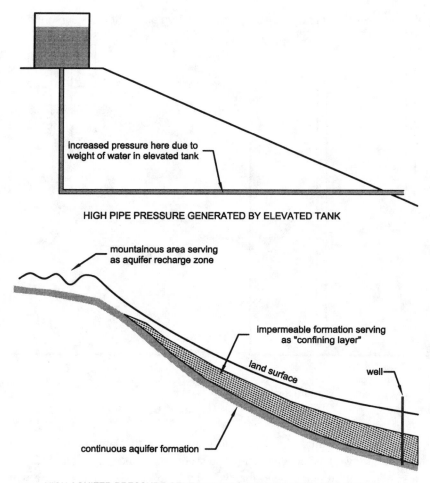

increased pressure here due to weight of water in elevated tank

HIGH PIPE PRESSURE GENERATED BY ELEVATED TANK

mountainous area serving as aquifer recharge zone

impermeable formation serving as "confining layer"

land surface

well

continuous aquifer formation

HIGH AQUIFER PRESSURE GENERATED BY ELEVATED RECHARGE AREA

FIGURE 5-1
Physical basis for an overpressured (artesian) aquifer

you penetrate into the aquifer? If your well samples water from just the top part of the aquifer, will you find the same level as you would if your well took water from the deeper parts of the aquifer? Think about this for a moment, perhaps consulting Figure 5-2 for some ideas.

This figure can give some useful insights. The left side shows two pipes extending into a tank of water. Obviously, the water level in each will rise to exactly the same level, namely, the level of the water in the tank. While not an earth-shattering revelation, look at the following aspect. The pressure at the bottom of the longer pipe is definitely higher than the pressure at the bottom of the shorter pipe. (Remember, the deeper you go in a body of water, the higher the pressure. If submarines go too deep, they can even be crushed by the pressure.) So the pipes are tapping into water at two different pressures. But the higher pressure at the bottom of the longer pipe is only enough to push the water column in that pipe back up to the same level as the lower pressure can for the shorter pipe. So you would measure the *same head* in both cases, even though you have *different pressures* at the points of measurement. So you

Two open pipes extending to different
depths in a tank of water

Two adjacent wells screened at different
depths in an aquifer

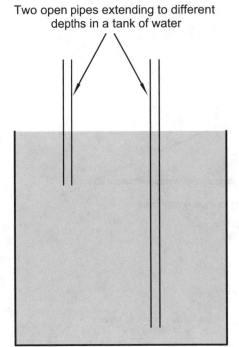
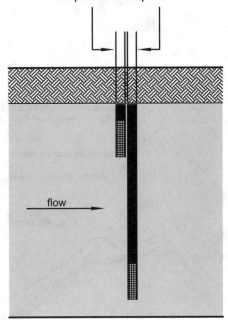

FIGURE 5-2
Measuring head by sampling at different depths

want to be careful not to think that head and pressure are the same thing! (Sometimes head is broken down into two basic components, "pressure head" and "elevation head," and in fact on some occasions a third component, the "velocity head" is also added in. We will not need these concepts here, although very analogous concepts will be seen in Chapter 7 in a different setting.)

Turning to the right side of Figure 5-2, two adjacent wells are shown "screened" at distinct depths in an aquifer. A screen is simply a specially perforated portion of a well casing that lets water in but (hopefully) keeps soil or rock particles out. Would the water level in these two wells be the same, even if the well were "overpressured" as in Figure 5-1?

The answer is a little more complicated than in the case of the tank: If the flow in the aquifer is essentially horizontal (as the sketch indicates), then the two wells will yield the same head value. But if there is a vertical component to the flow in the aquifer, then the wells would show different values for the head. This is basically because of the principle behind Darcy's law: you need a difference in head to move the water, so if it's moving vertically, there must be a vertical difference in head.

There are important cases where the site analysis or the remediation program for contaminated Superfund sites have had to take into account significant vertical flows, and there are many other similar situations. However, we will postpone consideration of this case until Section 5-5, when we treat two-dimensional flows. For the remainder of this section we will treat flows that are sufficiently horizontal that we can ignore any vertical component. Even for a slightly sloping water table aquifer with a gradient of 0.01 or 0.001 (the range encountered in most of the examples in Chapter 2), this assumption is well satisfied. It follows that the head value does not depend on the depth of the point where we sample the aquifer, but only on where we are located along the horizontal axis of flow.

The general framework within which our basic form of Darcy's law was developed is summarized by the left side of Figure 5-3, where the focus is on the meaning of the "head loss" Δh. You divide this head loss by the distance L to get the actual hydraulic gradient. Remember from a number of the earlier problems that the gradient itself can change from one location to another, just as the closer lines on a topographic map show areas of steeper slope. Suppose now that we want to calculate the hydraulic gradient i at the point P. An important observation for this problem is that in order to make an accurate estimate of the hydraulic gradient at the point, we would like to have a second point available whose distance from point P, represented by L in the figure, is relatively small. In this way, any changes in conditions farther along in the aquifer would not factor into the value calculated for P.

This observation is the basis for moving to the new notational scheme presented on the right-hand side of Figure 5-3. Here the physical situation is identical. We are considering the one-dimensional flow of fluid from left to right, and we want to calculate the value of the hydraulic gradient i at the point whose coordinate value is x. (Note that only one coordinate is needed because the flow pathway in this simplified diagram is assumed to be one-dimensional, namely, directly horizontal from left to right.) Using notation that the reader should recognize as similar to that in elementary calculus, we can convert the representation of hydraulic gradient

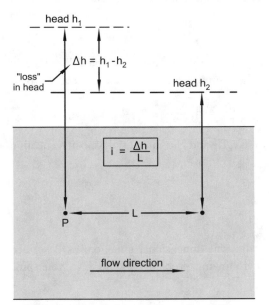

Discrete case, as discussed in earlier sections in which Δh refers to the loss in head (i.e., a decrease was treated as positive)

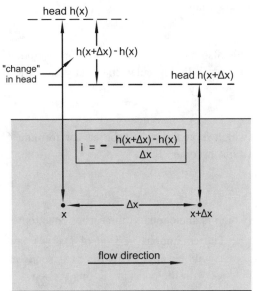

Continuous case, in which the net change in head $h(x+\Delta x)-h(x)$ is emphasized, corresponding to $-\Delta h$ in the left figure

FIGURE 5-3

Discrete and continuous notational schemes for hydraulic gradient

on the left-hand side of the diagram to the new notation from the right side as follows. First,

$$\Delta h = \text{head loss}$$
$$= h(x) - h(x + \Delta x)$$
$$= -\big[h(x + \Delta x) - h(x)\big],$$

and so

$$i = \frac{-\big[h(x + \Delta x) - h(x)\big]}{\Delta x}.$$

Note that the one symbol Δh is being used here as the negative of the way in which it is usually used in elementary calculus. This is because we will continue to use it to represent the loss in head rather than the net change in head.

Note that we will get a more accurate value of the hydraulic gradient i at the specific point x if we use relatively small values of Δx. In fact, the precise value of i at that point can be represented as the limit of this expression as Δx approaches 0; that is,

$$i = \lim_{\Delta x \to 0} \frac{-\big[h(x + \Delta x) - h(x)\big]}{\Delta x}$$
$$= -\lim_{\Delta x \to 0} \frac{h(x + \Delta x) - h(x)}{\Delta x}$$
$$= -\frac{dh}{dx}.$$

So the exact value of the hydraulic gradient in this continuous framework is given by the negative of the derivative of the head function! This leads to the following continuous form of Darcy's law:

$$q = Ki = -K\frac{dh}{dx}.$$

This also has two obvious corollary formulas based on the same basic principles discussed in Chapter 2. In particular, the total volumetric flow rate through a cross section of area A would be given by

$$Q = qA = -K\frac{dh}{dx}A$$

(which is analogous to the earlier formula $Q = KiA$), and the interstitial velocity equation would become

$$v = \frac{-K\dfrac{dh}{dx}}{\eta}$$

(which is analogous to the earlier equation $v = \dfrac{Ki}{\eta}$).

The continuous version of Darcy's law for the one-dimensional situation, as described above, will be applied in the next section. The following problems should give you some practice with the concepts in this section.

Exercise 1. Suppose on the basis of some data collection and curve fitting, a hydrologist produces the following equation to represent the approximate hydraulic head value as a function of distance x in an aquifer, measured in an easterly direction from a fixed location:

$$h(x) = 29 + 0.02x - 0.00000063x^2.$$

This is an acceptable approximation on the interval from $x = 0$ to $x = 12{,}000$ feet.

    a) Is the flow from east to west or west to east?

    b) What is the value of the hydraulic gradient at the point corresponding to $x = 6{,}000$?

    c) Where is the hydraulic gradient the greatest along the given interval?

Exercise 2. Similarly to the situation in the previous exercise, the hydraulic head in a different aquifer is modeled by the equation:

$$h(x) = 112 + 0.03x - 0.00000028x^2,$$

although in this case the x-direction is measured north from a fixed point. Identify what might be considered to be peculiar behavior in this aquifer and describe geologic conditions that would be a possible explanation. Can you think of a topographic analog involving surface water flow?

Exercise 3. Returning to the aquifer in Exercise 1, suppose that it is uniformly 20 feet thick, 700 feet wide in the north-south direction, and has a hydraulic conductivity of 6.2 ft/day. Why would this situation be impossible?

5.2 Applying the Continuous One-Dimensional Version of Darcy's Law*

When a hydrologist first goes out into the field to begin to try to understand the flow system in an underground aquifer, he or she is going to be looking for either directly measurable parameters or other parameters that can be reasonably well estimated. For example, the hydraulic head can be measured by looking at the water level in monitoring wells. The hydraulic conductivity can be estimated by studying the nature of underground material that is brought to the surface during the drilling process, and then either using reference values for the hydraulic conductivity of such materials or (if it is in the form of intact drilling cores) subjecting it to direct laboratory tests.

But these parameter values are generally relatively sparse because they require the expenditure of considerable time and money for their collection. Therefore, it is not uncommon for data to be collected in an iterative fashion, meaning that one starts with a relatively small data set, tries to develop a conceptual understanding of the underground situation based on this small sample, uses a mathematical model to make a number of predictions of measurable parameters not included in the initial data set, and then compares these predicted values with data from a new round of field investigation. If there is good agreement at this stage, it is likely (although not guaranteed) that the original conceptual understanding of the underground situation is reasonably accurate, and the new data might then be used simply to make slight adjustments or modifications. On the other hand, if the new data differ considerably from the model predictions, then some of the assumptions upon which the model is based must be incorrect. This requires a more serious reworking of the conceptual framework for the model with the hope that the new predictions will agree more successfully with the most recent data. It is

---

* This section uses first and second derivatives; it also discusses simple antiderivatives. One of the later (optional) exercises in the section uses the fact that $\ln x$ is the antiderivative of $1/x$ for positive x. There is also a derivation and application of the one-dimensional Laplace equation.

not uncommon for hydrologists to proceed through several rounds of iteration of this general type before achieving a high level of confidence that their understanding of the underground structure and flow regime is reasonably accurate. We will see examples of various parts of this process in this section.

Begin by considering the relatively simple case of a flow regime that is essentially one-dimensional, as illustrated in Figure 5-4. This figure shows a side or section view of an aquifer in which the flow is assumed to extend from left to right in essentially a horizontal line. (Note that the total head loss of 20 feet over a distance of 2,000 feet may correspond, at least in a water table aquifer, to an actual slope that would be imperceptible in the scale of the diagram.) Since the flow lines are assumed to be horizontal, the head values can be written as a function of the x-coordinate alone, as discussed in the previous section. In fact, two such head values are indicated on the diagram, namely, $h(1000) = 70$ and $h(3000) = 50$. We further make the important simplifying assumptions that the aquifer has constant vertical thickness and constant horizontal dimension perpendicular to the flow axis, and that it is made up of uniform material with a constant hydraulic conductivity K.

What are the possible values that we might encounter if we drill a monitoring well at the point corresponding to $x = 2,000$ and measure the corresponding value of h? For example, could this value of h be 65? Could it be 60? Could it be 55? Are there a wide number of possibilities for the value of $h(2000)$, or are there only a very limited number of possibilities?

The answer to these questions is that there is only a *single possible value* for $h(2000)$ in this situation, namely, $h(2000) = 60$. To see that this is the case, consider any other possibility, such as $h(2000) = 65$. In this latter case the hydraulic gradient along the stretch of aquifer from $x = 1,000$ to $x = 2,000$, calculated by the elementary methods of Chapter 2, would be

$$i_{\text{left}} = \frac{70 - 65}{1,000} = .005$$

whereas the hydraulic gradient on the next portion of the aquifer would have to be

$$i_{\text{right}} = \frac{65 - 50}{1,000} = .015$$

But this would lead to an imbalance or a violation of the "conservation of mass principle" at the portion of the aquifer corresponding to $x = 2,000$. That is, the flow from the left side

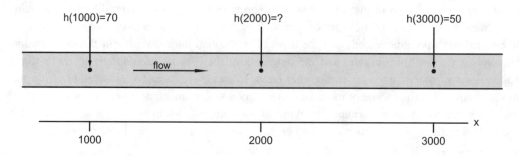

FIGURE 5-4
A section view through an underground aquifer in which the flow regime is essentially one-dimensional

through the aquifer cross section at $x = 2,000$ would be given by

$$Q_{\text{left}} = KiA = KA(.005)$$

whereas the flow through the same cross section but moving to the right would be given by

$$Q_{\text{right}} = KiA = KA(.015)$$

which is three times as much as the previous value. Since the aquifer has constant cross section and since we have assumed that the same value of K applies throughout the entire aquifer, these two Q values logically must be the same, but the equations show them to be different. Because of this inconsistency, we now know that one of our assumptions must be wrong. If we hold firm on all the assumptions about the aquifer and the measured head values, the only problem left can be the assumption that $h(2000) = 65$. Thus we must reject this possibility. In fact, it should now be fairly apparent that the only value for $h(2000)$ that would not lead to such an inconsistency would be the value

$$h(2000) = 60.$$

But what would happen if you went out into the field in this situation and actually measured $h(2000)$ and found that its value was really 65? What would this mean? Since this value is inconsistent with the previous assumptions that have been made about this situation, one of those other assumptions must have been wrong. Given the geometric constraints described by this problem, the only key assumption that was made was that the hydraulic conductivity K was uniform throughout the aquifer. At this stage, a hydrologist might be forced to hypothesize a more complex underground geologic environment with different values of K to the left and right of $x = 2,000$. This idea is investigated further in the following exercises.

Exercise 1. Assume in the context of the above discussion that we actually do measure a value $h(2000) = 65$. Assume further that we decide to assume a constant value of hydraulic conductivity to the left of $x = 2,000$ and a possibly different constant value of hydraulic conductivity to the right of $x = 2,000$. Without doing any computations but just thinking about the physical nature of this problem, would you expect the value of hydraulic conductivity on the left to be greater than or less than the value on the right? Explain your reasoning.

Exercise 2. Continuing the above discussion with the condition $h(2000) = 65$, can you determine specific values for K_{left} and K_{right} that must exist underground in order to lead to this distribution of head values? If so, find these values. If not, find the most detailed relationship you can between these values.

Exercise 3. Figure 5-5 shows a section view similar to a proposed nuclear waste repository site located deep beneath the surface in the southwestern United States. (A repository is a mine in which such wastes would be placed and then sealed off permanently.) The aquifer shown on the diagram is a dolomite bed that averages 25 feet thick, but whose characteristics gradually change to the west (i.e., towards the left side of the diagram). In particular, between Q and R, the average hydraulic conductivity is 0.03 ft/day with a porosity of .05, whereas from P to Q, the hydraulic conductivity increases to 0.3 ft/day due to increased fractures and solutioning, and the porosity increases to .10. The heads at P and R were measured at test wells and found to be 560 and 685 feet, respectively. Assume that the repository and the affected portion of the

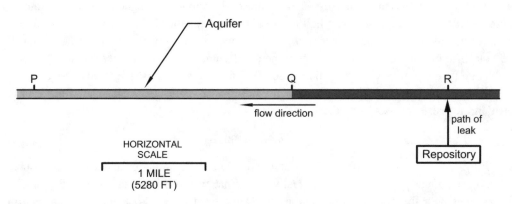

FIGURE 5-5
Conceptual hydrogeologic system near an underground nuclear waste repository in the Southwest

aquifer directly above it are 400 feet wide in the north-south direction (i.e., perpendicular to the section shown above), and use the scale on the diagram to determine distances.

a) What must the head be at Q?

b) What would be the ground-water travel time from point R to point P, just in case some radioactive material leaked from the repository and mixed with the water in the aquifer?

c) Find an equation for the head function $h(x)$ where x is distance measured west from point R.

d) Where is the hydraulic gradient the largest?

e) Is all of the given quantitative information necessary in order to answer the above questions?

On the basis of the foregoing discussion we can make two observations:

1) There are quite severe constraints on any function representing the hydraulic head distribution, especially in a situation where the hydraulic conductivity is assumed to be constant. (In our example, the value at simply two points was sufficient to determine a unique value for the head at every intermediate point.)

2) There is an interplay between assumptions that a hydrologist might make about hydraulic conductivity, and predictions that might result concerning hydraulic head values.

In fact, we will see later in this section that under the assumption of constant hydraulic conductivity, the hydraulic head function h, giving the hydraulic head at any point within the aquifer, must satisfy a famous equation called Laplace's equation:

$$\frac{d^2h}{dx^2} = 0.$$

This equation may not look very impressive, but it actually represents a very important step in our later development of similar Laplace equations for two- and three-dimensional problems. It is a "second-order differential equation." A *differential equation* is just any equation that has a derivative somewhere in it, as this one does. And it is a *second-order* differential equation because the highest (and in this case only) order derivative to be found is a second derivative.

The curious thing about differential equations is that their solutions are not numbers, but rather whole functions. For example, the function

$$h = 3x - 7$$

is a solution to the one-dimensional Laplace equation because when you plug it in, it works. That is, if you take the second derivative of this function, you do indeed get 0. In fact, you can probably easily determine the entire set of solutions to the one-dimensional Laplace equation. Try it with this exercise.

Exercise 4. Describe the complete set of solutions to the one-dimensional Laplace equation. Be sure to explain why no other function, no matter how strange, could possibly be a solution.

Sometimes it is useful to simplify the cumbersome notation used for derivatives, and you already are quite familiar with one way to do this. For example, you might write h' instead of $\frac{dh}{dx}$, or h'' instead of $\frac{d^2h}{dx^2}$. However, to set the stage for our upcoming use of "partial derivatives," we will often use the abbreviations h_x and h_{xx} for the first and second derivatives of h with respect to x. In this new notation, the one-dimensional Laplace equation would take the form:

$$h_{xx} = 0.$$

Now we want to derive this one-dimensional Laplace equation, and our method of reasoning may be quite new to you. A very important type of argument is being introduced here, one that should better prepare you for material in later parts of this chapter.

The typical one-dimensional situation is represented by Figure 5-6, where, as earlier, the flow lines are assumed to extend in a horizontal direction in the plane of the figure, and the head values can simply be written as a function $h(x)$. It is best for you to think of Figure 5-6 as a three-dimensional, not simply a two-dimensional situation. The third dimension of this aquifer is directly perpendicular to the paper. If the vertical direction on the paper represents the vertical direction within the aquifer, then the direction perpendicular to the paper represents the other horizontal direction in three space.

The box drawn in Figure 5-6 represents a *direct side view* of a *three-dimensional box* containing a small rectangular portion of the aquifer. The dimensions of this "incremental volume" are as follows: the length is Δx in the x-direction; the height is 1 in the vertical or z-direction; and the depth is 1 in the other horizontal or y-direction (this y-direction is the one that extends perpendicular to the plane of the paper and hence is not shown on the diagram).

The key physical principle that is used to derive Laplace's equation in this and higher-dimensional cases is a simple conservation of mass condition that may be stated as follows:

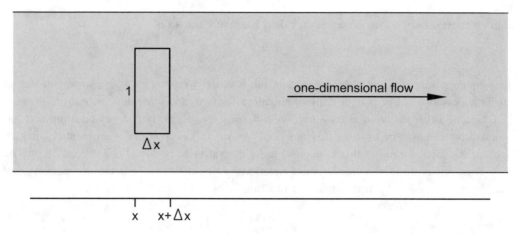

FIGURE 5-6
Aquifer flow system exhibiting essentially one-dimensional flow

the net amount (thinking in terms of mass) of fluid entering this fixed incremental volume at any moment must be 0. This follows from the simple observation that there is no mechanism for either creating or destroying fluid within the volume, nor is there any way to store it or take it out of storage. Therefore, everything that goes in one end must be balanced by the same amount coming out the other.

Furthermore, since water is essentially an incompressible fluid, saying that the net amount of *mass* entering the incremental volume must be 0 is equivalent to saying that the net amount of fluid *volume* entering the incremental volume must be 0.

Now let us calculate what the net volumetric flux must be through the boundary of this incremental volume. (Yes, you know that its numerical value must be 0, but we want to get an equivalent mathematical expression that we can then set equal to 0.) Since this volume is a rectangular solid, it has six faces. However, there can be no fluid flux through the top, bottom, front, or back faces, because the flow lines run directly along these faces, rather than through them. Therefore, in order to calculate the net fluid flow into this incremental volume, we need consider only the fluid flow through the left face and the right face. Keeping in mind our basic equation for fluid flow

$$Q = KiA$$

and calculating both such flows in a direction pointing *inward* toward the incremental volume through each of the two faces under consideration, we obtain the following equations:

$$0 = \text{flow in through left face} + \text{flow in through right face}$$

$$0 = K\big(-h_x(x)\big)(1 \cdot 1) + K\big(h_x(x + \Delta x)\big)(1 \cdot 1).$$

Here we have again used the subscript notation for differentiation. One can readily verify that the minus sign does belong on the first h_x expression, because on the left face of the incremental volume h_x is itself negative (since h is decreasing to the right), but the fluid flow into the incremental volume must be positive. Similarly, a plus sign is used with the h_x value at the right face, corresponding to points whose x-coordinate is $x + \Delta x$.

Now we perform some algebraic manipulations on this last equation in order to obtain Laplace's equation, as follows:

$$0 = -h_x(x) + h_x(x + \Delta x) \qquad \text{(divide by } K\text{)}$$

$$0 = \frac{h_x(x + \Delta x) - h_x(x)}{\Delta x} \qquad \text{(divide by } \Delta x\text{)}$$

$$0 = \lim_{\Delta x \to 0} \frac{h_x(x + \Delta x) - h_x(x)}{\Delta x} \qquad \text{(take limits of both sides)}$$

$$0 = h_{xx}(x) \qquad \text{(definition of the derivative of } h_x\text{)}.$$

Since x is an arbitrary point in the flow pathway, this equation must hold at all x values and thus can be written in the simple form

$$h_{xx} = 0.$$

This is the one-dimensional Laplace equation!

Exercise 5. Where would the derivation of the one-dimensional Laplace equation fail if the hydraulic conductivity were not constant, but rather were to vary along the aquifer?

Exercise 6. Suppose that in the same context as above we have an aquifer within which we hypothesize that the hydraulic conductivity varies linearly with x. That is, we assume that the hydraulic conductivity can be represented by an equation of the form

$$K = K_0 + \alpha x$$

for some constant α. What differential equation would h satisfy in this case?

These next exercises should bring you back from the theoretical world to the practical problem of figuring out what the head values are in an underground aquifer. At this point you know that on any interval of constant hydraulic conductivity K, the head function must satisfy Laplace's equation. Therefore it must be a straight line on that interval. Now apply this information to a situation where two or more such intervals occur in a row.

Exercise 7. Return to the situation of Figure 5-4 with the observation that $h(2000) = 65$ and the assumption that there are constant values of hydraulic conductivity on the intervals both to the left and to the right of the point $x = 2,000$. Draw a graph of the function $h(x)$ that would be the solution to this problem on the interval from $x = 1,000$ to $x = 3,000$.

Exercise 8. Figure 5-7 shows a still more detailed structure for a one-dimensional flow problem. In this case the interval from the 1,000 ft marker to the 3,000 ft marker is actually divided into three parts, and the head values are measured at all the nodes. These values are shown on the figure. The K values are assumed constant on each of the three sub-intervals. Furthermore, the hydraulic conductivity within the leftmost sub-interval has been extensively investigated and is known to have the value of 0.4 ft/day.

    a) Find the hydraulic conductivity in the other two sections of the aquifer.

    b) Find the head value at the point $x = 2,800$.

    c) Write an equation that will give the head value at any point in the interval.

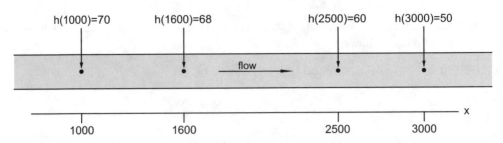

FIGURE 5-7
One-dimensional flow framework with three sub-intervals having different properties (Exercise 8)

d) If the porosity values of the three subintervals are .35, .3, and .2, from left to right, how long would it take ground water to travel from $x = 1,000$ to $x = 3,000$?

Exercise 9.   Figure 5-8 is a revised interpretation of the nuclear waste repository setting treated earlier in this section. However, the aquifer in this case is now hypothesized to be divided into three distinct sections, with parameter values as shown on the figure. Head values are known only at the extreme endpoints, as previously.

a) If you were to drill monitoring wells at points S and T, what head values would be predicted by this conceptual model of the aquifer?

b) What would be the ground-water travel time from R to P according to this model?

Before completing this section on applications of the continuous version of Darcy's law to one-dimensional flow problems, there is one more important situation to which this equation applies. Consider Figure 5-9, which shows a water well penetrating a slowly-moving but highly

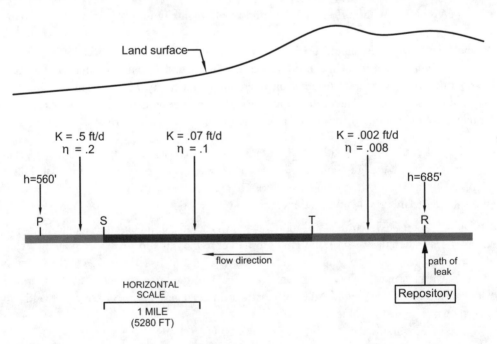

FIGURE 5-8
Three-section aquifer with specified hydrologic properties

productive confined aquifer, along with several head contours corresponding to the steady-state situation achieved when the well is pumping. These head contours are concentric circles whose corresponding head values obviously decrease closer to the well. Furthermore, the innermost head contour can be thought of as being right at the circumference of the well, with a head value corresponding to the head value measured in the well. You may wonder why the original natural head gradient is not reflected in the contours, and the answer is that it is essentially negligible compared to the "near-field" head gradients induced by the pumping process. Note that the actual head value is given only on the outermost contour. The radius of this particular circle is known as the *radius of influence* of the well, and beyond it there is minimal disruption of the original natural head values even when the well is pumping.

Although this looks like a two-dimensional problem, it is "essentially" a one-dimensional problem. This is because of the fact that there is complete symmetry around the circles, so that the head values along a given circular contour will depend only on the radius r of that contour line. Although Laplace's equation, in the form in which we derived it, does not apply to this case, the basic reasoning that went into it, involving conservation of mass and Darcy's law, certainly can also be applied here, the problem being to determine the head values on all the contour lines as well as the amount of water that can be pumped from the well under these conditions. This is the point of the following exercise, which may require some initial contemplation on your part as you plan a strategy for the new kinds of questions being asked.

Exercise 10. For the situation shown in Figure 5-9, assume the following parameter values. The confined aquifer is 20 feet thick and is fully penetrated and screened by a well that is 8

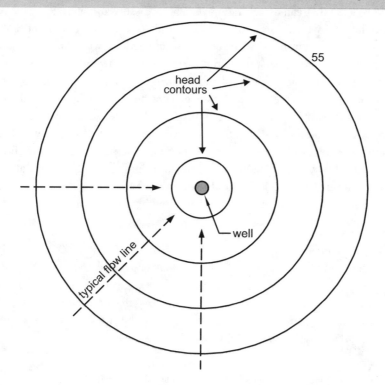

FIGURE 5-9
Head contours and flow lines in the vicinity of a water withdrawal well

inches in diameter. Its hydraulic conductivity is 28 ft/day. During pumping, the head at the well is maintained at 44 feet, whereas the head at the limit of the radius of influence (estimated to be 50 feet from the well) is 55 feet. Find the head function $h(r)$ and the quantity of water (in gallons per minute) being extracted from the well under these conditions. How sensitive is this latter value to the assumed radius of influence of the well?

5.3 Retardation Factors*

Essentially all the ground-water modeling discussion up to this point has been oriented towards the movement of the ground water itself, the implicit assumption having been made that any material that is dissolved in the ground water is simply going to be carried along at the same rate. Although this is usually quite a reasonable approximation, there are certain situations when it is important to pay greater attention to the physical processes that may impact the movement of the contaminant in a way that differs from the movement of the water itself. One process that can affect the transport of contaminant is diffusion, meaning that dissolved material may tend to diffuse or spread out in all directions as it is transported, the result being the lessening of maximum concentration but the spread of contaminant over a wider area. This process has been the subject of extensive modeling, much of which is quite analogous to the modeling for air problems in Chapter 3. Because of this close similarity, it has not been included for separate discussion here.

A second process that can affect the movement of dissolved material is called by the general term "retardation," which refers to the fact that under certain circumstances the dissolved material may undergo certain physical or chemical interactions with the surrounding geologic medium, the net effect being that these processes cause a slowing of the advance of a contaminant front well below the velocity of the fluid itself. (A *contaminant front* refers to the relatively compact region where the contaminant concentration builds from 0 to its peak or steady-state value.) A simple parameter R is used to characterize this retardation effect, this number referring to the factor by which the effective forward velocity of the contaminant front is reduced below the velocity of the fluid itself. For example, if the retardation factor R is 10, this is a mathematical way to say that the contaminant front is moving forward at a rate roughly 10 times slower than that at which the fluid itself is moving.

To get some initial computational familiarity with this concept, it would be useful to complete the following easy exercise.

Exercise 1. This problem is an extension of Exercise 3 in Section 5.2, to which you should refer for the basic figure and parameter values. You should have already calculated as part of that exercise the ground-water travel time from R to P, expressed in years. If a radioactive substance were to enter the ground water through a leak from the repository at R, how long would it take to get to P, assuming its retardation factor in the RQ portion of the aquifer is 50 and its retardation factor in the QP portion of the aquifer is 10?

---

* This section does not require any calculus background. While it does refer in one exercise to a problem treated in the previous section, that is only to provide some computational continuity. No knowledge of any groundwater topics other than in Chapter 2 is needed to understand the material in this section.

If some contaminant has just begun to enter the ground-water system and one wants to estimate how long it will take for the contaminant front to reach some critical off-site location, such as a drinking water well, then by combining the retardation factor with our usual estimates of ground-water flow velocity, as in the above exercise, it is easy to estimate the time within which some kind of remedial action might need to be taken. The trouble with the retardation factor is that it cannot be measured directly. Rather, it needs to be calculated on the basis of its relationship to one or more other parameters that can actually be measured for a given situation.

Some further discussion of the nature of the interaction between the dissolved contaminant, called the *solute,* and the surrounding water and soil or rock system, called the *substrate,* is necessary in order to develop useful equations for this factor R. In particular, consider the following hypothetical laboratory experiment. Take a sample of the material comprising the aquifer, which for convenience we may think of as a soil, and put it in a beaker into which you also add water containing some of the solute of interest. If you mix up the contents of this beaker and then pour off the water, it is possible that less solute may be leaving with the water that you pour off than you originally added to the system in the first place. This is because there may have been some kind of interaction between the solute and the soil that caused some of the solute to bind to the soil particles and remain with this solid phase in the beaker after the liquid was removed.

There are many different kinds of physical and chemical interactions between solute and soil that could lead to such binding, but we shall assume in this discussion that the reaction that takes place falls into the general category of *adsorption,* which refers to the binding of solute molecules to the surfaces of soil particles or rock pores. These reactions generally involve the establishment of an equilibrium between the amount of material in solution and the amount adsorbed onto the soil or rock. For example, if you were to increase the concentration in solution, more would then also attach itself to the surfaces. In fact, we will actually assume that the reactions discussed here follow a so-called *linear isotherm,* which simply means that if the temperature is kept constant ("isotherm"), the amount adsorbed is proportional to the amount in solution. (Further clarification of units for these concepts will be given shortly.)

These reactions generally reach equilibrium relatively quickly and are also easily reversible, meaning in the case of our experiment with the beaker that if we now pour clean water into the beaker, some of the solute that was bound to the soil will be released into the solution (where the concentration is less than it was at the beginning of the original experiment). In fact, if we repeatedly flush the soil with clean water, we will be able to remove more and more of the solute that has been bound to it.

For a given type of soil, some solutes may undergo no adsorption onto the soil particles, whereas other solutes may have much stronger affinity for the soil and bind in large part, practically removing them from the solution. Thus the degree of adsorption depends both on the solute and the soil. It may also depend on other chemical conditions in the ground water, such as pH, or physical conditions, such as temperature.

Now suppose that we knew that for a given solute and soil, roughly 99% of the solute would be adsorbed in an experiment of the type just discussed, regardless of the amount orig-inally present. If we then transferred the liquid successively through a sequence of beakers, as illustrated in Figure 5-10, the concentration of solute would very quickly be reduced to negligible levels. For example, after the first beaker, the fraction left would be only 0.01 of the original. After the second beaker, it would be $(0.01)^2 = 0.0001$, since it would be 1% of

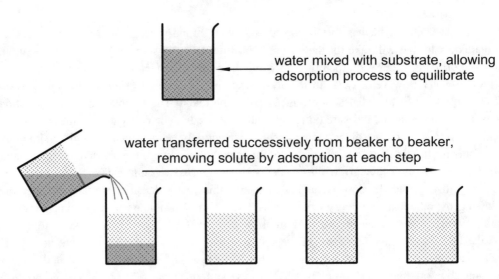

FIGURE 5-10
Laboratory experiment involving adsorption

the 1% getting through the first beaker. Similarly, after the third beaker, the fraction would be $(0.01)^3 = 0.000001$.

The natural analog of this process is quite similar and is illustrated in Figure 5-11. This figure shows a portion of an underground aquifer along the axis of flow and having a unit cross-sectional area perpendicular to the axis of flow. It has been divided into individual segments of length d, which correspond roughly to the beakers in the previous figure. We want to track the rough concentration of solute in the ground water as it moves from each segment to the next, and we will use the same hypothetical removal factor of 99% as we used above.

When contaminated ground water first enters the aquifer section shown in this figure (through the face at the left end), suppose that the concentration of solute is given by C. As the first segment is filled with this contaminated water, the adsorption process takes place essentially instantaneously so that approximately 99% of the solute is pulled out of solution and left attached to the soil matrix. The water continues to move on to the second segment but now its concentration is starting at a level approximately equal to $.01C$. But then 99% of the solute moving through that segment is removed from solution so that the concentration as it enters the third segment is going to be only $(.01)^2C$. By repeating this simplified analysis of events, it is

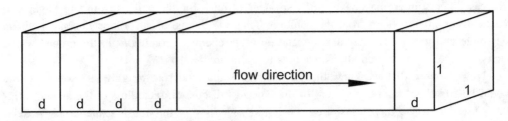

FIGURE 5-11
A portion of the flow pathway in an underground aquifer of arbitrary length and unit cross-sectional area, segmented into individual sections of length d

easy to see that the concentration of solute, by the time the ground water reaches the far right end of this aquifer segment, will be negligible. The main difference between this situation and the experimental setup in Figure 5-10 is that in this latter case we are observing a "continuous process" whereby more and more contaminated water keeps coming in through the left end.

This leads to an important proviso applying to our earlier conclusions. In particular, the adsorption process as described will only continue as long as we do not use up all the adsorption capacity of the soil. For example, if we were to fully saturate the soil with all the solute that could attach to it, no more solute could be removed from solution and so the contaminated ground water would begin to move through the system at essentially full strength. This would be expected to happen first at the leftmost portion of the aquifer, since it is receiving the highest strength solution. Then the following unit would become saturated, etc.

If this saturation process takes, say, 10 times as long to occur, as it takes for the ground water itself to traverse this section of the aquifer, then the solute would begin to appear at the right end of the aquifer segment only after a time delay 10 times as long as the time it takes for the ground water itself to traverse this segment. In other words, it would appear as though the contaminant front were moving at a velocity 10 times slower than the ground water. This would, of course, correspond to the statement that the retardation factor for this situation is $R = 10$.

You may, of course, notice that as the aquifer section gets very close to saturation, the concentration at the end begins to rise. For example, when only the last segment is unsaturated, you will have at best an exit concentration of $.01C$, rather than the truly negligible values earlier in the process. But there would still be quite a rapid rise in concentration as this last segment also reached saturation, so that the exit concentration would have the general shape shown in Figure 5-12. When we talk about the time it takes for the contaminant to pass through the aquifer section, we generally are referring to the time for breakthrough to occur, as illustrated in the figure, but since the concentration rise is generally quite steep and since the idealized adsorption process discussed here is only an approximation of reality, no great quantitative significance should be attached to the particular point where the concentration is 50% of the maximum. It is simply used as a representative of the region where breakthrough occurs.

The 99% estimate used in the previous discussion is only a heuristic concept. We did not, for example, define its units or explain precisely how it would be measured. But now we do want to make these aspects more precise. In particular, the actual experimental parameter that

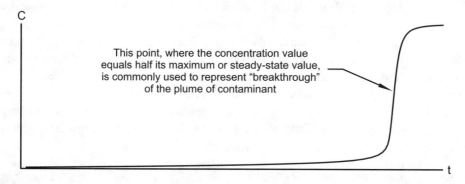

FIGURE 5-12
Concentration profile near contaminant breakthrough point

is typically measured in the laboratory is called the *distribution coefficient* K_d. It measures the distribution or point of equilibrium between dissolved solute and solute that is adsorbed onto the solid matrix. The following equation defines this parameter:

$$K_d = \frac{\text{mass of solute adsorbed on solid phase per unit mass of substrate}}{\text{concentration of solute in solution}}$$

This definition may initially seem somewhat awkward because the numerator is in units of mass of solute per mass of soil or rock, and the denominator, being a concentration, is in units of mass of solute per volume of solution. Thus the final reduced units are in units of volume of solution per mass of substrate, such as milliliters per gram, the most common reported units.

However, the expression may seem much more reasonable if it is rewritten in the form

mass of solute adsorbed on solid phase per unit mass of substrate

$$= K_d \times \text{concentration of solute in solution}$$

since this simply says that the mass adsorbed (per unit of substrate) is simply proportional to the concentration of the solution. For example, if you double the concentration of the solution, then twice as much will be adsorbed by the same amount of substrate. The K_d value is just the constant of proportionality!

It can be shown (see the exercise below) that the retardation factor R can be calculated once the distribution coefficient K_d has been measured. The relationship between these parameters is the following:

$$R = 1 + \frac{\rho_b}{\eta} \times K_d$$

where h is the usual porosity and ρ_b is the bulk soil or rock density described above. To clarify this latter parameter, suppose its value were 2.3 g/cm^3. This would mean that there would be 2.3 grams of solid material in a total volume, including any pore space, of 1 cubic centimeter of the soil or rock.

It turns out that for many dissolved contaminants the retardation factor is negligible and does not slow the spread of contamination. However, for certain special kinds of contaminants, such as a large family of radioactive materials, adsorption processes may lead to retardation factors on the order of 10 to 1,000. This has been an important area of research in connection with the national program to develop safe systems for burying radioactive wastes, because even if the wastes escape from their original location of burial into the surrounding ground-water system, one might find an additional layer of protection in the fact that they would likely be substantially absorbed in the aquifer itself to a sufficient extent that they would decay to harmless levels by radioactive decay before ever reaching accessible points of the environment. For critical applications of the retardation concept, such as here, extensive research must be carried out on appropriate values of the distribution coefficients for these substances, and, in addition, one must be sure that the underground geochemical environment is such that the sorption processes experienced in the laboratory really do occur to an equivalent extent in the heterogeneous and unpredictable geologic environment.

Exercise 2. Consider a system in which the bulk rock density is 2.3 g/cm^3 and where the porosity is 30%. In order for the retardation factor to turn out to be 100, what would the K_d value have to be? (Use the relationship between the retardation factor and the distribution coefficient given in the text.) Furthermore, make a simple statement about the percent of mass of solute that would be adsorbed onto the solid phase in this system (analogous to the hypothetical 99% figure that was used in the text).

Exercise 3.   Derive the equation given in the text for the retardation factor as a function of the distribution coefficient.

Exercise 4.   Give a direct physical interpretation of the distribution coefficient K_d in terms of its typical units of milliliters per gram.

Exercise 5.   Find the values of several retardation coefficients for specified solute/substrate combinations. (Hint: look in the radioactive waste literature, much of which is available on-line.)

5.4 Determining the Hydraulic Head Contour Lines and General Flow Directions Using Data Available From Moderate or Large Numbers of Wells*

Recall from Chapter 2 that a hydraulic head contour map is extremely useful in determining the direction of ground-water flow at any point (since flow will be perpendicular to the contour lines), the shape of the overall flow lines, and the hydraulic gradient along these flow lines. Many of our calculations in Chapter 2 began with such head contour maps. But where do such maps come from? How do you take raw data from a limited number of wells in the field and convert this to accurate contour maps?

We have encountered one situation where the details of this procedure were worked out. In particular, in Section 2.8 we investigated how one might use data from three wells to estimate how the contour lines and flow directions might look in the neighborhood of those three wells. But now we approach the problem on a larger scale, perhaps involving an area of several or even hundreds of square miles. Here we would hardly expect to get much useful information out of three wells, so we must assume that we have quite a few more. But if we do, how then would we put this information together to construct the actual contour lines?

There are many computer programs that do these kinds of operations. You would input the well locations and the head values, and the computer would generate some estimated contour lines. But how much faith should you have in these results? Could two different computer programs generate substantially different shapes for the contour lines? What would make one of these "contouring programs" better than another? And what actual procedure does the computer use anyway?

This section is different from most others because it offers many questions and few answers. It invites you to experiment on your own to try to answer some of these questions. It is important that you take your time and really think about the choices you are making at each step. You should have a logical rationale for everything that you do, but that doesn't mean that you can actually prove that it is correct. Some people think that there is no room for opinion or judgment in mathematics, but in some important applied areas this would not be correct, for a good supply of experience and physical intuition can often guide investigators to valuable results that there

---

* This section requires no calculus background. It reintroduces the two-dimensional framework from most of Chapter 2 and is largely open-ended and experimental in nature. If desired, this section can be skipped without loss of continuity.

would hardly be time to develop and prove rigorously under the real-time constraints in which the problems need to be addressed.

The general situation to be investigated in this section is typified by Figure 5-13, which shows a 5,000 ft by 5,000 ft geographic area within which a number of wells have been drilled and then measured to determine the hydraulic head at each of their locations. In particular, 24 wells are available and the measured head values range from a low value of 203.2 feet to a high value of 268.2 feet. This kind of well distribution would not be uncommon in this roughly mile square region; and, in fact, the wells might at least in part represent wells drilled in the past for a diversity of purposes, such as drinking water, irrigation water, contaminated plume sampling, or water supply investigation. They are all shown in the figure to be located at grid points on a 500-foot grid pattern, but this is only to simplify their use in later computations.

Through a sequence of exercises presented below, you are asked to provide some original thinking about intelligent ways to use the data shown in the figure in order to construct hydraulic head contour lines. There are a number of different approaches to this problem, and you may be surprised at the radically different approaches taken by people approaching the problem from

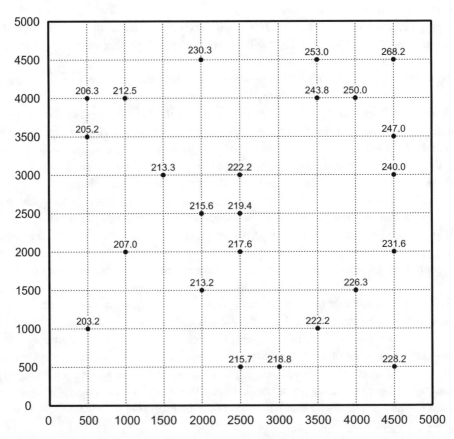

FIGURE 5-13
Scattered well data available early in a hydrogeologic investigation

TABLE 5-1

Tabular data showing location (in terms of x- and y-coordinates) and head values at monitoring wells shown in Figure 5-13

x	y	head	x	y	head	x	y	head
500	1000	203.2	2000	4500	230.3	3500	4500	253.0
500	3500	205.2	2500	500	215.7	4000	1500	226.3
500	4000	206.3	2500	2000	217.6	4000	4000	250.0
1000	2000	207.0	2500	2500	219.4	4500	500	228.2
1000	4000	212.5	2500	3000	222.2	4500	2000	231.6
1500	3000	213.3	3000	500	218.8	4500	3000	240.0
2000	1500	213.2	3500	1000	222.2	4500	3500	247.0
2000	2500	215.6	3500	4000	243.8	4500	4500	268.2

different points of view. To simplify the calculations you are asked to carry out within the exercise set, Table 5-1 contains a summary of the well data presented in the figure.

Exercise 1. Make a list of at least five distinct methods that a reasonable person might try to use to estimate the hydraulic head values at an arbitrary point in the geographic region shown in Figure 5-13. Be very specific in defining your methods so that they are completely "well-defined," meaning that a competent person should be able to carry them out for any given point in the region without asking you for further details about how the proposed method is supposed to work. (Such well-defined procedures are called *algorithms*.)

Exercise 2. For each of the methods that you listed in Exercise 1, comment very briefly on any limitations or weaknesses that you think it might have in trying to estimate reasonably the value of the hydraulic head throughout the region shown. After completing these comments, rank order the proposed methods based on your own judgment about which ones you think are likely to give the most accurate results.

Exercise 3. For your highest ranking method from Exercise 2, carry out the associated calculations to estimate the hydraulic head values at the following three points: $A(3000, 4000)$; $B(3500, 2000)$; $C(1000, 500)$.

Exercise 4. Repeat the previous problem for your second-highest ranked method, and see how the results compare with those using your highest ranked method.

Exercise 5. Based on your initial computational experience above, think once again, as creatively as possible, about what would be a good definition or criterion for a desirable method for estimating hydraulic head values in the present situation. See if you can formulate such a criterion in mathematical or quasi-mathematical terms. Based on your thoughts in connection with this problem, feel free to revise your list of methods originally proposed under Exercise 1. Furthermore, if you have ever studied any statistics, see if there are any lessons from that subject that you may want to take into consideration as you refine your list of possible methods here. (If you have not studied statistics, just ignore this suggestion.)

Exercise 6. Suppose that you now have an acceptable algorithm for generating estimated values of the head at any point within the region of interest (not just grid points). Design a conceptual algorithm that could then be used to generate a set of contour lines. Assume that the algorithm ultimately has to specify pairs of points between which straight-line segments are to be drawn.

Exercise 7. Find a commercial computer program to which you have access that you can apply to the problem of estimating the head contour lines corresponding to the original data in Figure 5-13 and Table 5-1. [Hint: many quantitative programs can be applied or easily adapted to this problem, including math packages, statistical packages, and even some spreadsheet programs. This will generally be a two-step process: finding a way to approximate the head value at a general point, and then drawing a contour plot of this function. You might even want to use different programs for these two steps, depending on what you are most comfortable with.]

Exercise 8. Sometimes the interpolation or extrapolation of given data can lead to incorrect results because of some key factor that is not represented by the limited set of data points provided. As an example of this situation, consider, within the context of Figure 5-13, the point $D(3500, 3000)$. A cursory review of Figure 5-13 suggests that the unknown hydraulic head value at that point should be somewhere in the range between 222.2 and 240.0. Based on your knowledge of the relationship between geologic, topographic, and hydrologic values, provide at least one plausible physical situation that is both consistent with the data shown in Figure 5-13 but in which the actual hydraulic head value at point D would be greater than 250.

Exercise 9. Suppose that the firm investigating the hydraulic head contours in the geographic region described by Figure 5-13 has sufficient funds available to drill one additional monitoring well. Identify on Figure 5-13 those areas where such a monitoring well would be most likely to reduce the amount of uncertainty associated with the contour lines. Explain your reasons. Could you define an objective quantitative criterion to capture your underlying logic? (If you completed Exercise 7, feel free to incorporate those results in your answer to this exercise.)

5.5 Exploration of the Relation Between Hydraulic Head Contour Lines and Ground-water Flow Lines*

We begin in this section to develop the two-dimensional mathematical framework that will let us "upgrade" our treatment of ground-water flow from the graphical methods of Chapter 2 to more precise two-dimensional versions of Darcy's law and Laplace's equation. This will be somewhat analogous to the one-dimensional continuous form of Darcy's law treated in Section 5.1 and 5.2, and will also shed further light on some of the contouring issues raised in Section 5.4. This development will extend through the remainder of this chapter.

Recall the basic approach presented in earlier sections for determining the direction and rate of ground-water movement: identify hydraulic head contour lines, assume that the flow

---

* A reader with a good understanding of the basic concepts of vectors as treated, for example, in either high school or college physics, should have no trouble following the basic ideas in this section, although a few of the exercises will not be accessible. Readers who have studied multivariable calculus should find in this section an interesting application of some of the concepts seen there. Early exercises are designed to review such background material. There is further reference to linear algebra for readers with such background.

direction is always perpendicular to these contour lines, sketch in the flow line beginning at any point of interest (such as a point where contamination enters the aquifer), and analyze distances and hydraulic gradients along the flow line. Key to this logical sequence has always been the assumption that ground water flows in a direction perpendicular to the hydraulic head contour lines. The purpose of this section is to further investigate this assumption and to understand when it applies and when it does not apply.

Here is the first physical situation we will investigate in this section. We have an aquifer made up of a type of geologic material that is *isotropic,* meaning that at any particular location its hydrologic properties in any direction are identical. For example, this assumption would imply that if we took a sample of this geologic material and subjected it to some experiments to measure its hydraulic conductivity, then the outcome of the those experiments would not depend on the particular way we picked the sample or the way we oriented it for the experiments. In other words, we are assuming that ground water can flow equally well in any direction through a sample of this material. We are also assuming that our aquifer is relatively flat, so that the only significant direction of flow of ground water is in the horizontal direction. Thus the flow is essentially restricted to two dimensions. Within this context, the basic question is this: *Why does ground water flow in a direction that is perpendicular to the hydraulic head contour lines?*

To answer questions like this, it is important to think in terms of vectors. The following exercises are designed to help you review various aspects of vectors that you may have encountered earlier, involving both simple physical reasoning as well as the manipulation of mathematical vectors in multivariable calculus or linear algebra. Ideally, one would like to be comfortable with both approaches, but either one would be sufficient for the main purposes of this section.

Exercise 1. (Review of key concepts from basic physics.) Recall that a vector is an object that has magnitude and direction. Complete the following:

a) Give an example of several quantities in physics that can be represented by vectors.

b) Given two vectors, illustrate by means of a diagram how you can combine them to find the "resultant vector." (Hint: maybe you are familiar with this under the name "parallelogram rule for combining vectors.")

c) Given a vector drawn in the xy-plane, illustrate by means of an example how you can reverse the above process and represent the vector as the sum of two individual vectors, one pointing in the x-direction and one pointing in the y-direction.

d) In the previous part of this problem, you were asked to break a vector down into components in two given perpendicular directions, namely the x-direction and the y-direction. Is it possible, in general, to break a vector in the xy-plane down into components that are in two given directions, even if those two directions are not perpendicular to each other? If so, illustrate by means of a nontrivial example.

e) If your answer to the previous part was "no," describe under what limited conditions you may actually be able to break down a vector into components in two non-perpendicular (but non-identical) directions. If your answer to the previous part was "yes," see if you can find a good reason why it may be preferable to break a vector down into perpendicular components rather than simply into general components in other directions.

f) Draw the following three vectors in the xy-plane: Vector 1, from the point $(1, -3)$ to the point $(5, -2)$; Vector 2, from the point $(1, 1)$ to the point $(3, 3)$; Vector 3, from the point

$(-1, 2)$ to the point $(-4, 5)$. Can you break down Vector 1 into two components along the two directions represented by Vectors 2 and 3? If so, do it, being sure to explain exactly what your logic is for each step. If not, explain why this is impossible. Be sure to work with good diagrams.

g) As in the previous part of this problem, arrows in the xy-plane can be used to represent two-dimensional vectors. Consider the following two vectors: Vector 1, from the point $(1, 1)$ to the point $(2, 4)$; and Vector 2, from the point $(4, -1)$ to the point $(5, 2)$. Do these two arrows represent different vectors or exactly the same vector? Be very precise in your explanation.

Exercise 2. (Review of selected multivariable calculus topics.) Answer the following questions in the context of a general function of two variables of the form $z = f(x, y)$ that is assumed to be "smooth" (meaning that it has continuous first partial derivatives on its domain):

a) What is meant by the "directional derivative" of such a function? Describe your answer in words, not equations, but be precise.

b) What is meant by the "gradient" of such a function?

c) Is the directional derivative at a point a scalar value or a vector?

d) Is the gradient at a point a scalar value or a vector?

e) Give a typical example of a nontrivial function of the type being discussed in this problem. (You will use this function for later parts of this problem.)

f) Give a typical example of the calculation of the directional derivative of this function, and also an example of the calculation of the gradient of this function, both of them at a given point. If you need any additional assumptions or input values for these calculations, make clear what assumptions you are making or what values you are assuming.

g) Use dot product notation to provide a formula for the directional derivative of a function at a point in terms of the gradient at that point and a vector pointing in the direction in which you wish to calculate the directional derivative.

h) What special property does the directional derivative at a point have when taken in the specific direction represented by the gradient? (Hint: you may wish to use the representation of the dot product in terms of the cosine of some angle in order to help you answer this question.)

i) Consider a point along an arbitrary level line (= contour line) of the function f. Prove that the gradient is perpendicular to the tangent to that level line at the given point. Use your previously defined function to present an actual numerical example to verify this as well.

j) The statement is occasionally made that "the gradient points in the direction of steepest ascent." Explain this statement. Furthermore, give a corresponding characterization of the direction of steepest descent.

Exercise 3. (Review of key linear algebra concepts for readers with some linear algebra background.) Refresh your memory on the situation addressed in parts c, d, and e of Exercise 1, above. Address these same issues in the language of linear algebra, especially the concepts of spanning set or basis.

Now we can proceed to investigate the question that was posed earlier, namely: Why does ground water flow in a direction that is perpendicular to the head contour lines? The best way to do this may be to think first about a simpler but analogous case, suggested in Exercise 4, and then to apply similar reasoning to the ground-water case in Exercise 5. Be sure in these two exercises to do your best to "hammer the nail in straight" with precise statements and convincing logic at each step. This can be quite challenging at first, but if you are successful, you will have developed a deeper understanding of some basic concepts and be better prepared for the material that follows.

Exercise 4.   Imagine that you are standing on the side of a mountain, with the only peculiar aspect of this mountain being that its surface is quite smooth rather than being covered with vegetation or rocks. In mathematical terms, you could of course think of the mountain as the surface representing the graph of the function f described at the beginning of these exercises. Now imagine that you take a hockey puck and lay it down at your feet on the side of the mountain so that it can begin to slide downhill. In what precise direction will it start to move? (Hint: don't just say "down" because there are many different directions it could go that would be generally downward. And don't even just say "in the steepest downward direction" unless you can provide a precise argument why this would be the case.) Give a convincing explanation of your answer. You may base it either on the physics-type reasoning represented in Exercise 1, or you may use some of that reasoning in combination with the terminology and concepts reviewed in Exercise 2.

Exercise 5.   Give a convincing argument why it is that in an isotropic geologic medium the direction of ground-water flow at any point would be perpendicular to the hydraulic head contour line at that point. Be careful not to make any assumptions in your analysis without providing justification. Furthermore, be sure to understand where you are making use of the assumption that the geologic medium is isotropic.

If you have been successful in working through the previous exercises, you should now have a better understanding of the forces causing ground-water movement. The key issue in the movement of ground water in an isotropic medium is the fact that the water moves in the same direction as the external force applied to it by the variation in hydraulic head.

However, experience tells us that sometimes things do not move in the exact direction in which you try to push them with an external force. For example, consider the situation described in Figure 5-14, which shows a block sitting on an inclined plane. Gravity is the key external driving force, and it is trying to pull the block directly down, but obviously the block does not move directly down. It moves down and to the right, constrained as it is to follow along

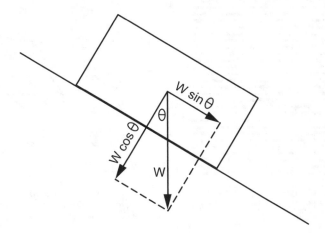

FIGURE 5-14
Resolution of the gravitational force acting on a block resting on an inclined plane

the surface of the inclined plane. This is a situation where the body is not moving along the line of the obvious external force. The underlying reason can be seen if we resolve the external force W, corresponding to the weight or gravitational pull on the block, into two perpendicular components as shown in the figure. One component is perpendicular to the inclined plane, and, in fact, is counterbalanced by an exactly equal force in the opposite direction which the plane exerts upon the block. (This latter force is not shown on the diagram.) Therefore, the *net force* on the block is simply the unbalanced component $W \sin \theta$, and this is the final force we must take into account to determine the motion of the block. Indeed, the block does move in the direction of this net force, which is in a direction going down to the right, parallel to the surface of the inclined plane.

Although a detailed analysis using vectors is more complicated, it is easy to understand that some geologic media might favor the flow of ground water in certain directions and impede it substantially in others. Two examples of such situations are illustrated in Figure 5-15. On the left side of the figure, there is the schematic representation of a sedimentary deposit (i.e., sediments that were laid down over long periods of time at the bottom of a lake, ocean, or other body of water) in which the individual soil particles that were settling out of the water column over time were not symmetric, but rather had a certain general "flat" or platy structure to them. These settled out in such a way that their orientation was basically horizontal, and they packed in much more closely in one direction than another because of these shape differences. The result is a geological formation that provides freer ground-water flow paths from side to side than from top to bottom. In rough terms, we might say that the hydraulic conductivity in the horizontal direction would be expected to be higher than the hydraulic conductivity in the vertical direction in this example. Furthermore, keeping in mind that over geologic time sedimentary deposits such as this might be tilted, folded, warped, or rotated extensively, it is easy to understand how the preferential hydraulic conductivity could turn out to be in almost any direction.

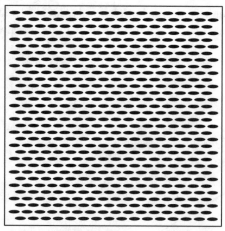

Side view of flat particles in a sedimentary formation, resulting in preferential flow from side to side

Top view of interconnected fractures system in a crystalline rock formation, showing preferential flow between upper right and lower left

FIGURE 5-15

Conceptual representation of two anisotropic geologic formations

The other side of Figure 5-15 shows a similar situation for a crystalline rock formation, wherein the predominant ground-water flow pathways are through interconnected fractures and joints. It is not uncommon for such fracture systems to display certain preferential directions because they often result from an external stress field provided by geologic forces, which themselves have very specific directions of action (e.g., continental drift, vertical pressures during periods of deep burial, regional slipping along faults, etc.). In this figure, it is clear that it would be easier for fluids to move from the upper right to the lower left (or vice versa) than for fluids to move from the upper left to the lower right.

The situations typified in Figure 5-15 contain geologic media that are not isotropic with respect to their fluid flow properties; they are called *anisotropic*. Does Darcy's law work for these media? What are the basic equations governing fluid flow?

A quick preview of the answers to these questions is contained in the following statements:

a) Darcy's law itself does not apply to these anisotropic media, but a matrix equation analogous to Darcy's law does apply wherein the hydraulic conductivity is actually represented by a matrix of various values.

b) In general, the direction of fluid flow in these situations is not perpendicular to the hydraulic head contour lines.

These facts will be discussed further in the next section, but a brief introduction to the underlying ideas should be obtained first by working out the following exercise.

Exercise 6. In this exercise you are asked to sketch in a conceptual way the general flow lines you would expect to encounter in an anisotropic situation.

a) Begin by drawing a large square filling out most of a page, and within it draw a typical set of hydraulic head contour lines. (You have seen many examples of these in previous sections.)

b) Go along each of the individual contour lines and sketch in a small arrow pointing in the direction of "steepest descent," which, as you know (if you have worked the multivariable calculus exercises in this section), is the negative of the direction of the gradient vector. In drawing in these arrows, make the length of the arrow roughly proportional to the magnitude of the gradient vector. (Hint: if you have locations where the contour lines are closer together, meaning that the surface is somewhat steeper, the gradient is larger so the arrows should be somewhat longer.)

c) Assume that in the two-dimensional region shown by your diagram ground water can move roughly twice as easily in the x-direction as in the y-direction. Taking this fact into account, with a different-color writing instrument from the one with which you drew the previous set of arrows, draw an additional arrow at the base of each of the previous arrows to represent a vector of estimated fluid flux at that point. Do your best to take into account both the magnitude of the driving force and the anisotropic nature of the medium in drawing these new vectors.

d) Using the fluid flux vectors as a guide, sketch in a set of estimated flow lines for this situation.

5.6 The Continuous Version of Darcy's Law in Two and Three Dimensions For Isotropic and Anisotropic Media[*]

Earlier in this chapter, we developed the so-called continuous version of Darcy's law for one-dimensional flow, which involved the calculation of the flow at any point in terms of the derivative of the head function there. In particular, Darcy's law took the form:

$$q = -K\frac{dh}{dx} \qquad \text{or} \qquad q = -Kh_x.$$

Note that the second version shown here simply uses the subscript notation for the derivative, which from this section on will become our standard method of writing derivatives.

Now we turn to the two-dimensional setting, and we will see at the end of this section that the three-dimensional setting is exactly analogous. In this case, we are looking at flow that we can consider to be restricted to a plane. The most common case of this that we have encountered is flow that is essentially horizontal, but can move in either of the two horizontal directions. This is the context within which we have generally drawn hydraulic head contour lines on map-like figures, and we have calculated flow directions and travel times between different points on these maps. While most of our diagrams will continue to be interpreted in this same sense, keep in mind that we have also seen another distinct two-dimensional case, namely the case of a side or "section" view of an aquifer in which flow may occur either in the horizontal or vertical directions shown on the diagram, although it is known not to occur in the other horizontal direction, namely, the one perpendicular to the drawn section. For example, the left side of Figure 5-15 shows such a view of an aquifer, one that actually has different vertical and horizontal flow properties.

For notational consistency, we shall use x and y to represent the two dimensions of the flow field, whether these are two horizontal dimensions or one horizontal and one vertical dimension. Thus the hydraulic head will now be a function of these two variables, and we can write it as $h(x, y)$ or simply h when there is no ambiguity about where it is to be evaluated. Since this is a function of two variables, it can have derivatives with respect to each, and these are called its partial derivatives. The partial derivative with respect to x gives the rate of change of h with respect to x as y stays unchanged; it can be denoted by either h_x or $\frac{\partial h}{\partial x}$, although we will generally use the shorter subscript notation. Similarly, the partial derivative with respect to y gives the rate of change of h with respect to y as x stays unchanged; it can be denoted by either h_y or $\frac{\partial h}{\partial y}$.

If you take the two partial derivatives of h and put them together to form a vector, namely, (h_x, h_y), then this vector is called the *gradient vector* of h, denoted ∇h. That is,

$$\nabla h = (h_x, h_y).$$

Since it is a vector, it has a magnitude and a direction. It is shown in multivariable calculus that this direction turns out to be perpendicular to the contour line at any point. This is illustrated in

---

[*] This section assumes that the reader is familiar with partial derivatives and other concepts from multivariable calculus, especially the gradient vector. There is a brief review of these concepts at the beginning of the section. Matrices (2×2) are used to represent linear equations, and further optional linear algebra topics, such as coordinate transforms based on eigenvectors, are also mentioned.

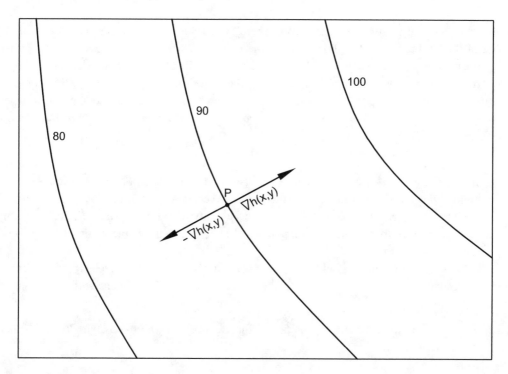

FIGURE 5-16

Two-dimensional flow field showing hydraulic head gradient vector at point P

Figure 5-16, which also shows the negative of the gradient vector, which points in the opposite direction. The magnitude or "Euclidean norm" of the gradient vector, which is simply

$$\|\nabla h\| = \sqrt{h_x^2 + h_y^2},$$

turns out to be the same as the value of the "directional derivative" of h taken in the direction of the gradient vector. Furthermore, it is greater than or equal to the directional derivative (or its absolute value) in any direction whatsoever, which is why the direction of the gradient vector is often called the "direction of steepest ascent." Similarly, the negative of the gradient vector is the "direction of steepest descent."

Now we will combine this mathematical framework with Darcy's basic physical principle and the work carried out in the previous section in order to develop two-dimensional versions of Darcy's law for two distinct situations: an isotropic medium and an anisotropic medium. Let us begin with the isotropic medium.

In the isotropic case, the medium exerts no preferential force to try to move the fluid other than in the direction in which it would ordinarily want to go. Therefore, as seen in the exercises of the previous section, it will move in a direction perpendicular to the contour lines, the same linear direction as the gradient vector and its negative. Since flow is from areas of higher head to lower head, obviously the flow is in the *negative* gradient direction. Therefore the flow vector q at any point is a positive scalar multiple of the negative gradient vector; that is,

$$q = K(-\nabla h) = -K\nabla h.$$

But is this constant multiple the same everywhere, independent of the point (x, y) or even the value of the gradient ∇h? Yes, this is really the point of Darcy's experiments and our earlier one-dimensional version of Darcy's law. For a given medium, the flow rate at any point is proportional to the rate of change of head along the flow path at that point, and this constant of proportionality, K, depends only on the medium and not on the particular point or other factors. Therefore, to summarize, *Darcy's law for a two-dimensional isotropic medium* takes the form

$$q = -K\nabla h.$$

Keep in mind that this is a vector equation because we are obtaining a two-dimensional vector q by multiplying the scalar value K by the gradient vector ∇h. It is the isotropic nature of the medium that preserves the essential one-dimensionality of the underlying physical problem because it guarantees that the flow direction will be along the same line as the driving force given by the gradient vector.

The first exercise below should help make some of these abstract concepts more concrete, and the second should help clarify some theoretical subtleties for readers with an adequate background.

Exercise 1. Suppose that the head function in the vicinity of a well that is withdrawing water is given by the function

$$h(x, y) = 8 \ln \left(10 + \sqrt{x^2 + y^2}\,\right).$$

Assume that the hydraulic conductivity of the aquifer is 30 ft/day, and consider the three points $(10, 15)$, $(-20, 30)$, and $(5, -5)$. (Hint: Begin by drawing the contour lines and the expected flow lines. Minimal or no computation should be necessary to do this.)

a) Calculate the hydraulic gradient vector at each of these points.

b) Calculate the flux vector at each of these points and specify the units for each of its components.

c) Calculate the directional derivative of the head at each of the points in the direction of flow.

Now evaluate whether each of the following statements is true or false, giving a logical rationale for your answer, and using your calculated values at all three points to illustrate the situation:

d) The magnitude of the flux vector is proportional to the magnitude of the gradient vector.

e) The magnitude of the flux vector is proportional to the directional derivative in the gradient direction.

Give a precise physical interpretation of the following:

f) The magnitude of the flux vector.

g) Each individual component of the flux vector. (Make sure that there is clear meaning to what you are saying. Use a diagram to illustrate if necessary.)

Exercise 2. The quantities represented by q and by ∇h in the above are referred to as vector fields. In other terminology, they might be called functions from R^2 to R^2, or vector-valued functions of two variables. The key concept is that for any input value (x, y), each of these expressions has an output value which also contains two components. The arrow representation used in Figure 5-16 is a useful vehicle for drawing such vector fields. The question you are going to be asked here is somewhat subtle, and you should attempt it only if you have a good grounding in multivariable calculus and mathematical derivations. In developing Darcy's law for the isotropic case and for two dimensions as above, we assumed that the vector q was proportional to the gradient vector ∇h at every point. Suppose we were to replace that assumption with the possibly weaker assumption that the magnitude of the vector q at any point (x, y) is proportional to the magnitude of the gradient vector. If we combine this weakened assumption with the other assumption made above, namely that q and ∇h always point along the same line, does it still follow that q must have the form of a constant K times the negative of the gradient vector?

Now we turn to the case of a geologic medium that might be anisotropic. In this case, once again the driving force for fluid flow is in the negative gradient direction, but because the medium favors flow in one direction over another, the actual direction of fluid flow will in general not necessarily be along the same line and may need to be represented by a vector pointing in a different direction. You may actually have worked out a sample case of this in the last exercise of the previous section. What would Darcy's law look like in this situation?

Before reading further, try the following exercise:

Exercise 3. Make your best guess as to how Darcy's law would need to be reformulated for the anisotropic situation just described.

There is a good chance that in working out the previous exercise, you may have concluded that Darcy's law could be written as follows:

$$q_1 = -K_1 h_x$$

$$q_2 = -K_2 h_y.$$

This pair of equations can be rewritten in vector/matrix form as follows:

$$q = - \begin{pmatrix} K_1 & 0 \\ 0 & K_2 \end{pmatrix} \nabla h.$$

If this is what you provided as your solution, it was indeed a good effort, but in general it is not quite right. The problem is a bit more complicated, and we begin to approach its solution by summarizing some basic observations about the form of the relationship we are looking for between input driving force ∇h and the output vector q:

1. When the hydraulic gradient is 0, the corresponding fluid flux should be 0.
2. The fluid flux q should be proportional *in magnitude* to the gradient vector ∇h. That is, even though the direction of the fluid flux may be bent in one direction or another from the direction of the negative gradient factor, certainly if we double or triple the actual magnitude of that gradient vector, we should reasonably expect a corresponding doubling or tripling of the fluid flux vector's magnitude. (This condition actually implies the previous one, but it can be useful to think about the previous condition first separately.)

These two observations are satisfied by a general "linear transformation" that would map two-dimensional vectors ∇h to two-dimensional vectors q according to the matrix equation:

$$q = - \begin{pmatrix} K_{11} & K_{12} \\ K_{21} & K_{22} \end{pmatrix} \nabla h.$$

This equation, also known as the *two-dimensional anisotropic form of Darcy's law,* can be rewritten in more compact form as

$$q = -K \nabla h$$

with the understanding that K now stands for the *hydraulic conductivity matrix*

$$K = \begin{pmatrix} K_{11} & K_{12} \\ K_{21} & K_{22} \end{pmatrix}.$$

It can also be written in more expanded form by writing it out component by component:

$$q_1 = -K_{11}h_x - K_{12}h_y$$

$$q_2 = -K_{21}h_x - K_{22}h_y.$$

In these equations q_1 and q_2 are the components of the vector q representing fluid flux.

Exercise 4. Verify that this more general anisotropic form of Darcy's law does indeed satisfy the two requisite conditions listed just previously. (Hint: depending on the extent of your background in matrix theory, you may want to work with either the matrix or scalar form of the equations.)

Exercise 5. [Recommended only for readers with a good background in mathematical proofs.] Is the linear transformation given above the *only* functional form that satisfies the requisite two conditions listed?

Referring to the scalar form of the equations, the second term on the right-hand side in the first equation and the first term on the right-hand side of the second equation appear to most people to be the "unexpected terms" because they imply that the rate of change of hydraulic head in the direction of one of the coordinate axes actually influences the rate of flow of fluid in the direction of the other coordinate axis. These are precisely the terms that are the most likely to have been overlooked in solving Exercise 3. Why must they be included? What is the physical reason for which you actually should expect such terms to be needed in Darcy's law?

The answers to these questions are somewhat complex, and we shall be content here to discuss them qualitatively. Note that for the two-dimensional flow system under consideration, there must be certain overall physical constraints on the amount of fluid passing through any given subregion of the region under consideration. For example, consider Figure 5-17, which shows certain hydraulic head contour lines and flow lines in a portion of an anisotropic aquifer. Note that because the region is anisotropic, the flow lines are not drawn perpendicular to the contour lines. Within this diagram two specific regions R_1 and R_2 are also indicated.

An example of a physical constraint on the system is the fact that at least under long-term steady state conditions the *net* amount of fluid passing through the boundary of region R_1 must be 0, since there is no way to create or destroy fluid within R_1 (unless there were a well there for pumping water in or out continuously) nor is there any way to continuously store it or take it out of storage. A similar physical constraint would exist for the region R_2.

While this principle applies equally well to R_1 and R_2, it is more useful when applied to a region such as R_2, because in this case it is fairly easy to see that fluid flow can occur through only two of the four boundaries of this region, as you are asked to verify below.

Exercise 6. Identify at which of the two boundaries of the region R_2 in Figure 5-17 fluid flow through the boundary can actually occur, and through which two boundaries fluid flow cannot occur. Explain your reasoning.

Constraints of the type just discussed suggest an important interconnection between what is happening in the x-direction and what is happening in the y-direction. After all, if more

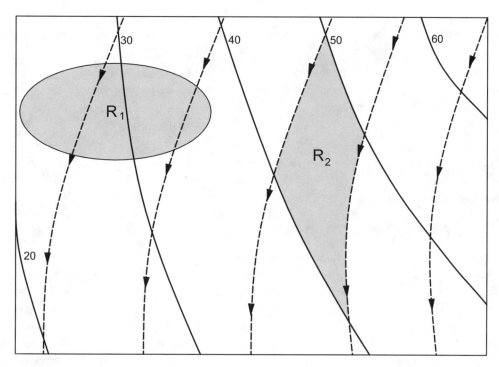

FIGURE 5-17

Hydraulic head contour lines and ground-water flow lines in an anisotropic situation

fluid is tending to move in one direction because of a higher change of hydraulic head in that direction, then less fluid might be available to flow in another direction. This is the key reason why the two-dimensional form of Darcy's law for anisotropic media includes the mixed terms that cause the fluid flux in each direction to be affected by the change in hydraulic head in both of the coordinate directions.

Exercise 7. [Requires knowledge of eigenvalues and eigenvectors of matrices.] The hydraulic conductivity matrix K discussed above is generally "symmetric," meaning that $K_{12} = K_{21}$. (The reasons for this are too complex for discussion here.) Assuming that this is the case, describe a coordinate transformation such that if the entire flow problem is reformulated in terms of these new coordinates u and v, the off-diagonal elements of the new but corresponding hydraulic conductivity matrix are actually 0. Explain why, in this new coordinate system, one might comment that "the flow in the u direction and the flow in the v direction are essentially decoupled." (These new directions for u and v, if indicated on the diagram of the original problem in x and y, would be referred to as the "principal directions" for the problem.) Describe the nature of the new coordinate system. For example, are u and v perpendicular in the xy-plane?

Exercise 8. With reference to the previous problem, now consider a situation in which the original x- and y-directions are already the principal directions, so that the hydraulic conductivity matrix in Darcy's law for this particular situation already has 0's as the off-diagonal entries. Find an expression for the angle between the flow direction and the negative gradient direction in terms of the directional hydraulic conductivities and/or the hydraulic head function. Simplify as much as possible so as to identify the factors that affect this angle.

Exercise 9. In the same context as the previous problem, consider an arbitrary unit direction vector $u = (u_1, u_2)$. Show that a one-dimensional generalized Darcy-type law applies to the flow rate, with the general form

$$q_u = -K_u \frac{dh}{ds},$$

where q_u in this equation is to be interpreted as the fluid flux in this direction and K_u is an *equivalent one-dimensional hydraulic conductivity* (which may not be constant) for that direction. Find an expression for this hydraulic conductivity K_u in terms of the directional hydraulic conductivities and/or the hydraulic head function, and discuss when it would be constant throughout a region.

One of the most important situations in which anisotropy has to be taken into account in ground-water modeling is in a two-dimensional "section view" of an underground aquifer in which the flow is known to be in the plane of the figure so that the two directions of flow are vertical and horizontal. In this case, the principal directions are generally horizontal and vertical, so that the problem is already formulated in a way that has the simplifying components of a principal directions problem. Thus if you consult a hydrology text, you might find a discussion of horizontal and vertical hydraulic conductivity as decoupled quantities, which would be correct in this particular context.

An additional comment that should be made at this point is that the restriction to two dimensions was only to make it easy to visualize the problems, but it was not at all essential to the mathematical formulation. Thus Darcy's law in three dimensions (as well as the process of changing coordinates to principal directions) can all be written down almost immediately for this case, as is suggested in the following exercise.

Exercise 10. State the three-dimensional version of Darcy's law for both the isotropic and anisotropic cases.

Just as in Chapter 2, Darcy's law leads immediately to a velocity equation for the ground water in the two- or three-dimensional case, at least in the case of an isotropic medium. Such an *interstitial velocity equation* would take the form

$$v = \frac{-K\nabla h}{\eta}$$

and could be used in connection with line integrals to calculate the total travel time along a flow path. The problem with trying to apply this same concept to the anisotropic case is the issue of whether the porosity itself needs to be considered as a multidimensional property because a cross section of a flow tube in an anisotropic medium is likely to look quite different as its direction changes. This latter aspect will not be pursued further here.

Certainly this section has included far more theoretical discussion than numerical calculations, and you may be wondering about applications. It would be relatively uncommon to solve numerical problems based on the equations in this section except in very special cases where the head function could be specified by a neat analytical equation. It is more common to know the head function at a number of discrete points and then to use some numerical approximation scheme to estimate the flow paths, flow rates, velocities, and travel times. For implementing such numerical schemes, one needs to understand the concepts discussed in this section. This theme will be further investigated in subsequent sections, although restricted to the isotropic case. An important value of the material in this section is that it should give you a deeper understanding of the nature of fluid movement through a porous medium and of the problems one needs to be aware of before too quickly applying or accepting the results of standard models.

The following analytical problem, although somewhat challenging, integrates a number of the concepts in this section, at least for the isotropic case, and would be valuable for you to spend some time on before proceeding to the next section.

Exercise 11. Suppose that the head function in the vicinity of a well (located at the origin), which is withdrawing water, is given by the function

$$h(x, y) = 55 + 12 \ln \left(10 + \sqrt{x^2 + 2y^2}\right).$$

Suppose further that the aquifer is isotropic with a hydraulic conductivity value of 15 ft/day and a porosity of 0.25.

a) Sketch the head contour lines and describe them geometrically.

b) Find the equation of the flow line along which water would travel from the point $(100, 150)$ to the well.

c) Find the travel time required by the ground water to move along this flow line from the point $(100, 150)$ to the well.

5.7 Laplace's Equation and Inverse Problems*

The previous section on Darcy's law discussed how ground water moves in two and three dimensions under the influence of a gradient in the hydraulic head function h. In this section, we learn more about the possible distribution of values of this function h, such information being very helpful in trying to estimate h values over a whole region on the basis of only limited data values. You may have considered this question previously in Section 5.4, where you explored methods of interpolating or extrapolating from limited well data in order to estimate head values and contour lines throughout a region. It will be interesting to see how the results compare.

We have considered this problem in one dimension in Section 5.2. In particular, for the situation considered there, the unknown head function had to satisfy a differential equation called Laplace's equation, which took the form

$$h_{xx} = 0$$

and hence could have only straight line functions for solutions. By combining this observation with the measured head values at the endpoints of the interval (= "boundary conditions"), the head value could be calculated throughout. If we had additional measured head values that did not fit the predicted line, then this would tell us that one of our underlying assumptions about the aquifer must be wrong. In this way, we could gradually improve our understanding of the underground situation.

---

* This section continues to use the multivariable calculus concepts introduced previously. In addition, the divergence theorem is introduced and used for an alternative derivation of Laplace's equation near the end of the section.

The analogous situation applies to two and three dimensions except that the solutions to the resulting Laplace equations can be far more complicated than simple straight lines. This is the material that we will pursue in this section. To begin, take a look at the Laplace equations in one, two, and three dimensions:

$$h_{xx} = 0 \qquad \text{(one-dimensional case)}$$

$$h_{xx} + h_{yy} = 0 \qquad \text{(two-dimensional case)}$$

$$h_{xx} + h_{yy} + h_{zz} = 0 \qquad \text{(three-dimensional case)}.$$

These equations apply to the case of isotropic aquifers with constant hydraulic conductivity, a standing assumption we shall make throughout this section except for one exercise where a variation is pursued. The subscripts in these equations are used to denote derivatives, although they are partial derivatives in the two- and three-dimensional cases. For example, h_{xx} stands for the second partial derivative of h with respect to x.

Exercise 1. Rewrite the two- and three-dimensional versions of Laplace's equation using the common del (∂) notation for partial derivatives.

You do not need any prior experience with partial differential equations to continue with this section. Partial differential equations are simply equations containing partial derivatives and for which you would like to find a function that works when plugged into the equation (meaning that the equation is satisfied at every point for this function.) The usual situation, as you will see soon, is that one tries to find a solution to a partial differential equation (e.g., Laplace's equation) such that that solution also satisfies certain auxiliary conditions called *boundary conditions*. In the example treated earlier in Section 5.2, the boundary conditions consisted of the requirements:

$$\left.\begin{array}{l} h(1000) = 70 \\ h(3000) = 50 \end{array}\right\} \text{ "boundary conditions".}$$

Now we continue our discussion by beginning the derivation of Laplace's equation for the two-dimensional situation. Recall that there are various typical situations where the flow problem is essentially a two-dimensional problem. One such situation is when a vertical section of the aquifer is chosen so as to correspond to the direction of the flow lines, so that the two available dimensions for flow would be the horizontal direction and the vertical direction. The other two-dimensional situation, the one that we have emphasized more in previous discussions, is the situation where vertical flow is assumed to be negligible, and so we are looking at the flow pattern in the two horizontal x- and y-directions. Although the mathematical structure of the two problems is identical, we shall develop Laplace's equation for the latter situation because it is more familiar.

This general situation is sketched in Figure 5-18, which shows a top or plan view of an underground aquifer, in which are sketched a few hydraulic head contour lines simply to remind us that this is the same situation that we have looked at many times before. We really want to think of this diagram as a depiction of a three-dimensional situation, where the third dimension, or z-dimension, is exactly perpendicular to the plane of the paper. Flow in the z-direction is assumed to be negligible, but nevertheless the aquifer does have thickness and z is the variable that characterizes position through that thickness. This aquifer is assumed to be isotropic and uniform, so that there is a single K value that characterizes hydraulic conductivity throughout.

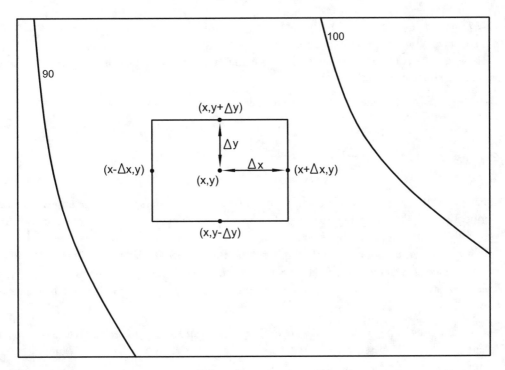

FIGURE 5-18
Volume increment in a ground-water flow field exhibiting two-dimensional flow

Figure 5-18 also shows an "incremental volume," which we shall use to balance flows and ultimately to derive the two-dimensional Laplace equation. This incremental volume is the shape of a rectangular solid whose center is at the point (x, y) and which has length $2\Delta x$ in the x-direction, with $2\Delta y$ in the y-direction, and depth 1 in the z-direction.

By reasoning analogous to that used earlier for several situations, the net fluid flow into this incremental volume must be 0. In this case, the fluid flux q is a vector quantity, having components q_1 in the x-direction and q_2 in the y-direction. Furthermore, fluid may be entering the incremental volume through four of its six faces. There is no flow possible through the top and bottom faces (since flow has been assumed to be horizontal), but flow is certainly possible through any of the other four.

We make one additional approximation as follows. Consider the left face of the incremental volume. We do not need to consider the third or z-coordinate for this face or any other points, because everything is essentially constant in the z-direction. Therefore we can write variables such as q and h as functions of only the two location variables x and y. Our key assumption is that along this left face, the inward fluid flux, which may vary as the y-coordinate varies, can be represented by the one specific fluid flux value at the center of the left face, namely $q_1(x - \Delta x, y)$. Note that this is the first component of the flux vector q, and it is the only one relevant to the left face since any y-direction component, q_2, does not contribute any flow through the face and into the incremental volume. We make similar assumptions along the other three faces by again using their midpoints as typical points at which to take q or h values. This kind of approximation is valid because near the end of our derivation we will be letting Δx

and Δy approach 0, so that all the points along an individual face will converge on the point we have selected for our approximation.

Now we are ready to write down the basic equation that states that the net fluid flow *into* our incremental volume must be 0:

$$0 = \text{left face flow in} + \text{right face flow in}$$

$$+ \text{bottom face flow in} + \text{top face flow in}$$

$$0 \approx q_1(x - \Delta x, y)(2\Delta y \cdot 1) - q_1(x + \Delta x, y)(2\Delta y \cdot 1)$$

$$+ q_2(x, y - \Delta y)(2\Delta x \cdot 1) - q_2(x, y + \Delta y)(2\Delta x \cdot 1).$$

We have used the "approximately equal" symbol ($\approx$) to remind ourselves of the approximation described in the previous paragraph, and thus when we eventually take limits as Δx and Δy go to 0, this will be replaced by exact equality. We have also used the fact that the total flow through a face can be obtained by multiplying the flux perpendicular to the face by the total area of the face. The above mass balance equation can be easily reduced to the two-dimensional Laplace equation, and the reader is asked to complete this part of the derivation in the next exercise.

Exercise 2. Beginning with the above mass balance equation, derive the two-dimensional Laplace equation. (Hint: if you are not sure how to proceed, review the derivation for the one-dimensional situation in Section 5.2.)

While the one-dimensional Laplace equation might have seemed practically trivial because its only solutions were straight lines, the two-dimensional Laplace equation has been the subject of extensive investigations for over two centuries and is the subject of many advanced mathematical treatises. A function satisfying this equation is called a *harmonic function*. Much of the study of such harmonic functions is included under the general title of *potential theory*.

Recall once again the situation of the one-dimensional Laplace equation, treated in Section 5.2. We needed to solve this equation in combination with some additional constraints called boundary conditions. In other words, we were not just looking for "any old solution" to the one-dimensional Laplace equation, but rather we were trying to find the particular solution also satisfying the boundary conditions implied by the specific hydrologic situation we were trying to understand.

The analogous situation applies as well to the two-dimensional situation. In this case, we know that the hydraulic head in the region under consideration must satisfy the Laplace partial differential equation (as long as our basic assumptions about the aquifer are correct), and we also know that the hydraulic head in our calculations must be consistent with whatever measured values are available to us from field investigations. These values, which are actually quite often found around the edges or the boundary of the region under study, are a typical form of boundary conditions for Laplace's equation in two dimensions. In fact, just to illustrate how important problems such as this have been over the years, the problem consisting of finding a solution to the Laplace equation that meets specified boundary conditions around the border of the given region has a special name of its own: the Dirichlet Problem. If we find ourselves in the situation where we need to find the hydraulic head by finding a solution to the Laplace equation, but for which problem the boundary conditions have a somewhat different form, some of these other variations also have special names of their own and have been the subject of quite elaborate investigations. We will encounter some examples of these situations momentarily, and this should help to clarify the ideas in this paragraph.

Exercise 3. Find three distinct, non-trivial examples of functions that satisfy the Laplace equation in all or some portion of the xy-plane.

Exercise 4. For one of the three functions you identified in the previous exercise, pick a relatively simple region of the plane in which it is defined, such as a square, rectangle, or circle, and determine the actual boundary conditions that it exhibits on the boundary of this region. (The more common situation would be to begin with the boundary conditions and try to find the unknown harmonic function satisfying those conditions, but this exercise simply asks you to look at the situation in reverse in order to see some typical boundary conditions.)

Exercise 5. The derivation of the two-dimensional Laplace equation was based on the assumption of constant hydraulic conductivity throughout the region of interest. Assume instead in this exercise that the hydraulic conductivity gradually increases to the right within the region shown in Figure 5-18. In particular, assume that the hydraulic conductivity function can be written in the following form:

$$K(x, y) = K_0 + \alpha x.$$

Derive the partial differential equation that would need to be satisfied by the hydraulic head $h(x, y)$ in this case. (The aquifer is still assumed to be isotropic with respect to flow properties.)

Exercise 6. Return to the derivation of the two-dimensional Laplace equation, but assume that the aquifer is anisotropic, with the x- and y-directions being principal directions with respective hydraulic conductivities K_1 and K_2 that remain constant throughout the region of interest. What partial differential equation would need to be satisfied by h in this case?

Exercise 7. Draw an appropriate diagram and provide a complete derivation of the three-dimensional Laplace equation for the head function $h(x, y, z)$ in an isotropic aquifer with uniform hydraulic conductivity K and in which flow is three-dimensional.

It is interesting that there are many other physical situations aside from ground-water modeling which lead naturally to the Laplace equation. Looking back on our derivation of Laplace equation and on the governing flow equation given by Darcy's law, we can see that there are two key characteristics to this physical situation:

1) The flow of some quantity, in this case ground water, at a rate that is proportional to the gradient of some "potential," which for us has been hydraulic head h.

2) A conservation condition requiring that throughout the flow regime no material spontaneously appears or disappears, or is stored or taken out of storage.

These two characteristics are common to a number of other situations, and the net result is that in all of these situations the potential function can be shown to be a solution to Laplace's equation. Examples of these situations are given in Table 5-2.

It is important to note in connection with this table that Laplace's equation does not *always* apply to the variable in the second column, namely, the variable representing the potential, even in the case of ground water, which we have been discussing in some detail. There is one additional key assumption that we have made somewhat implicitly up to now, but which at this point should be discussed more explicitly. This assumption is that the systems we are discussing are exhibiting *steady-state behavior.*

TABLE 5-2

Examples of physical phenomena controlled by potential fields that are solutions to Laplace's equation under steady state conditions

Variable or quantity subject to flow or movement	Variable or quantity whose gradient determines the rate of movement	Brief statement of the basic physical principle
ground water in a porous medium	hydraulic head	Darcy's law: fluid flow is proportional to the head gradient.
heat flow in a solid or other conducting body	temperature	Fourier's law: heat flows in body by conduction at a rate proportional to the temperature gradient.
dissolved material in a liquid solution	concentration	Fick's law: solute diffuses throughout the liquid at a rate at any point proportional to the concentration gradient.
molecules of a gaseous substance in an environment consisting of inert gas	concentration	Fick's law: molecular diffusion into the inert gas environment is proportional to the concentration gradient at any point.
electric current (i.e., flow of electric charge)	electric potential	Ohm's law: electric current at any point is proportional to the voltage drop or the gradient in electrical potential.

To take the case of ground water, there are certainly situations in ground-water flow where this assumption does not apply. For example, consider an aquifer into which a town has drilled a water supply well that is used only some of the time. When the well is not pumping, the aquifer is likely to reach a certain steady-state condition, but as soon as the well pump is turned on, the situation begins to change over time and the water table in the vicinity of the well may gradually drop and form a "cone of depression." During this transient period, the system is not at steady-state and the head distribution would not be expected to obey the Laplace's equation. To relate this to the two conditions above, the second condition would be violated because there would be a net withdrawal of water from regions of the aquifer within the cone of depression. Even if the aquifer were not a water table aquifer, but rather were a confined aquifer where the pumping would lower the head but not actually empty parts of the aquifer, the lower head would interact with the elasticity properties of the bulk rock matrix (as well as perhaps the fluid). In particular, lower fluid pressure would allow the rock or soil particles to expand, slightly reducing the pore spaces and total pore volume, hence leading to a net reduction in the amount of water in a given incremental volume of the aquifer. Once again, this would violate the second condition above. Naturally, the degree to which this factor is significant depends on the variation in head over time. If the variation is quite small, then Laplace's equation might still be a good approximation for each time increment even though the boundary conditions for which it would be solved might need to be changed from one time increment to the next.

The same restriction applies to the other physical phenomena listed in Table 5-2. In particular, some of them can violate Condition 2 above quite easily, as heat or solute concentration, for example, can easily build up or diminish in a portion of the medium as time goes on. Nevertheless, if the boundary conditions are such that these systems remain at some steady state, then Laplace's equation will still apply.

The investigation of situations that depart from the steady-state assumption is well within the reach of the same kinds of methods we have been using in this chapter to investigate steady-state situations, but the addition of these additional cases would take us beyond our current objectives. Suffice it to say that if condition 2 is violated, we can still set up an analogous mass balance condition, but we need to add a term to account for the rate at which material is being created, destroyed, stored, or taken out of storage. In the two-dimensional case, this leads to a new partial differential equation of the form

$$ah_t = h_{xx} + h_{yy}$$

where a is a constant. Note that this reduces to Laplace's equation in the case of steady state conditions, since the partial derivative with respect to t would be 0 in this case. This partial differential equation is at the foundation of our diffusion models for air pollution, and it will be investigated in some detail in Chapter 6.

Exercise 8.   Consider a two- or three-dimensional steady-state problem involving one of the sets of quantities in Table 5-2 other than those relating to fluid flow. For this situation, draw appropriate diagrams and provide a direct derivation of Laplace's equation. (Hint: the key issue here is to convert the flow equations and the conservation condition from the ground-water derivation to physically meaningful statements in terms of the new variables.)

At the beginning of this section on Laplace's equation, we briefly discussed the iterative approach that might be taken by a hydrologist in attempting to develop a reasonable conceptual model and understanding of the flow parameters in an underground aquifer. Now we shall continue that discussion.

Consider the situation shown in Figure 5-19. Here we have a geographic region 6,000 feet square, around the borders of which hydraulic head values have been determined from monitoring wells on a regular grid spacing. (For the moment ignore the test point shown in the interior of the region.) A cursory examination of the boundary values for hydraulic head suggests that the general flow direction is from the upper right towards the lower left, although it can easily be seen that the flow lines are not uniform, and so you would expect them to follow some curvilinear structure. A hydrologist attempting to understand the features of the underground aquifer might initially approach this data-set with the additional assumption that the underground aquifer has constant thickness and hydraulic conductivity K. On the basis of these assumptions, our earlier derivation of Laplace's equation for the two-dimensional case applies, and thus one might try to solve the corresponding problem consisting of Laplace's equation in combination with conditions specified along the boundary.

Suppose you knew how to solve Laplace's equation in this framework, a topic we will treat further in the next section, and upon carrying out this solution, you determined that the value of hydraulic head h at the test point P should be 270. To test the validity of your analysis, you now collect some additional field data (or you look at some additional field data that you collected previously but held back to use as a check against your model), and you see that the measured

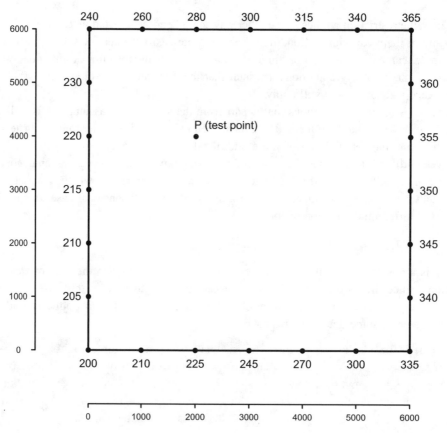

FIGURE 5-19
Typical boundary conditions and conceptual framework early in the efforts to analyze subsurface flow conditions

value of hydraulic head at the test point is actually 300, which differs substantially from the value predicted by the model. This means that something is wrong with your conceptualization of the aquifer, and the most likely culprit is the assumption of constant hydraulic conductivity. You go back and review all the available geologic information on this region, and you see that it is at least possible that the hydraulic conductivity in the lower left region of the diagram might tend to be somewhat lower than in the upper right, and you reason physically that this would tend to raise the hydraulic head value at the test point P.

Exercise 9.   Explain the basis for this last sentence.

In particular you divide the large region up into two subregions as shown in Figure 5-20, and you assume that the hydraulic conductivity in the triangular region to the lower left is twice as high as the hydraulic conductivity in the remainder of the larger region. Now you wish to recalculate the predicted values of h to see how they compare both at your test point and at any other points for which you may have additional data available.

However, now there is another catch. You must treat the two subregions separately because Laplace's equation only applies to regions in which the hydraulic conductivity is constant. However, if you apply Laplace's equation to the region at the lower left, you do not have a

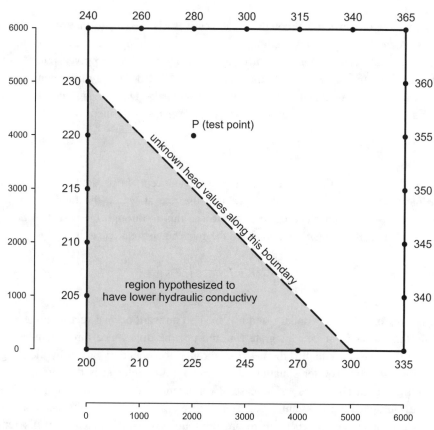

FIGURE 5-20
A revised conceptual understanding of the ground-water flow regime in the previous figure, hypothesizing different hydraulic conductivity values in two subregions

complete set of boundary conditions because you don't have hydraulic head values all along the boundary between the two sub-regions. The same limitation applies if you were to try to solve Laplace's equation for the region towards the upper right. So here is a situation where you would want to work with Laplace's equation, but the boundary conditions are not quite those covered by the Dirichlet Problem. What kind of boundary conditions would you need to be prepared to deal with for this new situation? The answer to this question is investigated in the following exercise.

Exercise 10. Let $h(x, y)$ and $g(x, y)$ represent solutions to Laplace's equation in the two subregions under discussion in connection with Figure 5-20, and assume that these functions do satisfy the relevant boundary conditions for hydraulic head along their respective portions of the exterior border of the overall square region. What additional consistency condition(s) must apply to h and g along the slanted border between the two subregions in order for these solutions to Laplace's equation to form a physically meaningful solution to the overall unknown hydraulic head distribution throughout the large region?

This paragraph summarizes certain more advanced topics from multivariable calculus with which the reader may be familiar and that can add to one's understanding of flow systems. We will work in three dimensions, although sometimes it is also useful to interpret them for two dimensions. A *vector field* in this setting is a function that assigns to each point in 3-space a three-dimensional vector, often thought of as an arrow based at the given point. Examples from our flow system context include the head gradient vector (for 3-dimensional flow) and the actual flux vector calculated from the gradient vector by Darcy's law. For a vector field $F(x, y, z) \equiv (f(x, y, z), g(x, y, z), h(x, y, z))$, its *divergence* is defined by:

$$\operatorname{div} F \equiv f_x + g_y + h_z$$

This may look strange at first, but it is a very useful concept. In particular, its name derives from the fact that it represents, loosely speaking, the rate at which the field is "expanding" or "diverging" at each point. The easiest way to see this is through the so-called *divergence theorem*, which is called *Green's theorem* for the two-dimensional case. It says

$$\iiint\limits_{R} \operatorname{div} F = \iint\limits_{S} F \cdot n$$

where R is a region, S is its boundary, and n is the unit outward normal vector on the boundary. In English, this says that if you integrate (or sum up) the divergence throughout any region, large or small, it must be equal to the net flow through the boundary, given by the integral on the right. For our ground-water situation, the vector field of interest is the flux q, or, almost the same thing, at least in the isotropic case, the gradient of the head function, which differs only by the multiplicative constant $-K$. Thus, since water is essentially incompressible, if we are working in a region where there are no sources or sinks (such as wells removing water), the net flow through the boundary of any closed region should be 0. This observation can be used to give a very simple derivation of Laplace's equation, as the reader is asked to work out in the next exercise. In fact, our methods for deriving Laplace's equation earlier in this section are quite similar to the methods usually used to derive the divergence theorem itself.

Exercise 11.   Use the divergence concept to derive the Laplace equation for the case of steady-state ground-water flow.

The general class of boundary value problems considered in this section are often called "inverse problems." Rather than having as a given a fixed set of hydrologic properties, one is asked to figure out what hydrologic parameters and flow assumptions would lead to the observed output values or boundary conditions. In fact, often a number of alternative assumptions about the flow regime might also be consistent with given boundary conditions, so that such inverse problems are fraught with the additional complication that their solutions may not be unique! A general principle for this kind of situation (as well as for most mathematical modeling) is that one should try to use the simplest possible set of assumptions or the simplest possible model that is consistent with the given data.* Even with this general guidance, the solution of inverse problems in the field of hydrology is still partly a science and partly an art.

---

* This principle is sometimes referred to as the principle of "Occam's razor."

5.8 Introduction to Numerical Modeling[*]

The purpose of this section is to provide a brief introduction to the question of how to solve Laplace's equation (and other partial differential equations) "numerically," meaning that we are not looking for an exact equation for the solution function at any or all points (practically impossible to find in most cases in two and three dimensions), but we will be satisfied to find approximate values for the unknown solution at least at a specified finite number of specific "grid points" or "nodes." In general, the strategy for the numerical solution of such equations is to find a set of simultaneous linear algebraic equations that provide a reasonable approximation to the partial differential equation, and then to solve these simultaneous linear equations by standard methods of algebra, with which the reader is no doubt familiar, or by more exotic shortcut methods.

First we need to see how derivatives can be approximated by simple algebraic expressions. To do this, we return to one of the most basic concepts from elementary calculus, namely, the definition of the derivative of a function $f(x)$:

$$f'(x) = \lim_{\Delta x \to 0} \frac{f(x + \Delta x) - f(x)}{\Delta x}.$$

Remember that the derivative of a function can be interpreted geometrically as the slope of the tangent line to the graph at a given point. Since the derivative is given by a limit expression, we might also write down a corresponding approximation equation for the derivative in the form

$$f'(x) \approx \frac{f(x + \Delta x) - f(x)}{\Delta x},$$

and this approximation should be quite an accurate approximation for Δx values that are relatively small. This concept of approximation, which the reader has surely seen earlier at least in an elementary calculus course, is illustrated in Figure 5-21.

The left-hand side of the approximation equation is simply the slope of the tangent line to the curve at the point $(x, f(x))$, and the right-hand side of the equation is the slope of the secant line shown connecting the points $(x, f(x))$ and $(x + \Delta x, f(x + \Delta))$. As is discussed in elementary calculus, as Δx gets smaller and smaller, the slope of the secant line approaches the slope of the tangent line. From a different point of view, generally not stated in elementary calculus courses, even if Δx is not small, as long as the function f does not have a high degree of curvature, the slope of the secant line should also be a reasonable approximation to the slope of the tangent line. This latter viewpoint is the one that we shall adopt here. Since this approximation involves the original point x and a second point farther to the right, it is called a *forward difference approximation* to the derivative.

This is not the only reasonable way in which we might approximate the derivative at the point $(x, f(x))$. For example, as is also shown in Figure 5-21, we could also approximate the desired derivative as the slope of the secant line connecting the point $(x, f(x))$ and $(x -$

---

[*] This section uses derivatives and their approximation by difference quotients. Although applied to Laplace's equation in two dimensions, the concepts are basically one-dimensional. The simultaneous solution of linear equations is required, and there is discussion in certain exercises of Newton's method from calculus. Some advanced concepts pertaining to iterative or fixed point methods, as well as harmonic functions, are introduced in the final exercises for readers with adequate background.

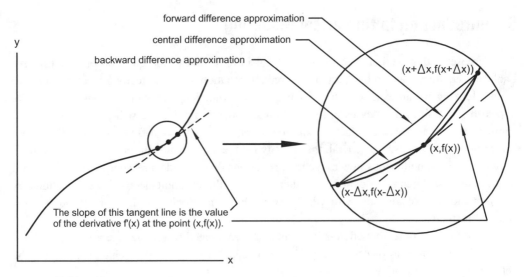

FIGURE 5-21

A graphic representation of three approximation schemes for the derivative of a function at a point

$\Delta x, f(x - \Delta x))$, as indicated below:

$$f'(x) \approx \frac{f(x) - f(x - \Delta x)}{\Delta x}.$$

This would logically be called a *backward difference approximation* to the derivative.

But further examination of the figure suggests what may appear to be an even better approximation to the derivative, namely, the slope of the secant line connecting the points at the far left and the far right. This approximation would take the form

$$f'(x) \approx \frac{f(x + \Delta x) - f(x - \Delta x)}{2\Delta x},$$

and this would be called a *central difference approximation* to the derivative. In fact, this one looks so good that you might even be wondering why anyone would bother with the others!

All three of these approximations have a place in the numerical solution of differential equations, but you would probably agree that the central difference approximation would probably be the preferred method as long as it is feasible to use it in a particular case.

Exercise 1. Show that the central difference approximation to the derivative is simply the average of the forward and backward approximations.

Exercise 2. [For readers who are familiar with Taylor Series and Taylor's Theorem.] Use Taylor's Theorem with Remainder or Taylor Series to expand the expressions $f(x + \Delta x)$ and $f(x - \Delta x)$ around the value x itself; and then, by plugging these expressions into the formulas for the forward, backward, and central difference formulas, demonstrate that the central difference formula has a "higher order of convergence." (This means that as Δx approaches 0, this approximation would be expected to converge to the true derivative value much more rapidly.) Construct a numerical example for a specific function and use it to demonstrate this conclusion.

We will now use the same approach to approximate the second derivative of f. In particular suppose we are trying to approximate $f''(x)$. We could use a central difference approximation to this value based on values of the function f' at equally spaced points to the left and the right of x. Furthermore, the distance we move to the left or the right to form this central difference formula can be any convenient distance; it does not have to be the same Δx as has been used above. So, in fact, we will take our points a distance $\Delta x/2$ to the left and the right of x. That is,

$$f''(x) \approx \frac{f'\left(x + \frac{\Delta x}{2}\right) - f'\left(x - \frac{\Delta x}{2}\right)}{\Delta x}.$$

Now we observe that the original forward difference approximation for $f'(x)$ could actually also be interpreted as a central difference approximation for $f'(x + (\Delta x)/2)$, with an analogous observation about the original backward difference approximation as well. Using these approximations in our approximate equation for $f''(x)$, we obtain

$$f''(x) \approx \frac{\frac{f(x+\Delta x)-f(x)}{\Delta x} - \frac{f(x)-f(x-\Delta x)}{\Delta x}}{\Delta x}$$

which reduces to

$$f''(x) \approx \frac{f(x + \Delta x) - 2f(x) + f(x - \Delta x)}{(\Delta x)^2}.$$

We will call this our central difference approximation to the second derivative.

Exercise 3. [Involves Taylor's Theorem or Taylor series.] Analyze the quality of the above approximation to $f''(x)$ by using Taylor expansions, as earlier, and determine what the lowest power of Δx is in an expression for the difference between $f''(x)$ and this approximation. (This power is called the *order* of the approximation, and the error being evaluated here is called the *truncation error*.)

With the above background on how we can easily approximate first and second derivatives using algebraic expressions, we now proceed to the context of the two-dimensional Laplace equation to see how the partial derivatives in that equation can also be represented in a similar fashion. Therefore, let us see how for a general function $h = h(x, y)$, we might approximate the partial derivative values by algebraic expressions. Consider Figure 5-22. Suppose that we want

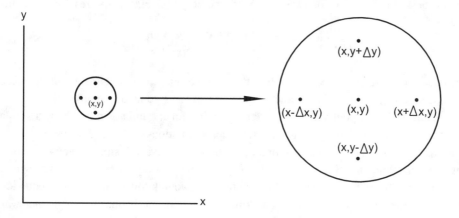

FIGURE 5-22
Grid points in the domain of a function $f(x, y)$ that are used to construct numerical approximations to the partial derivatives at a given point (x, y)

to evaluate the partial derivatives of $h(x, y)$ at the point (x, y) shown in this figure, and that we know the function values h at the additional grid points shown on the right side of the figure. That is, we know the value of $h(x, y + \Delta y)$ and the values of h at the other points surrounding (x, y), as well as the h value at the actual point (x, y) itself. By direct application of the central difference formula for the second derivative in both the x-direction and the y-direction separately, we can write the following approximations:

$$h_{xx}(x, y) \approx \frac{h(x + \Delta x, y) - 2h(x, y) + h(x - \Delta x, y)}{(\Delta x)^2}$$

$$h_{yy}(x, y) \approx \frac{h(x, y + \Delta y) - 2h(x, y) + h(x, y - \Delta y)}{(\Delta y)^2}$$

We could also use similar logic to evaluate other partial derivatives of h at the given point, as suggested by the following exercise.

Exercise 4. Construct reasonable approximations to the partial derivatives h_x, h_y, and h_{xy} at the point (x, y). If for one or more of these approximations you feel you need to assume that the h values are also available at additional grid points analogous to those in Figure 5-22, you may state and use such an assumption. (Note that these partial derivatives do not show up in Laplace's equation; however, they do show up in many other important partial differential equations governing ground water and other environmental problems.)

Now we are ready to apply these ideas to an actual numerical solution of a ground-water problem, namely, the problem first depicted in Figure 5-19. Recalling that figure, the objective was to find the distribution of hydraulic head values within the region, given the availability of measured head data at selected grid points along the boundary. For convenience, Figure 5-19 has been slightly modified in Figure 5-23, where we have identified specific internal grid points at which we would be satisfied to find values of h, and we have numbered the unknown head values at those interior points consecutively.

Here is our strategy: we replace the second partial derivatives in Laplace's equation, which h must satisfy, by their central difference approximations, using Δx and Δy values equal to the grid spacing of 1,000 feet. So, for example, at the point $(1000, 5000)$, the resulting equation would be:

$$0 = \frac{h_6 - 2h_1 + 230}{(1000)^2} + \frac{260 - 2h_1 + h_2}{(1000)^2}$$

This equation could certainly be simplified, which we shall do in a moment, but the reader can see immediately that it is a simple linear equation involving some of the unknown h values. In fact, for each of the 25 points at which we are looking for the unknown h values, a similar equation could be written, again using numerical approximations to the partial derivatives in Laplace's equation. At the end of this process, we would have 25 simultaneous equations involving the 25 unknown values $h_1, h_2, \ldots, h_{25}$. Although 25 equations in 25 unknowns is certainly more complicated to solve than two equations in two unknowns, say, or three equations in three unknowns, it is essentially the same kind of algebraic problem. One might not want to solve it by hand, but there are readily available computer methods for solving simultaneous linear equations, and these methods are available in standard mathematical packages and in everyday spreadsheet programs.

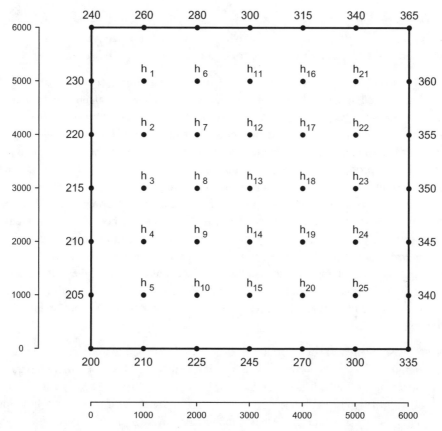

FIGURE 5-23
Region of unknown hydraulic head discussed previously, except with specific points labeled at which head values are to be calculated

But now let us simplify the above equation by multiplying through by the common denominator and then collecting terms. It reduces to:

$$h_1 = \frac{230 + h_6 + 260 + h_2}{4}.$$

But this equation has a simple interpretation: the value at a grid point is simply the average of the four neighboring values! The same is true at any point of the grid. This observation will make it easier for us to organize some of our calculations. But it also gives an important insight about the nature of the head function and, presumably, any function that satisfies Laplace's equation. It suggests that the value at any point is, at least roughly speaking, the average of the values at the surrounding points. This issue will be investigated further near the end of this section, but for now we return to the numerical problem at hand.

We have 25 equations, each of which can be simplified to the general form described above. The reader is asked to complete the solution to these equations in the following exercise.

Exercise 5.   Find the unknown head values at the grid points in Figure 5-23.

Exercise 6.   With respect to your solution for the previous problem (even if you never got to the point of actually calculating numerical values), reiterate in simple physical terms what you would expect a typical h value, say h_{17}, to represent. Did you need to make any assumption about the specific value of hydraulic conductivity K to carry out the above solution? Summarize the assumptions of the model on which your calculated h value are based.

There are many approaches for solving the simultaneous linear equations derived above. Some of them proceed along the lines ordinarily learned in high school algebra and are generally referred to mathematically as "Gaussian elimination." These methods are called *direct methods*. They are based on a sequence of well defined operations that in principle (i.e., neglecting round-off problems when long decimals occur) lead to the exact solutions to the linear equations. You (or your computer) probably used such a method to solve the equations in Exercise 5. Methods involving matrix inverses also fall into this general category.

However, there is also an interesting category of very powerful methods called *iterative methods* that are somewhat analogous to the way Newton's Method is used in elementary calculus to provide successively better approximations to the solution to an equation. These are called *indirect methods*. The subject of numerical analysis is devoted in part to evaluating different approaches to solving equations, taking into account the development of various kinds of errors (e.g., truncation error and round-off error) and efficiency (i.e., leading to solutions without excessive use of computational resources). One such iterative method is introduced in the following exercises.

Exercise 7.   Use a spreadsheet program or other computer program to carry out the following iterative method for solving the same problem as in Exercise 5. Make an initial guess at all 25 unknown values in the grid. (It does not have to be a good guess. In fact, feel free to try this problem with a crazy guess. You will still get the right answer, but it may take a little longer.) Your first iteration involves replacing each value with the average of its four nearest neighbors. Your next iteration repeats that process on the new values. Continue this process over and over until the values seem to be "converging" (that is, there is little change from step to step). (Hint: Try to automate this process with your computer so that you can perform a large number of iterations conveniently.) How do your answers compare with those from Exercise 5?

Iterative methods are not only a fascinating concept, but they are very important for solving linear systems with very large numbers of variables. If such equations are solved by the usual methods, the small round-off errors generated in each computation can start to add up and affect the validity of the results. On the other hand, any such round-off errors generated at a step of an iterative process simply translate into a slightly different "guess" or starting point for the next iteration; thus they are not cumulative. The only error is the "truncation error" resulting form the original finite difference approximation to the derivatives in the equation. While the iterations you performed in Exercise 7 probably converged to the solution rather slowly, be assured that there are clever ways to rewrite the equations so as to yield iterative methods that will converge very rapidly. A brief initial introduction to the underlying strategy is given in the next two exercises.

Exercise 8.   Most iterative methods involve rewriting an equation as a "fixed point problem," $x = g(x)$, and then generating successive approximations to the solution by the formula $x_{n+1} = g(x_n)$, starting from some initial guess x_0.

a) When Newton's method from elementary calculus is interpreted in this way, what is the corresponding function g?

b) When g is an ordinary (scalar) function, not necessarily the one from Newton's method, show that if the derivative of g at the unknown solution is close to 0, the method will converge faster, at least if you start close enough to that solution.

c) For Newton's method, what is the value of the derivative of g at the unknown solution? (This should give an idea of why Newton's method converges so fast.)

d) [For readers with a strong linear algebra background.] The same iterative concepts can be applied to matrix equations of the form $Ax = b$, where A is square matrix and x and b are vectors. This is just the case of n equations in n unknowns. If you were to find several ways to rewrite such a system in the fixed point form $x = Dx + c$, so as to try iterations or successive approximations, what qualities do you think D should have so as to encourage rapid convergence?

Exercise 9.   Show that the method used in Exercise 7 can be interpreted in the framework of Exercise 8d. In particular determine the corresponding matrix D. Also find an alternative iterative scheme, using a different matrix D, although you need not carry out the actual iterative calculations to test its convergence.

We made an important observation earlier in this section, namely, that our numerical scheme for approximating a solution to Laplace's equation simply involves finding values that are the average of their four closest neighbors in the grid. This is actually very similar to a fundamental property of exact solutions to the exact partial differential equation, a fact that is explored in the following challenging but hopefully thought-provoking exercise.

Exercise 10.   Let (x_0, y_0) be a point in a region in which the head function satisfies the two-dimensional Laplace equation, and let C be a circle centered at (x_0, y_0). C can be of any radius r, large or small, as long as it and its interior are completely contained in the portion of the aquifer to which Laplace's equation applies. Show that the head value at the center of the circle, $h(x_0, y_0)$, is simply the average of the head values all around C, represented by the line integral

$$\frac{1}{2\pi r} \oint_C h(x, y).$$

This property is called the *mean value property of harmonic functions*. (Hint: show that this integral is actually independent of r, from which the result then follows easily.)

There are many important aspects of Laplace's equation and its numerical solution that we do not have room to treat in this brief introduction. For example, how would you approximate the derivatives if the boundary or grid values you were given were irregularly spaced? Could the averaging principles mentioned above be extended to these cases, and could they be used to improve the interpolation and contouring approaches you may have experimented with in Section 5.4? How would you deal numerically with more complex boundary conditions, such as those encountered in connection with Figure 5-20? If you decide to pursue any of these topics, you will find a wealth of fascinating ideas.

5.9 Guide to Further Information

From the practical, computational aspect of solving real ground-water problems in the real world, you might find it interesting to explore ground-water resources on the Internet or other computer network. Such resources are extensive, related both to interesting individual cases, where people often post their model results, as well as to software available for downloading. For example, you would have no trouble finding some standard packages for modeling ground-water flow, and you might even try to gain some experience with one of these by applying it to a problem as simple as the one that you solved earlier in this chapter in connection with Figure 5-23.

With respect to some of the mathematical topics raised in this chapter, you may wish to review a book on calculus, multivariable calculus, or advanced calculus, especially if you were somewhat uncomfortable with the extensive use of concepts such as gradient and directional derivatives, and you may also have identified some areas in linear algebra where some further study or review would be valuable. Standard text books on these topics abound. Although there are many books on ground water and ground-water hydrology, a particularly good reference is *Groundwater* by Freeze and Cherry, which, although first published in 1979, is still in print and is beginning to be regarded as a classic. If you have been troubled by any questions related to the rigorous mathematical basis for some of the concepts discussed in this chapter, one of the best sources would be books by Jacob Bear of the Technion University in Israel, such as his book *Dynamics of Fluids in Porous Media*. Few hydrology texts achieve its precision in their statement of mathematical concepts. For modern methods of fitting surfaces to data values, along the lines of Section 5.4, one might want to read material on multivariate splines and specifically on "NURBS," or nonlinear rational B-splines. Although this is a difficult topic, a particularly good reference is The NURBS Book, by Les Piegl and Wayne Tiller. For the subject of numerical methods, there is certainly a wide range of numerical analysis texts available, many of which are quite good, but for the specific issue of indirect or iterative methods for solving simultaneous systems of linear equations one of the classic texts is *Matrix Iterative Analysis* by Richard Varga. For a good discussion of the mathematical models associated with the transport of contaminants over and above the calculation of the flow regime for the ground water itself, a good recent text is *Mathematical Modeling of Groundwater Pollution* by Ne-Zheng Sun. For the subject of partial differential equations and potential theory, you will find numerous texts listed under these topics in a mathematical library.

6

Additional Topics in Air Modeling and Diffusion Processes

The purpose of this chapter is to provide a brief introduction to a number of additional aspects of air pollution modeling or the underlying diffusion-related principles, ranging from elementary to advanced levels. At the beginning of each section, some discussion will be given in a footnote as to the level of mathematical background required for the section. The exercises for this chapter are interspersed throughout the text to be sure that you master each concept before moving on to the next. Working through such exercises is essential for gaining an in-depth understanding of the material. Nevertheless, the continuity of the subject matter can be followed at a more general level by reading the text portions, and at least reading all the exercises and thinking briefly about each one, even if not all are pursued to their full level of detail.

6.1 Using Calculus to Obtain Further Information From the Diffusion Equation*

Recall the one-dimensional diffusion equation as presented earlier:

$$C = \frac{M}{\sqrt{4\pi Dt}} e^{-\frac{x^2}{4Dt}}.$$

Let us think of this equation in its simplest physical interpretation, namely, as an expression for the concentration of material at any point in an imaginary, infinitely long, thin tube filled with a substrate through which a mass M of the material is diffusing. At the initial moment, the mass M is deposited in the tube at the point corresponding to $x = 0$, and then it starts to diffuse out both in the negative and positive x-directions over time. The longer the period of time that elapses, the more the mass would be expected to spread out.

Before starting our discussion of this particular equation, let us discuss exactly what we mean by the concentration "at a point." After all, in the one-dimensional case, concentration

---

* This section uses concepts from elementary calculus, namely, max-min theory for functions of a single variable and the definite integral as a summation process. The last portion of the section makes use of the concept of partial derivatives, although it does not make use of any special properties of these derivatives or of functions of several variables. (One exercise, which can be skipped, uses iterated integrals.)

would be measured in grams per centimeter or some other units of mass per length, and a single point has no length. To answer this, just think about what you would want to do to measure, as best you could, the concentration at a point. You might take a small interval around the point, measure the total amount of mass within that interval along the diffusion tube, and then divide that mass by the length of the interval. If you took quite a small interval, then you would indeed be getting a good estimate of the concentration at the point of interest. Using this idea, you could actually make up a definition of the concentration at a point as the limiting value of these kinds of measurements as the size of the interval approaches 0. Stated in symbolic terms, this would be:

$$C(x) \equiv \lim_{\Delta x \to 0} \frac{\text{amount of mass in the interval from } x - \Delta x \text{ to } x + \Delta x}{2\Delta x}.$$

Here we have written the concentration as a function of x, the variable under discussion, but if the dependence on time t were also under discussion, we would write it as $C(x, t)$. We will indeed have occasional use for this more precise expression for concentration as a limit, but its main value right now is just to assure ourselves that this concept of concentration at a point is logical and well-defined.

We now return to our discussion of the diffusion equation. When we worked with this equation in Chapter 3, much of that work was carried out using numerical experiments. In particular, you were asked to construct a calculator or computer program that you could use to evaluate this equation with different values of the input parameters and variables, and based on those experiments you were encouraged to develop some physical intuition about the way the diffusion process works. Before proceeding further, let us clarify our use of these words "variable" and "parameter." In general we would say that the above equation involves two independent variables, x and t, and one dependent variable C. In other words, x and t are the input values, and C is the output value. However, before you can carry out this transformation from input values to output values, you also need the values of additional quantities shown in the equation, such as the diffusion constant D and the initial mass M. In a certain way, these are both variables and constants. They are variables in the sense that you can put in a range of values for them, this range corresponding to different physical situations that you might encounter in different diffusion experiments. For example, as has been mentioned earlier, the diffusion constant D for dissolved sugar in water would likely be different from the diffusion constant for ozone in air. But this quantity is also a constant in the sense that for a given physical situation or experiment, the value of D remains fixed throughout. Such "variable constants" are called *parameters*.

With an equation such as the one-dimensional diffusion equation, which has two independent variables, it is often very instructive to keep one variable fixed and look at the variation of the output value with respect to the remaining input value. For example, if we keep x fixed, the resulting function will give us the concentration as a function of time at the fixed location specified by x. The following group of quite easy exercises pursues this concept.

Exercise 1. Without performing any numerical calculations but basing your answer solely on your physical intuition, what would be the general shape of the graph of the function just described, namely, C as a function of t for an arbitrary fixed value of x?

Exercise 2. Does the general shape of the graph that you drew in connection with Exercise 1 hold for every single x value, or can you find one or more x values where the shape of the graph is qualitatively different? Once again, do this problem simply on the basis of your physical intuition.

Exercise 3. Now consider the one-dimensional diffusion situation for the following parameter set: $M = 200$ grams, and $D = 40$ cm^2/sec. For the point corresponding to $x = 100$ cm, use the standard max-min methods from calculus to find the value of time t at which the concentration at that point reaches its maximum. What is this maximum concentration? (Be sure to specify your units.)

Exercise 4. Consider the general one-dimensional diffusion equation. For an arbitrary but fixed value of x, can you find a general expression for the time t at which the concentration at x will temporarily achieve its peak value? (The answer to this should obviously depend on x, as well as on one or more other parameters.)

Exercise 5. Explain by physical reasoning why it was reasonable that the answer to the previous problem did not depend on the mass M.

Exercise 6. Show mathematically that the concentration value at the point corresponding to $x = 0$ is always decreasing with time, and then explain on physical grounds why this is precisely the result you would have expected.

In the previous set of exercises the focus was on the variation of concentration with time at a given point. You might have been thinking of such a point as the location of a receptor (such as a person's residence) that is subject to air pollution from a nearby power plant. However, there is a key qualitative difference between this situation and the way we modeled power plants and other pollution sources in Chapter 3 because in the earlier cases we assumed a constant release of material from the stack over a relatively long period of time, so that the concentration at any given receptor location would always stay the same. However, if there were a momentary "puff" type release from the stack, then the resulting physical situation would not be unlike the kind of situation that we have been analyzing here. Certainly these kinds of situations do occur, and, in fact, we will further investigate such puff releases later in this chapter.

Now let us look at the one-dimensional diffusion equation from a different point of view. In particular, we will now consider the value of time t to be fixed, and we will look at the

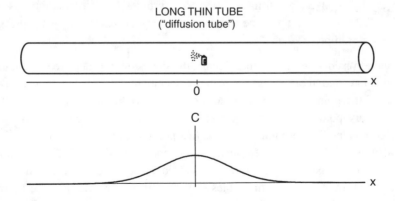

LONG THIN TUBE
("diffusion tube")

FIGURE 6-1
The underlying physical model for one-dimensional diffusion and the associated graph of concentration as a function of distance x for a fixed point in time

variation of concentration C with respect to spatial location x. This would be like taking a "snapshot" of the entire concentration profile along the tube at a fixed point in time. As you know from our previous discussions, the general shape of this concentration function will be that of a classic bell-shaped curve. It would be good to keep in mind two images simultaneously, namely, that of the conceptual one-dimensional diffusion situation involving a long narrow tube, as well as the graph of the corresponding concentration function C as a function of position. These two images are shown in Figure 6-1. If you were to redraw the concentration graph in Figure 6-1 for various points in time, you would always have the same general bell-shaped curve, the only difference being that for later values of time (corresponding to a longer period allowed for diffusion to take place), the curve would be flatter. To develop some familiarity with the use of derivatives to examine properties of this concentration function, we begin by verifying mathematically the fact that is obvious from the graph, namely, that the concentration function has a relative (or local) maximum at the point corresponding to $x = 0$, and that this relative maximum is also an absolute maximum.

Exercise 7.   As described above, consider t to be a fixed point in time so that the concentration C may be regarded as a function of the single variable x. Demonstrate mathematically that this function has a relative maximum at the point $x = 0$.

Exercise 8.   To continue with the previous situation, demonstrate mathematically that the point $x = 0$ also corresponds to an absolute maximum.

The last exercise in this series related to the diffusion equation pertains to the question of how one might actually measure the diffusion constant D in a laboratory experiment. You should be able to use your knowledge of mathematics to refine the proposed experiment so as to reduce the associated experimental error.

Exercise 9.   Suppose once again that you have a one-dimensional diffusion apparatus of the type discussed previously, and you are trying to plan an experiment to let you determine the value of the diffusion coefficient D. You plan to inject one gram of solute into the diffusion tube at the initial time. Assume that you have the ability to measure the concentration of diffusing material at any point and at any time, but you are permitted to carry out only one such measurement. Using such a measurement and assuming that all experimental errors are negligible, would you have enough information to determine D? Explain your answer. If your answer is "yes," illustrate with a numerical example. If your answer is "no," illustrate with a numerical counterexample.

Once again consider the one-dimensional diffusion situation depicted in Figure 6-1. As time goes on, more and more diffusing material is moving within the tube away from the neighborhood of the point $x = 0$ and off to the left and right sides of the tube. If you were to pick an arbitrary point along the tube and watch with an imaginary microscope how the diffusing material is moving through the tube at that point, then you would observe that material is moving past this point at some given rate. This is depicted in Figure 6-2. For example, to use purely hypothetical numbers, if you carry out such observations at the location corresponding to $x = 10$, say, you might observe molecules of the diffusing material passing by from left to right at a net rate of 1,000 molecules per second, or, you might measure this in other units such as 10^{-6} grams per second. This rate of material transfer by an observation point is called a *flux*. This is analogous to the concept of flux in the ground-water chapters. However, in this

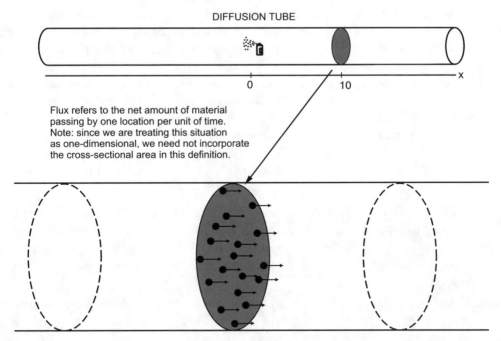

FIGURE 6-2

A close-up view of the transport of material by diffusion past a given observation point

current idealized one-dimensional case, we do not need even to talk about a unit cross-sectional area. (However, one could also think of the tube as having a unit cross section, in which case the flux could be interpreted in its earlier sense as the rate of material movement through a unit cross-sectional area.)

The concept of flux can be used to restate the basic *diffusion principle* as follows: *the flux at any point is proportional to the gradient (or the slope, or the derivative) of the concentration function.* We can rewrite this principle in the form of an equation by agreeing to let the letter q stand for the flux and recalling that proportionality is equivalent to multiplication by a constant, in the following sense:

$$q = -K \frac{dC}{dx}.$$

K is simply whatever constant of proportionality applies to the particular situation, and the equation is written so that K will be positive. In particular, the reason for the minus sign is that the flux is moving in the opposite direction from the concentration gradient or derivative. For example, if the derivative has a positive value, meaning that the concentration is increasing to the right, then the flux will be to the left, towards the region of lower concentration. In this equation we have written the derivative $\frac{dC}{dx}$ because we have previously agreed to treat the other input variable t as a fixed value during this part of the discussion. If you are familiar with the concept of "partial derivatives," then you would know that we could just as well write the above equation as follows:

$$q = -K \frac{\partial C}{\partial x}.$$

When we use the partial derivative symbol "∂" instead of "d", that is equivalent to automatically assuming that any variables other than x are to be treated as constants. (This is basically the definition of a partial derivative.) We will now consistently stick with this partial derivative formulation of the principle.

Note that we used the term "gradient" in the previous paragraph synonymously with derivative, whereas in Chapter 5 on ground water we reserved this term for the vector of partial derivatives. We shall not have use for the vector concepts in this chapter, and therefore we will often use the gradient in this same simpler sense in this chapter, namely, as a rate of change or hence as any regular or partial derivative.

The next exercises pertain to the concept of flux or to the rate of diffusion in such a process.

Exercise 10. For the one-dimensional diffusion situation, is the flux the same at all locations, if you consider them all at the same fixed instant in time, or does it vary from location to location? Provide two distinct lines of reasoning to answer this question: one based on your physical intuition as applied to this situation, and the other based on mathematical analysis of the appropriate equation.

Exercise 11. For a fixed time t, find any and all points of inflection of the concentration considered as a function of x. Give a physical interpretation of these x locations in terms of a special property that the flux has at those points.

Now we are ready to investigate the flux, the constant of proportionality, and the diffusion coefficient more thoroughly. To begin, let us recall the logical framework within which we have been operating. We have defined a diffusion process as one in which the transfer of some material or quantity takes place at a rate that is proportional to the gradient or derivative of its actual level. We have used this concept since Chapter 3, but we recently put it in mathematical form with the equation

$$q = -K \frac{\partial C}{\partial x}.$$

This flux q is of course a function of both x and t also, just like the concentration, because we have seen that the flux may vary from one point to another and from one time to another. It was stated, but not proved or derived, in Chapter 3 that for the one-dimensional diffusion situation with an initial mass M injected at the point $x = 0$ at the time $t = 0$, the actual concentration function satisfying this diffusion condition turns out to be given by the following equation:

$$C = \frac{M}{\sqrt{4\pi Dt}} e^{-\frac{x^2}{4Dt}}.$$

This equation contains a parameter D that we have called a *diffusion coefficient*. How did we figure out in the first place that this equation really is the solution to the diffusion problem? Even simpler, how can we even verify that this equation satisfies the basic diffusion principle stated above? These are not easy questions to answer, and if you are wondering about them, you are indeed wondering about very central issues in this subject area.

To address the second question, for example, we would like to use the one-dimensional diffusion equation to demonstrate that it yields a flux q at each point that is indeed proportional to the concentration gradient at that point. In other words, we would like to take the one-dimensional diffusion equation as a given, and use it to derive the proportionality equation given earlier. In fact, in carrying out that process, it will actually develop that the diffusion coefficient D *is* the constant of proportionality K in the diffusion relation!

While much of this calculation will be contained in an exercise below, there is one form of reasoning that you probably need an introduction to before you would be ready to undertake it on your own. In particular, consider the short portion of the one-dimensional diffusion tube illustrated in Figure 6-3. The illustrated portion of the diffusion tube is centered at an arbitrary but fixed point x and extends a distance Δx both to the left and to the right. Assuming that this value of x is to the right of the point $x = 0$ where the mass M has been injected, the general movement of mass is from left to right. In particular, mass is entering the short portion in question through its left boundary and mass is exiting through its right boundary. The rate at which mass is passing through those boundaries is simply the flux value at each of the corresponding x values, namely at $x - \Delta x$ and $x + \Delta x$. Therefore, we can write this "mass balance" condition as an equation in the following form:

$$\text{Net mass flow rate into section} = q(x - \Delta x, t) - q(x + \Delta x, t).$$

This might be in units of molecules per second or grams per second, or other units of mass per time.

We can also keep track of the inventory or amount of mass within the given section by noting that its concentration value is $C(x, t)$ at the center of the section, and that this will be a good approximation for the average concentration throughout the section as long as the length Δx is relatively small. In fact, as was mentioned earlier, for the one-dimensional diffusion

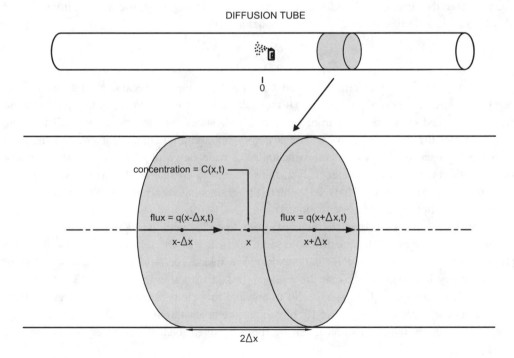

DIFFUSION TUBE

FIGURE 6-3

Selected short portion of the one-dimensional diffusion tube illustrating components of mass balance

situation, concentration is specified in terms of mass per unit length. Therefore, if we multiply this concentration $C(x, t)$ by the length of the section in question, which is $2\Delta x$, we will obtain the total amount of mass in this section at the given time. We can also write this as an equation:

$$\text{Total mass in section at time } t \approx C(x, t) \cdot 2\Delta x.$$

We have used the approximately equal symbol ($\approx$) because of the approximation we made in using the C value at the center of the section in order to approximate its value throughout. But now, examining the quantities in the previous two equations, we can see that the relation between them is that the net mass flow rate into this section should simply be the rate of change (or derivative with respect to time) of the total mass in the section. After all, the total mass within the section is changing only because there is a net mass flow rate into the section resulting from an imbalance between the fluxes at the left side and the right side. Therefore we can write

$$q(x - \Delta x, t) - q(x + \Delta x, t) \approx \frac{\partial}{\partial t}[C(x, t) \cdot 2\Delta x].$$

Since $2\Delta x$ does not depend on the t variable on the right-hand side of the equation, it behaves essentially as a constant from the standpoint of differentiation with respect to t and we can take it out of the derivative. In fact, we can even then divide both sides of this approximate equality by this quantity $2\Delta x$, leading us to

$$\frac{q(x - \Delta x, t) - q(x + \Delta x, t)}{2\Delta x} \approx \frac{\partial C(x, t)}{\partial t}.$$

We now take the limit of both sides as Δx goes to 0, enabling us to use an equality sign and also leading to a derivative on the left-hand side, as follows:

$$-\frac{\partial q}{\partial x} = \frac{\partial C}{\partial t}.$$

We have not written the arguments x and t in this last equation because everything is now being evaluated at the same point (x, t). (It may take a few moments to verify that the left-hand side does indeed approach the partial derivative indicated, and the reader is asked to show this in one of the exercises below.) This is a very important equation in itself, and the kind of reasoning used to derive it is fundamental to a wide range of physical problems arising in the environment and elsewhere. In fact, the reader who completed the previous chapter on ground-water modeling would have encountered a very analogous argument there in connection with the derivation of Laplace's equation.

Now we are ready to complete the verification that the one-dimensional diffusion equation really does satisfy the diffusion principle, and, in fact, that the constant of proportionality, previously written as a generic K, is actually the diffusion coefficient D from that equation. This is a very important result that the reader is asked to derive in Exercise 13, below. The other exercises are related to the above discussion and to various aspects of the concept of flux. If you elect not to complete Exercise 13, please be sure at least to understand the result, namely, that for the concentration function given by the one-dimensional diffusion equation, the flux satisfies:

$$\text{flux} = -DC_x.$$

Exercise 12. Verify that the limit process described above does indeed lead to the partial derivative $\partial q / \partial x$ as indicated.

Exercise 13. [This is a particularly challenging exercise.] As anticipated earlier in the text, show that the one-dimensional diffusion equation does indeed satisfy the fundamental diffusion property and that the corresponding constant of proportionality is actually the diffusion coefficient D. (Hint: it may be helpful to use the relationship just derived above between the x partial derivative of q and the t partial derivative of C.)

Exercise 14. It is not difficult to attach intuitive meaning to the statement "diffusion occurs faster in experiment A than in experiment B." We would think of two one-dimensional diffusion tubes A and B where the combination of solute and substrate in tube A is such that the solute tends to spread out through the tube faster than in the corresponding tube B. We also assume that we begin with the same mass M in each case. Which of the following are logically equivalent to the above statement?

a) For every pair of values of x and t, the magnitude of the flux value in A is larger than the corresponding value in B.

b) The diffusion coefficient for A is smaller than the diffusion coefficient for B.

c) For every time t greater than 0, the magnitude of the drop rate in the concentration at the origin, given by the time derivative $C_t(0, t)$, is greater for A than for B.

d) For every time t greater than 0, the magnitude of the concentration at the origin, given by $C(0, t)$, is less for A than for B.

e) For every time t greater than 0, the positive point of maximum flux in A is farther to the right than the point of maximum flux in B.

Exercise 15. We have been talking loosely and intuitively about the concept of a quantity of mass M injected at the point $x = 0$ at the initial time $t = 0$. There is a subtle logical problem associated with this concept. Can you identify this problem, and can you suggest a mathematical approach to this concept that shows promise in getting around the problem?

Suppose that we were conducting a one-dimensional diffusion experiment and that we wanted to keep track of the total amount of mass included within a given section of the tube. For example, suppose we were to consider the portion of the tube ranging from $x = a$ to $x = b$. At any given point in time there is a certain total amount of mass (whether measured in grams or molecules or some other mass units) located within this portion of the tube, and obviously this amount varies with time. For example, at the very beginning of the experiment, in the first few instants of time, as long as this interval does not contain the center point $x = 0$, there will be practically no diffused mass within this portion of the tube. Then gradually, material will begin to diffuse through it and the inventory within this section will grow. Then, over the long run, as material continues to diffuse out towards the left and the right directions, once again the concentration within this section will drop and thus the total inventory within it will drop as well.

Given that we have a complete solution to the one-dimensional diffusion situation in the form of an explicit function giving the concentration at any point at any moment in time, we should certainly be able to develop a corresponding expression for the total amount of mass in such a section of the tube. In fact, the answer involves the simple concept of the definite integral, taking the form:

$$\text{Total mass between } a \text{ and } b = \int_a^b C(x,t)\,dx.$$

The t within this integral should not be at all troubling because we are performing all our calculations at a fixed (although arbitrary) point in time. So t essentially behaves like a constant during the process of integration. Why is this equation true? The answer to this involves two concepts: our definition of the concentration at a point, and the definition of the definite integral as a limit of Riemann sums. (In fact, if you have a strong background in integration, you can probably see this immediately.) This equation will be explored in the next few exercises.

Exercise 16. In the context of the preceding discussion, imagine that you divide the interval from a to b up into n subintervals each of length Δx, so that $\Delta x = (b-a)/n$. Use the concentration function to write expressions for the approximate value of the total mass contained with the first subinterval, the second subinterval, etc. Then use the definition of the definite integral as the limit of sums to demonstrate that the total mass between a and b can be represented by the integral given in the text.

Exercise 17. Consider a specific one-dimensional diffusion problem in which three ounces of material are injected at the center of a long thin tube for diffusion within the substrate, and assume that the diffusion constant D has the value 5 in^2/sec. At the time given by $t = 80$ sec, find the total amount of mass in that portion of the tube between the points corresponding to $x = 3$ and $x = 25$. If you encounter some portion of your calculation where you cannot find an exact value, use tools at your disposal to approximate the numerical value.

Exercise 18. Find the value of the following improper integral:

$$\int_{-\infty}^{\infty} \frac{M}{\sqrt{4\pi Dt}} e^{-\frac{x^2}{4Dt}}\,dx$$

(Hint: there is both an easy way and a very hard way to answer this question.)

Up until this point, this section has focused on the properties of the one-dimensional diffusion equation. But recall that Chapter 3 also discussed two-dimensional diffusion equations, which had the general form

$$C = \frac{M}{4\pi t \sqrt{D_1 D_2}} e^{\left(-\frac{x^2}{4D_1 t} - \frac{y^2}{4D_2 t}\right)}$$

where D_1 and D_2 were the diffusion constants in the x- and y-directions respectively. (Note: if you have studied Chapter 5 and are concerned about the complexities of anisotropy, in this case we are assuming that the x- and y-directions are the principal directions. In fact, the diffusion case is less complicated than the ground-water case for reasons that we shall not discuss here.)

Much of what we have done earlier in this section could have been done in the context of the two-dimensional diffusion problem, although the computations would have been more difficult. If the reader has worked through Chapter 5, the nature of these complications has already been experienced, including issues such as gradient, directional derivative, contour lines, and others. There is no intention to reintroduce these concepts here. However, as a very brief introduction to some relatively elementary aspects of the two-dimensional diffusion equation, consider the following exercises.

Exercise 19. Consider the two-dimensional diffusion equation in the situation where the diffusion constants D_1 and D_2 are equal to a single value D. Show that the value of the concentration at any point can be represented as a function of a single new spatial variable. (Hint: you should be able to anticipate in advance what this variable is even before you apply some simple algebraic manipulations to obtain the required form.)

Exercise 20. For this same two-dimensional diffusion problem with equal diffusion coefficients, describe the family of geometric curves in the x, y-plane along each of which the resulting values of concentration will be constant.

Exercise 21. For the two-dimensional diffusion situation in which the diffusion constants are different in the x- and y-directions, describe the precise class of curves in the x, y-plane that represent points of constant concentration. These are sometimes called isopleths, contour lines (as in the case of ground water), or level lines.

Exercise 22. [Involves multiple integration.] For the two-dimensional diffusion situation in which the diffusion constants are different in the x- and y-directions, what should the value be of the double integral of the two-dimensional diffusion equation over the entire plane? Verify, using the techniques of integration for such integrals, that you do get the value that you expect on physical grounds. (Hint: this problem is not so hard as it may sound.)

Exercise 23. [Partly involves multiple integration, but the equation should be able to be guessed at without one's being experienced with such integrals.] Write down what you would expect to be the obvious form for the *three-dimensional diffusion equation* based on analogy with the equations for the one- and two-dimensional cases. Assume the possibility of different diffusion constants in each of the three principal directions. Use the integration constraint, of the type that you investigated for the two-dimensional case in the previous exercise, to determine the correct constant(s) for the three-dimensional diffusion equation.

6.2 Relation Between the Diffusion Equation and the Gaussian or Normal Distribution from Statistics*

If you have any familiarity with elementary statistics, you have no doubt encountered the famous *normal distribution,* also known as the *Gaussian distribution.* It is represented by a so-called probability density function

$$f(x) = \frac{1}{\sqrt{2\pi}\sigma}e^{-\frac{1}{2}\left(\frac{x-\mu}{\sigma}\right)^2}$$

that has two parameters, a mean μ and a standard variation σ, that may vary from one situation to another. This is the equation of the classic bell-shaped curve, and its use in statistics is essentially the result of the following fact. The probability distribution of many important quantities X in

---

* This section will be of greatest interest to readers who have used the Gaussian or normal distribution for elementary statistical calculations. However, it also summarizes these concepts in a reasonably self-contained fashion, and so could also serve as an introduction. The mathematics includes notions from elementary calculus (integrals) and statistics (the probabilistic interpretation of integrals of the normal distribution). One problem, which can be skipped, uses multiple integration.

real life can be represented by or well approximated by this probability function in the sense that

$$\text{Prob}\{a < X < b\} = \int_a^b \frac{1}{\sqrt{2\pi}\sigma} e^{-\frac{1}{2}\left(\frac{x-\mu}{\sigma}\right)^2} dx.$$

Therefore, in order to calculate some desired probability of something, all you need to do is find the integral of this function over the appropriate interval, a process that is usually carried out by using tables, such as are found in many statistics books.

One cannot help but notice the similarity between the normal probability density function and the one-dimensional diffusion equation. Here once again, for comparison, is the one-dimensional diffusion equation:

$$C = \frac{M}{\sqrt{4\pi Dt}} e^{-\frac{x^2}{4Dt}}.$$

It would be valuable to begin by working out a precise comparison between the equations.

Exercise 1. Compare the corresponding terms from the normal distribution and the one-dimensional diffusion equation to identify the correspondence between the parameters and variables in one equation and those in the other.

Exercise 2. Based on probability theory, find the value of the integral

$$\int_{-\infty}^{\infty} \frac{1}{\sqrt{2\pi}\sigma} e^{-\frac{1}{2}\left(\frac{x-\mu}{\sigma}\right)^2} dx.$$

Describe the relationship between this result and the answer to a very analogous exercise in the previous section.

Exercise 3. Give a heuristic (i.e., intuitive) explanation, referring to the comparison developed within Exercise 1, above, of why the counterpart for the standard deviation σ from the normal distribution does indeed appear reasonable. Be sure to consider both components, D and t, of this counterpart in your explanation.

Exercise 4. [This problem presumes that the reader has studied basic multivariable calculus.] Prove from the basic principles of calculus that the integral of the normal distribution from $-\infty$ to $+\infty$ is equal to 1, as asserted in terms of its probability interpretation above. (Hint: this is a standard problem from multivariable calculus wherein it is usually suggested that one square the given integral, using two different letters for the dummy variables in the two integrals, and then convert the result to a double integral written in polar coordinates.)

On the basis of the above discussion and calculations, it is very easy to see an obvious correspondence between the normal distribution and the one-dimensional diffusion equation. The real question is this: why is there such a correspondence? What is it that causes these two situations to result in essentially the same equation? In this section we will actually derive the one-dimensional diffusion equation by means of probabilistic arguments involving the normal and other probability distributions.

We will, of course, first need to consider the normal distribution itself. Where does it come from, and why does it turn out to be such a key tool in answering so many probabilistic questions?

Just as a basic course in statistics usually begins with a discussion of "discrete distributions," which are generally simpler than continuous distributions such as the normal distribution,

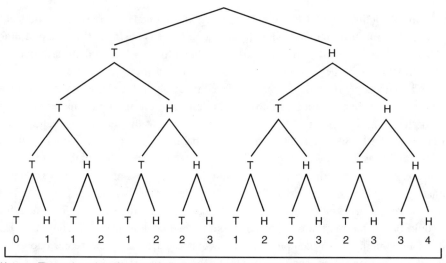

Notes: The total number of heads is indicated above for each of the 16 possible sequences.
         The probability of a certain number of heads is the fraction of the 16 sequences leading to that number.
         The points on the graph are connected simply to suggest the general shape.

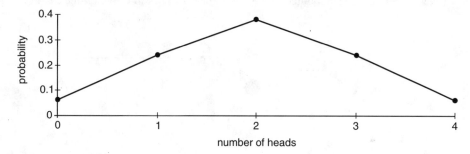

FIGURE 6-4

Tabular and graphical analysis of the probability of different outcomes when tossing a fair coin four times

that is also an appropriate starting point for this discussion. Consider the experiment of tossing a fair coin four times, keeping track of the total number of heads obtained during those four tosses. There are five possible outcomes for this experiment, namely, 0 heads, 1 head, 2 heads, 3 heads, or 4 heads. These five outcomes are not all equally likely, because there is only a single way to get a total of zero or four heads, whereas there are quite a few different ways to obtain the other outcomes. These results are tabulated and graphed in Figure 6-4.

The probability of each of the five possible outcomes can be calculated from the formula for the *binomial distribution* for the number of "successes" in four "independent trials":

$$\text{Prob}(k \text{ successes}) = \binom{4}{k}\left(\frac{1}{2}\right)^4 = \frac{4!}{k!(4-k)!}\left(\frac{1}{2}\right)^4.$$

(Almost any book on probability, statistics, or finite mathematics will provide a discussion of this, if desired.)

There is another useful physical model for an experiment that is "probabilistically equivalent" to the experiment of tossing a coin four times in a row. This experiment involves rolling a marble down an incline on which a number of pegs have been placed so as to deflect the marble either to the left or to the right when it hits them. (This is basically the idea of a pinball machine.) Such a device is shown in Figure 6-5. As shown in this figure, there are four levels of pegs. A marble starts by being rolled out of the opening at the center top of the incline. It rolls down and contacts the front peg, which is placed precisely in the center, so that the marble has an equal chance of being deflected either to the left or to the right. No matter which way it has been deflected at the first step, it will reach another peg at the second level and again be deflected either to the left or the right with equal probability. This process continues through two more levels of pegs, and then at the end of the incline there are five containers to catch the marble depending on where it finally ends up. As in the case of tossing the coin, it is relatively unlikely that the marble will wind up at the far left or the far right, because there is only one sequence of the four individual events that can place it at those extreme ends. The most likely outcome would be for the marble to wind up in the middle container, with a somewhat lesser probability of its winding up in the containers immediately adjacent to the center one.

Consider the movement of one single marble through this device. At each of the four levels, it can go left or right with equal probability, corresponding to the tossing of the coin earlier. In fact, if you let the final outcome be the number of the bin in which the marble winds

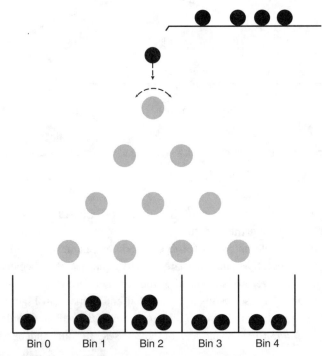

Collection bins for marbles, showing hypothetical results after initial few trials.
Bins above are numbered to correspond to equivalent number of heads in coin tossing experiment.

Note: alternative bin numbering, shown below, would record net number of steps to left or right.

Bin -4   Bin -2   Bin 0   Bin 2   Bin 4

FIGURE 6-5

Experimental apparatus to simulate several probabilistic events as a marble rolls down an incline

up, this corresponds to the number of steps on which the marble moved to the right. If you identify the movement of the marble to the right with getting a head on a coin toss (since they have equal probability), then the number of the bin in which it winds up corresponds exactly to the total number of heads. Since the outcomes can be made to correspond, we regard these experiments as probabilistically equivalent. Even if we were to label the bins differently, say by marking them with the net number of steps on which the marble moved to the right, the events would still be regarded as probabilistically equivalent. This is shown as the "alternative numbering system" in the figure.

There are two aspects of this experiment that we might vary: the number of levels of pegs on the pinball device and the number of marbles we pass through the device. For reference, let us define:

$$n = \text{number of levels of pegs in the pinball device}$$

$$m = \text{number of marbles processed through the device}$$

We will want to keep track of the effects of increasing both n and m, but it is important for us to follow the effects one at a time.

First, let us imagine that we increase the number n of levels of pegs. (This would be like increasing the number of tosses of the coin in the original experiment.) There would then need to be a total of $n + 1$ bins at the bottom, and the probability of a single marble's winding up in any individual one would be given by the general binomial distribution for n independent trials:

$$\text{Prob(bin } k) = \binom{n}{k} \left(\frac{1}{2}\right)^n = \frac{n!}{k!(n-k)!} \left(\frac{1}{2}\right)^n.$$

If we calculate these probabilities for larger values of n than the number 4 represented in the graph in figure 6-4, say $n = 10$ or $n = 30$, then we obtain graphs which have a very familiar shape. See Figure 6-6. The curves shown, which are for the binomial distribution, look very much like the bell-shaped curve we have associated with the normal distribution, especially for the case $n = 30$!

In fact, it can be proven mathematically that as n increases, the probability values of the binomial distribution approach the shape of a normal distribution. To put this differently, the normal distribution (which is usually easier to use at least for large n values) can be used as an approximate replacement for the binomial distribution for large values of n. You may have encountered this in studying probability and statistics, although very likely you have not seen or thoroughly assimilated the actual mathematical proof.

The following exercises provide further practice and investigation concerning the probability distributions with which we are working. They are recommended for an in-depth understanding of this section and may help you review material you might have studied in statistics.

Exercise 5. When the binomial distribution for large values of n is approximated by the corresponding normal distribution as described above, what are the values of the parameters σ and μ needed to specify the normal distribution? (You do not have to derive the answer. The value for μ should be easy to see, but the value for σ may be a bit more difficult if you have not come across it before. You may wish to consult a reference to find it.)

Exercise 6. The probability function for the binomial distribution with equally likely outcomes has been given in the text above. Write the corresponding probability function for a

binomial vs. normal approximation for n=10

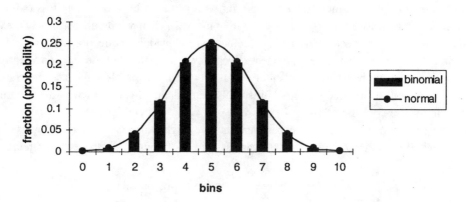

binomial vs. normal approximation for n=30

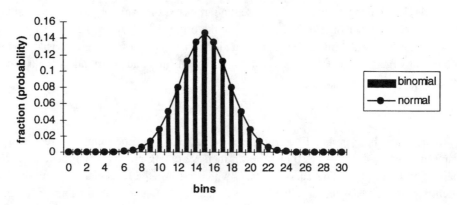

FIGURE 6-6

Theoretical histograms for results of the pinball experiment with an increasing number n of levels of pegs, and corresponding normal curve approximations

binomial distribution where the outcomes are not necessarily equally likely, so that they must be characterized by probabilities p and $1 - p$.

Exercise 7. When the value of p is not equal to $\frac{1}{2}$, the histogram that you obtain when graphing the probabilities of various possible outcomes is no longer symmetric, unlike the histograms shown in the figures in this section. Nevertheless, the normal distribution is always symmetric. Does the binomial distribution converge to the normal distribution only for the case of equally likely outcomes, or does it converge in the more general case? Explain your answer, perhaps illustrating with a numerical experiment. (Hint: you may wish to consult some statistics reference material to answer this question.)

Exercise 8. [Difficult problem.] As has been asserted in the text, for large values of n, the binomial distribution follows very closely the shape of the bell-shaped curve given by the normal distribution. Prove this fact mathematically for the case of equally likely outcomes, as applied in the text. (Hint: the key fact you will need is Stirling's formula, which gives an approximation to factorials in the form:

$$n! \approx \frac{n^n}{e^n} \sqrt{2\pi n}.$$

The approximation is an asymptotic approximation, meaning that as $n \to \infty$, the ratio between the exact and approximate values approaches 1. You may also want to keep in mind the following limit for e:

$$e = \lim_{n \to \infty} \left(1 + \frac{1}{n}\right)^n,$$

which can also be written: $e = \lim_{\varepsilon \to 0} (1 + \varepsilon)^{1/\varepsilon}$.)

Exercise 9. Show that the product of a normal random variable and a constant is itself a normal random variable. (Hint: show that there are appropriate mean and standard deviation values so that the probability function for the new random variable fits the general form of a normal distribution.)

Exercise 10. Show that if X is a normal random variable, then so too is $aX + b$, where a and b are constants. (Exercise 9 was a special case of this.)

Exercise 11. Relate the conclusion of the previous exercise to the relationship between the following two random variables for the pinball device: X, the original bin-number (e.g., number of steps to the right) described in Figure 6-5, considered as a normal random variable; and Y_{net}, the bin number based on the alternative bin numbering scheme in the same figure.

So now we know that for a pinball device with a large number of levels of pegs, the probability of the various outcome bins can be approximated by the normal distribution, and we would further expect this approximation to be excellent for large values of n. In fact, the graph shown in Figure 6-6 suggests that even $n = 30$ yields quite a good approximation.

Now we look at the effect of increasing m as well. That is, we start to process a large number of marbles through the pinball device. Let us enumerate some actual experimental results, keeping track of the *fraction* of marbles showing up in each of the bins at the bottom. (We would get this fraction simply by counting the number of marbles in the bin and dividing by the number m of marbles being run through the device.)

We begin by looking at the case where there are only four levels of pegs, as shown earlier in Figure 6-5. If we were to keep a running tally of our results after processing various numbers of marbles through this device, by the time we had run only five or ten marbles through the system, the results could be quite variable. However, as you know from the principles of probability, if you repeat the process many times, the experimental results will be much closer to the actual predicted probability values. This is called the *law of large numbers*. For example, as shown in Figure 6-7, if we were to process 100 or 1,000 marbles through the device, then the fractions in the different containers at the end would much more closely approximate the pattern predicted by the theoretical probability calculations for the binomial distribution or for its normal approximation.

Let us then summarize here the effects of increasing both n and m:

- An increase in n, the number of levels of pegs, causes the *theoretical* distribution of the results to approach a normal distribution.

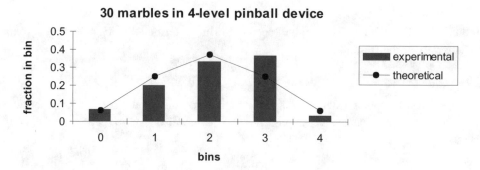

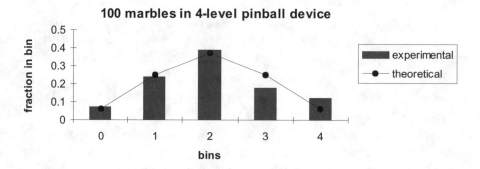

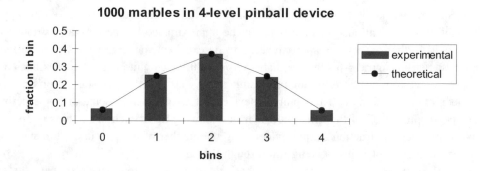

FIGURE 6-7

Tally of simulated experimental results from pinball device for different numbers of marbles processed through the system

- An increase in m, the number of marbles, causes the *average* of the *experimental* results to follow the theoretical prediction more closely.

With the above experiments in mind, we are now prepared to think of the diffusion process as it really is, namely, a probabilistic process. In particular, we have a large number of molecules being injected into a substrate at the point $x = 0$. The molecules are like the marbles in the above experiment, and so let us take a single one and watch what happens to it. Every individual molecule contains kinetic energy (energy of motion) and is constantly moving around, colliding with, and bouncing off the molecules around it, whether those are molecules of the diffusing material or molecules of the substrate. There are so many molecules and they are moving so rapidly, that these motions may be considered to be completely random.

So now, when the first molecule is injected into the diffusion tube at the point $x = 0$, we look at what might happen to it during a sequence of successive very small intervals of time. In the first time interval, it might move to the left or it might move to the right, and these possibilities are equally likely. This is like the marble contacting the peg on the top level. (For the moment, we assume that it always moves in the same size steps during each time step.) Then in the second interval of time it might again decide to move to the left or to the right, depending on the interactions it has with the surrounding molecules. This process is repeated over and over again as time goes on. Theoretically, it might always move to the left at each time step or might always move to the right, but you know that this is really extremely unlikely, just as it was unlikely to obtain all heads or all tails in flipping the coin or unlikely to obtain very many marbles at the far left and right extremes of the pinball machine, especially if there were a lot of pegs.

After a certain fixed number of time steps (corresponding to n, the number of levels of pegs), this molecule could be to the left of the starting point or to the right. Let us use the symbol X to stand for the location of this molecule. As long as we allow many time steps, this will be a normal random variable. (And since molecular motions are so short and fast, even one second would correspond to a very large number of steps.) This means that there are some appropriate parameters, μ and σ, such that the probability distribution of the molecule's final location X can be described by the probability density function for this normal distribution. In fact, it is easy to see that for our particular diffusion problem the value of the mean μ must be 0 since the molecule is injected at the point $x = 0$ and always has an equal chance of moving left or right.

The final probability that the molecule is in some particular interval would simply be given by the integral of the probability density function over that interval, as discussed earlier:

$$\text{Prob}\{a < X < b\} = \int_a^b \frac{1}{\sqrt{2\pi}\sigma} e^{-\frac{1}{2}\left(\frac{x-\mu}{\sigma}\right)^2} \, dx.$$

Therefore, if we were now to inject m molecules simultaneously into the diffusion tube, the expected number of molecules in the same interval would be:

Expected number in interval = total number times probability of interval

$$= m \times \int_a^b \frac{1}{\sqrt{2\pi}\sigma} e^{-\frac{1}{2}\left(\frac{x-\mu}{\sigma}\right)^2} \, dx.$$

Now we apply this principle to calculate the concentration at any point x, which, as you may recall from Section 4.1, is the limit of the amount of material in a small interval around x, divided by the length of that interval, as the length of the interval approaches 0:

$$C(x) = \lim_{\Delta x \to 0} \left(\frac{\text{molecules in interval}}{\text{length of interval}} \right)$$

$$= \lim_{\Delta x \to 0} \frac{m \int_{x-\Delta x}^{x+\Delta x} \frac{1}{\sqrt{2\pi}\sigma} e^{-\frac{1}{2}\left(\frac{x}{\sigma}\right)^2} \, dx}{2\Delta x}$$

$$= \lim_{\Delta x \to 0} \frac{m \cdot \frac{1}{\sqrt{2\pi}\sigma} e^{-\frac{1}{2}\left(\frac{x}{\sigma}\right)^2} \cdot 2\Delta x}{2\Delta x}$$

$$= \frac{m}{\sqrt{2\pi}\sigma} e^{-\frac{1}{2}\left(\frac{x}{\sigma}\right)^2}$$

In going from the second to the third equation in this sequence, we replaced the integral by a product that approximates it well on a short interval (since we are going to be taking the limit as $\Delta x \to 0$ anyway), but you are asked to look at this step in greater detail in Exercise 12 below. However, for now, just look at what we did. We derived the concentration function based on our probabilistic model, and it did indeed turn out to have the form of the one-dimensional diffusion equation! (To make the equations correspond, of course we need the identification $\sigma = \sqrt{4Dt}$, which is explored further below. We also measured the source and the concentration in units of molecules here, but it would only take multiplication through by a constant to convert them to mass units, such as grams or pounds.)

So now we have the theoretical concentration function based on this probabilistic model for the movement of the molecules. Should it be very accurate? The answer is yes because of the "m factor." That is, we saw earlier that large values of m, whether applying to marbles in the pinball device or to molecules in the diffusion tube, should cause the actual experimental or physical system to fit the theoretical model.

It is important to remind ourselves that one important simplification was made in this derivation: We assumed a very simple probabilistic model for molecular movements. Naturally, the movement of real molecules is more complex, even if we restrict attention to their net movement in the x-direction alone during each fixed time step. There is certainly going to be much more of a continuum of possible distances moved at each step in the real case.

Let the random variable Y correspond to the amount of movement of a molecule in the x-direction in a single time step. Earlier, we essentially thought of Y as having two outcomes, 0 and 1, each with probability $\frac{1}{2}$. The movement of a marble down the pinball device, our basic experiment, would then consist of n independent trials of this variable Y, and the overall outcome would be the sum of the individual results, where movements to the right (corresponding to heads on the coin toss) were being counted with the value 1 and movements to the left with value 0. These individual trials could be thought of as n independent random variables Y_i, each with the same probability distribution, and our previous random variable X was just the sum of these.

To incorporate the symmetry of the diffusion system right from the start, we could also have thought of the outcomes as having the values -1 and $+1$ depending on whether movement was to the left or the right. (This value of unity would correspond to whatever distance the molecules were all assumed to cover in a single time step.) This would not change the probabilistic structure of the system, which would still consist of binomial probabilities, but it would have the following advantage: The net distance to the left or right covered by a single molecule moving through the device would be given by the composite random variable

$$Y_{net} \equiv Y_1 + Y_2 + \cdots + Y_n.$$

This was the "alternative bin numbering" scheme shown in Figure 6-5. Here was the key point in our development: we showed that Y_{net} was approximately or essentially normal because it was a binomial random variable for a relatively large number n of levels of pegs, which can be well approximated by a normal distribution. This whole argument would collapse if the Y_is were other than the simple two valued variables that had been assumed because then Y_{net} would not have a binomial probability distribution.

So what can we do in the real case where the Y_is are indeed more complicated? Fortunately, there is a very important theorem in statistics, called the *central limit theorem,* that still applies to the more general situation. It basically says that if Y_{net} is the sum of *any n independent, identically distributed, random variables, $Y_1, Y_2, \ldots, Y_n$,* then, no matter what kind of values and probability distribution the Y_i's have, their sum Y_{net} can still be approximated as a normal distribution. Furthermore, the standard deviation σ of Y_{net} is just the standard deviation of any of the Y_i's multiplied by $\sqrt{n}$. This is quite a powerful theorem, and it applies to exactly the situation we have here, with whatever more complex probability distribution is needed to portray more accurately the movement of real molecules at each time step. Thus our final concentration calculations above still apply to this more general case, and the diffusion equation can indeed be regarded as a deterministic model for the underlying probabilistic process.

In fact, we can even use the statement about the dependence of σ on n to incorporate the specific terms involving t into our solution to the diffusion equation. So far we know that the concentration function has the form

$$C = \frac{M}{\sqrt{2\pi}\sigma}e^{-\frac{1}{2}\left(\frac{x}{\sigma}\right)^2}$$

where we have converted from molecule to mass units as mentioned earlier. But the derivation was for a single value of time t. For a different value, the appropriate value of σ would of course be different. In fact, as time goes on you would expect this σ to increase because the molecules are expected to be spreading out more. Now recall that the number n, when applied to the molecule situation, was the number of time steps being considered, each of fixed length, and hence also the number of Y_i's used to get the sum Y_{net}. Therefore n is proportional to t. (If t is twice as long, for example, there will be twice as many time steps.) And by the central limit theorem, σ is proportional to $\sqrt{n}$ and hence to $\sqrt{t}$. Therefore σ has the general form $\sigma = K\sqrt{t}$, which puts our concentration function in the form

$$C = \frac{M}{\sqrt{2\pi}K\sqrt{t}}e^{-\frac{1}{2}\left(\frac{x}{K\sqrt{t}}\right)^2}$$
$$= \frac{M}{\sqrt{2\pi K^2 t}}e^{-\frac{x^2}{2K^2 t}}$$

which is exactly our one-dimensional diffusion equation as long as we rewrite it with a new constant D defined as $D = K^2/2$.

Exercise 12. In the four-step sequence of equations above, leading to the expression for the concentration function $C(x)$ as a limit, provide a more rigorous justification than in the text for going from the second to the fourth and final step. (Hint: you may wish to apply the average value concept for a function, or you may find the mean value theorem for integrals helpful.)

Exercise 13. In the expression for the concentration $C(x)$ derived in this section, it was asserted in the text that multiplication through by a constant would convert everything from molecule units to mass units, as we are accustomed to in the diffusion equation. Demonstrate this process.

Exercise 14. [For readers with a strong statistics background.] Find a precise statement of the central limit theorem in a statistic reference, and use it to clarify the nature of the normal approximation that has been applied in this section. Explain also why some people say that "the central limit theorem rescues many people who apply statistical tests without checking that the distributions of their variables meet the assumptions of those tests."

Our probabilistic description of the diffusion process is also useful because it enables us to have a more detailed understanding of the concept of mass flux. Consider Figure 6-8, which shows a one-dimensional diffusion tube, of the type we have discussed previously, although of course it is blown up in the vertical direction to show details. (Thus, while it may look two-dimensional, you should still think of it as a one-dimensional situation.) The molecular distribution at some time after all the molecules have been initially injected at the point $x = 0$ is shown, and it is clear that they have begun to spread out through the tube. Furthermore, the individual molecules are annotated with small arrows to remind you that they are all moving in random directions, each independent of the other. In fact, even though the predominant movement of material is away from the point of injection and out towards the extremities, where the concentration is lowest, along the way individual molecules may spend some of their time moving to the left and some moving to the right. This is just like the situation with the pinball machine, where a marble might wind up in one of the containers on the right side even though at some of the pegs it encountered moving down the incline it might have randomly gone to the left.

So now consider the point on the diagram where we wish to measure the flux. Since this point is to the right of $x = 0$, we expect the flux to be in the positive direction, namely, moving

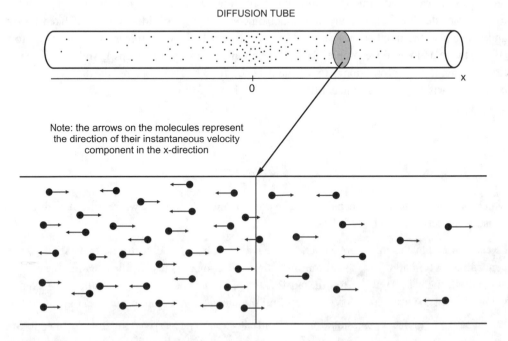

FIGURE 6-8

Exploded side view of a one-dimensional diffusion experiment, showing individual molecules and a location at which the flux is to be measured

from left to right. However, we are also reminded by the diagram that because of the random movement of the molecules, in a given time step some of the molecules may be moving from right to left past the measuring point whereas in the same time step others might be moving past that measurement point from left to right. This reminds us that when we talk about the flux at any given point along the tube, we are actually talking about the *net flux,* meaning that on the molecular level, we would look at the number of molecules passing the point from left to right minus the number of molecules moving from right to left in the same time step. In fact, all the fluxes we have been discussing earlier as being represented by "deterministic" mathematical equations are really net fluxes from the kind of random probabilistic molecular motions being discussed here. The deterministic equations being used for these fluxes have a high level of validity because the very large numbers of molecules involved and the very large number of time increments (since individual molecular movements are very short) cause this probabilistic system to converge very closely to a deterministic system represented by the one-dimensional diffusion equation and by its counterparts in higher dimensions.

Exercise 15. Consider the situation shown in Figure 6-9 where for the first time we are considering a diffusion situation in which mass is being injected at more than a single point. In particular, in this case equal masses M are injected at the two points shown, and we wish to analyze the flux passing by the flux measurement point shown on the figure.

(a) In a given time step, do you expect molecules to be passing by the flux measurement point in either direction?

(b) What is the flux value q at the flux measurement point?

A very important conclusion to be drawn from the above discussion is this. We have talked about a diffusion process as being characterized by a flux at any point that is proportional to the concentration gradient there. Using the probabilistic model, you can see why this really makes sense. If the concentration to the left of a given point is higher than the concentration to the right, based on the laws of probability you would expect that in a given time step a larger number of the molecules from the left are going to be moving toward the right than molecules on the right are going to be moving toward the left. Therefore you would expect the

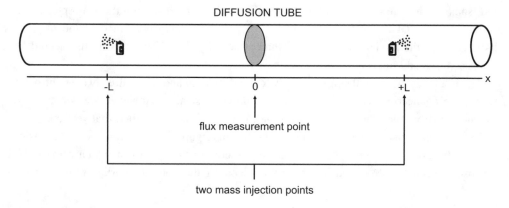

FIGURE 6-9

One-dimensional diffusion experiment with two points of mass injection and the flux measurement point halfway between them

net flux to be to the right. Furthermore, if you doubled, say, the difference in concentrations on the two sides of the point in question, you would be doubling the difference in the number of molecules available for movement on both sides, and since their random motions are essentially all identical, you would be doubling the difference in the numbers that would be moving from left to right and from right to left, and hence doubling the flux. This is the real physical basis for the fundamental principle we have used to characterize the diffusion process throughout.

6.3 Modeling No-Flow Boundaries Using the Reflection Technique*

Recall the Gaussian plume equation that was used extensively in Chapter 3:

$$C = \frac{Q}{2\pi\sigma_y\sigma_z u} \left[e^{-\frac{y^2}{2\sigma_y^2}} \right] \left[e^{-\frac{(z-H)^2}{2\sigma_z^2}} + e^{-\frac{(z+H)^2}{2\sigma_z^2}} \right].$$

This equation differs from all the other diffusion-type equations encountered so far in that it contains the sum of two distinct exponential terms in the factor on the far right. In this section we will see how the use of additional terms such as these can be used to account for the existence of "no-flow boundaries" such as the ground.

As you know, if we were to use a typical two-dimensional diffusion model (in one horizontal and in the vertical direction) in order to model diffusion of material away from a plume, the model would predict the gradual diffusion of material downward vertically an indefinite distance, and hence effectively right through the surface of the ground and downward. Not only would this be somewhat unrealistic, but it could not even be justified under our philosophy of conservatism, under which we have occasionally made slightly inaccurate but simplifying assumptions in the mathematical model when we were sure that the results they would lead to would tend to overestimate the level of air pollution experienced at the locations of interest. In this case, if we were to use a model that theoretically has material moving downward without bound, some of the mass of the pollutant would effectively be lost from the atmosphere above the land surface, and hence the total amount of pollutant being contained in the physically realistic zone by the model would be less than is actually there. That is why this would be a non-conservative and unacceptable approach.

Here is a fairly clever and simple remedy to this situation, and it has already been introduced in concept in connection with Exercise 9 of the previous section. However, we will begin our treatment of it from the beginning here in case the reader has not studied that section.

Consider the situation depicted in Figure 6-10. This figure shows a typical one-dimensional diffusion tube except for one difference, namely, there are two points at which a mass M is being instantaneously introduced into the tube. Consider the point P marked in the center of

---

* This section does not use any calculus or more advanced mathematics, although it does require reasonable comfort with the diffusion equation (the solution to the partial differential equation, not the partial differential equation itself) and the Gaussian plume equation.

SIDE VIEW OF DIFFUSION TUBE

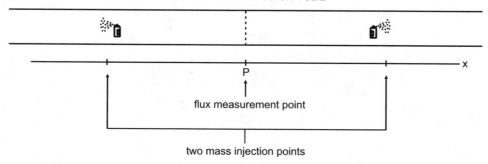

FIGURE 6-10
Side view of diffusion tube with two mass injection points

the tube halfway between the two points where the mass is being injected. Can you tell by a relatively cursory examination of Figure 6-10 what the flux must be at the point P?

The answer to this question is quite simple. The flux at the point P must be exactly 0. This is the case because there is a perfect symmetry between the part of the tube to the left of P and the part of the tube to the right of P. Therefore, there could be no more driving force for moving material by diffusion from the left injection point over to P and through it to the right than there would be for moving diffusing material from the right injection point over to the left to P and through it in that direction. Therefore the flux at P could neither be positive (net flux toward the right) nor negative (net flux towards the left), so that the only feasible value for the flux at that point would be 0, as stated.

We would refer to the point of the tube corresponding to P as a "no-flow boundary," since there can be no net flow through or past this point in either direction under the given conditions. In fact, this means that if we were actually to put a physical boundary or wall across the tube at that point, it would have no effect whatsoever on the pattern of diffusing material. This has been shown in Figure 6-11. Since the boundary in this figure serves to completely "decouple" the left side of the experiment from the right side of the experiment, we could really consider these two diffusion experiments as being carried out completely separately and independently.

SIDE VIEW OF DIFFUSION TUBE

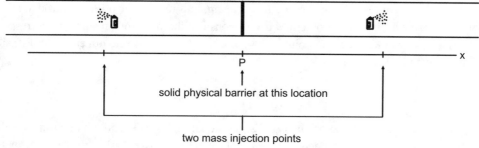

FIGURE 6-11
Diffusion situation described in the previous figure, except that a physical boundary has been incorporated in the tube at point P

SIDE VIEW OF DIFFUSION TUBE CLOSED AT ONE END

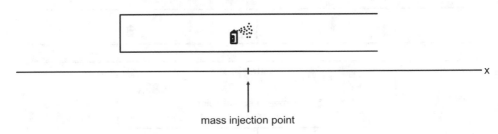

mass injection point

FIGURE 6-12
The right-hand side of the diffusion experiment from the previous figure shown completely separated from the left-hand side of that figure

In fact, we might be interested in the results of only one of the two of these experiments. For example, Figure 6-12 depicts a situation in which the right side of the tube is isolated by itself.

We could actually reverse the logic connected with the three figures in this sequence. For example, if we were to begin with a one-ended problem like that shown in Figure 6-12, we could then say that the desired diffusion pattern would obviously be the same diffusion pattern as we would have if we conducted both sides of the experiment, as in Figure 6-11. And then we could argue that the physical boundary shown in Figure 6-11 has absolutely no effect on this diffusion pattern at all, so that the diffusion pattern we are looking for could just as well be gotten from analyzing diffusion through the infinitely long tube shown in Figure 6-10. So if we can manage to solve the problem depicted in Figure 6-10, we can solve the one-ended problem in Figure 6-12 as well.

So now we will look more closely at the solution to the diffusion problem in Figure 6-10. Previously, we had agreed that the flux at the point P would have to be 0. Remember that the flux at a point corresponds to the "net flux" at that point, meaning that although there may be molecules moving both to the left and to the right through the point, the flux is simply the *net* rate of movement, subtracting the rate at which they are flowing in one direction from the rate at which they are flowing in the other direction. To analyze the diffusion pattern in Figure 6-10, we will break it up into steps as follows. Consider Figure 6-13 and for the moment focus on the first part of this figure. This shows the basic diffusion situation of Figure 6-10 but where we have included the injection of diffusing material only on the right-hand side of the point P. We have also added a coordinate system, and unlike our original introduction of the one-dimensional diffusion problem, we will consider the point of injection to correspond to the point $x = L$ rather than the point $x = 0$. The solution to this one-dimensional diffusion problem is given by the equation

$$C = \frac{M}{\sqrt{4\pi Dt}} e^{-\frac{(x-L)^2}{4Dt}}$$

where the only change from the original one-dimensional diffusion equation is that the x-value from the original equation has been changed to $(x - L)$ in this equation to account for the movement of the injection point L units to the right of the origin. The graph of this equation is simply the original bell-shaped curve except that its peak is now located at the point $x = L$. This is shown below the diffusion tube in the top section of Figure 6-13.

Similarly, the second part of Figure 6-13 shows another one-dimensional diffusion problem, this time with a single source now located L units to the left of the origin. By reasoning exactly analogous to the above, the solution to this problem is given by the equation:

$$C = \frac{M}{\sqrt{4\pi Dt}} e^{-\frac{(x+L)^2}{4Dt}}.$$

Notice that this equation differs from the earlier one only in the $(x+L)$ factor in the exponent, which now implies that the peak of this bell-shaped curve is achieved at the point where $x = -L$. The graph of this function is also shown in Figure 6-13.

As our final step, we want to combine the solutions to the first two parts of Figure 6-13. The key to this is the *principle of superposition*, which is a physical principle applicable here that states that the solution to the diffusion problem as shown in the third part of Figure 6-13, namely, involving two distinct sources, can be obtained as the sum (or "superposition") of the solutions to the individual decoupled problems shown in the first two parts of the figure. A good way to think about it is as follows. The molecules from the source on the left side "do their own thing" and don't care what the molecules from the right-hand source are doing. Similarly, the molecules from the right-hand side "do their own thing" without paying any attention to what the molecules from the left-hand side are doing. (In fact, no solute molecule pays any

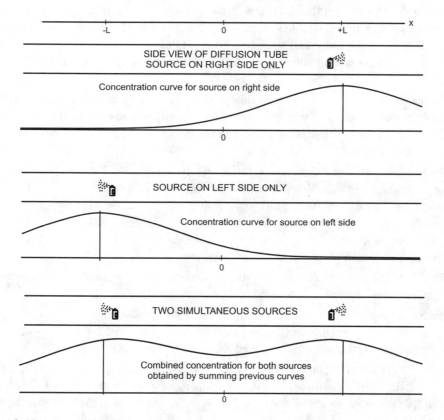

FIGURE 6-13

Gradual development of a framework for solving the diffusion problem with two sources, based on the treatment of the two sources individually and then combined

attention to any other solute molecule. They all move randomly and it is only the result of relative numbers in different places that gives the effect of net systematic migration.) When we observe the system as "outside observers," we can count only the total number of molecules of diffusing material at any point and at any moment in time, without knowing on which side they might have originated. The number that we observe (and that corresponds to the total concentration of material at any point) is simply the sum of the number that came from each of the two individual sources, and hence is given by the sum of the solutions to the two individual diffusion problems. Therefore, the solution to the diffusion problem shown in the third part of Figure 6-13, namely, the part with both diffusion sources, is given by

$$C = \frac{M}{\sqrt{4\pi Dt}} e^{-\frac{(x-L)^2}{4Dt}} + \frac{M}{\sqrt{4\pi Dt}} e^{-\frac{(x+L)^2}{4Dt}} .$$

Remember this key point: The solution to the diffusion problem with two sources, as shown in the third part of Figure 6-13, is exactly the same as the solution to the two individual decoupled parts of this problem (such as in Figure 6-11 or Figure 6-12) on their respective intervals.

This key observation leads to the following general strategy for solving any kind of diffusion problem in which there is some physical barrier or "no-flow boundary":

1. Add an imaginary source (of the same quantity and same type of diffusing material) in some imaginary location in space so that the symmetric nature of the two sources would imply a no-flow or zero flux condition at the location of the original no-flow boundary.
2. Find the solution to this double system, using the principle of superposition. (That is, find the solution to each individual source and then add the results.)
3. Use this resulting sum of solutions as the solution to the original problem on its original interval.

The reader is asked to try to apply this strategy for solving several additional diffusion problems in the exercises. The title of this section applied the name "reflection technique" to this method for treating no-flow boundaries. The reader is asked to investigate the basis for this terminology in Exercise 6.

Exercise 1. Consider a one-dimensional diffusion tube that is closed at one end. Suppose that a mass of 3 grams of material is injected instantaneously into the tube at a point 5 centimeters from the end and that the diffusion coefficient has the value $0.1 \mathrm{cm}^2/\mathrm{sec}$. Determine the equation for the concentration profile as a function of location and elapsed time. In addition, through a sequence of graphs, illustrate the evolution of the shape of this profile over time. Pay particular attention to the evolution of any local maxima that may exist.

Exercise 2. In the situation described by the previous problem, at what precise time would a local maximum concentration cease to exist anywhere within the tube?

Exercise 3. Consider the two-dimensional diffusion layout shown in Figure 6-14. Find the concentration function for all points (x, y) for all points in time. You need not assume that the diffusion constants are the same in both the x- and the y-direction.

Exercise 4. Consider the following two-dimensional diffusion problem with a no-flow boundary. As usual, the xy-plane will describe the two-dimensional region available for diffusion, but assume in this case that an initial mass M is injected at the point $(2, 5)$, and that a physical boundary along the line corresponding to $y = -3$ causes this line to function as a no-flow boundary. Draw a figure for this situation and find the resulting solution function. Describe the region within which your solution function applies.

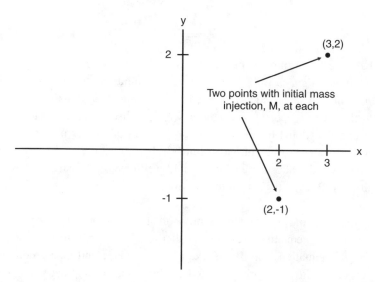

FIGURE 6-14
A basic two-dimensional diffusion problem where mass M is initially injected at two distinct points

Exercise 5. Provide a clear explanation of how the principles described in this section are represented in the Gaussian plume equation.

Exercise 6. Investigate the use of the term "reflection principle" for the technique described in this section. In particular, consider the one-dimensional diffusion situation of Figure 6-12 and assume that every time a diffusing molecule wanted to move to the left through the boundary at the left end, it was reflected instead back towards the right. Show that the resulting solution actually equals the solution obtained in this section.

Exercise 7. [This exercise assumes that you completed the Exercise 23 on three-dimensional diffusion modeling in Section 6.1.] Imagine that a mass M of air pollutant is emitted into the atmosphere H meters above the ground. For simplicity in this particular problem, let us ignore all atmospheric effects (such air movement, stability classes, temperature, etc.), and let us consider this as a pure three-dimensional diffusion problem. Since this is not a continuous plume but only a momentary release, it is referred to as a "puff" release, and, in fact, you were asked to develop the general form of its solution in Section 6.1. However, in this case the ground behaves as a no-flow boundary. Taking this one additional factor into account, derive the appropriate solution to this problem.

6.4 The Basic Partial Differential Equation for Diffusion Processes*

Much of both this chapter and Chapter 3 have revolved around the so-called *one-dimensional diffusion equation,* given by

---

* This section makes free use of partial derivatives, but does not presume any prior background in differential equations. Certain specific exercises do require reasonably advanced mathematical sophistication, but these are indicated as such.

$$C = \frac{M}{\sqrt{4\pi Dt}} e^{-\frac{x^2}{4Dt}}.$$

Certainly we have treated a number of variations on this equation, including its counterparts in two and three dimensions as well as the related air pollution model given by the *Gaussian plume equation*. This diffusion equation is so fundamental that it will be very valuable to investigate it further at this point. For example, we have discussed the analogy between mass transport at a rate proportional to a concentration gradient and other physical processes, such as heat flow, ground-water movement, etc. Although the diffusion equation was not introduced in the chapters on ground-water movement, it could well have been had we chosen to take that subject just a little bit further. In fact, in this section, some brief comments will be included on this topic.

In the previous section, we adopted a probabilistic viewpoint and used it to derive the one-dimensional diffusion equation. There is another totally different line of reasoning that can be used to derive this equation, one that is totally deterministic and much more similar to our earlier discussion of the diffusion principle. Indeed, both lines of approach represent important lines of thought, and in this section we shall pursue the latter as well.

Even though we have used the term "one-dimensional diffusion equation" to describe the above equation, the reader should know that many texts would not use this terminology, reserving the term "one-dimensional diffusion equation" for a fundamental "partial differential equation" (that is, an equation that contains some partial derivatives) from which our version of the equation can be derived. Now that the reader has some experience with diffusion processes and the nature of their corresponding equations, we shall also investigate this basic partial differential equation.

No previous background in partial differential equations is necessary to follow this discussion, and, in fact, this would be a good way to learn how such equations typically arise. Partial differential equations were actually invented about two centuries ago specifically to model physical problems such as heat flow which are, as noted earlier, basically diffusion processes. If you have read Chapter 5, you have already encountered an important partial differential equation, namely, Laplace's equation.

Remember that for a one-dimensional diffusion situation the concentration of diffusing material is represented by a function $C(x, t)$ that has two independent variables x and t. Recall that there are several common notations for indicating partial derivatives. For example, the first partial derivative of C with respect to the first variable x may be represented by any of the following:

$$\frac{\partial C}{\partial x} \qquad \frac{\partial C(x, t)}{\partial x} \qquad C_x \qquad C_x(x, t).$$

Similarly, the first partial derivative of C with respect to t could be indicated by any of the following:

$$\frac{\partial C}{\partial t} \qquad \frac{\partial C(x, t)}{\partial t} \qquad C_t \qquad C_t(x, t).$$

With respect to second partial derivatives, our interest will be primarily in the second partial derivative of C with respect to x, which may be represented by any of the following expressions:

$$\frac{\partial^2 C}{\partial x^2} \qquad \frac{\partial^2 C(x, t)}{\partial x^2} \qquad C_{xx} \qquad C_{xx}(x, t).$$

Earlier in this chapter, we primarily used the ∂ notation for partial derivatives, whereas in Chapter 5 on ground water, we frequently made use of the subscript notation. Each notation has its place in presenting mathematical constructs in a way in which they can be easily understood, and sometimes it is advantageous to shift from one to another during the course of a set of calculations. As mentioned above, a *partial differential equation* is nothing more than some kind of an equation that contains at least one partial derivative. Partial differential equations are frequently referred to as "PDE's."

Here is what we shall refer to as the *one-dimensional diffusion PDE,* written two different ways in accordance with the different systems for denoting partial derivatives.

$$C_t = DC_{xx} \qquad \frac{\partial C}{\partial t} = D\frac{\partial^2 C}{\partial x^2}$$

Remember that both of these equations say exactly the same thing, and we shall use the two different systems of notation interchangeably. This equation involves partial derivatives both with respect to t and with respect to x. As will be seen shortly below, this equation can be derived directly from the physical framework of the diffusion problem, considered once again as a deterministic situation governed by the diffusion principle; and this derivation will be quite similar to one that was carried out in Section 6.1.

Given this basic PDE to describe the physical situation, the next logical step would then be to "solve" it, meaning to find a function $C(x, t)$ that works when it is substituted into this equation. The only catch here is that the one-dimensional diffusion PDE has many solutions (in fact, an infinite number) so we shall also have to specify certain auxiliary conditions called initial conditions or boundary conditions in order to pin down the exact solution that we want. (This is similar to evaluating the constant of integration to find the particular solution to an antiderivative problem.) Not surprisingly, the one-dimensional diffusion equation that we have been discussing all along is one of the solutions to the one-dimensional diffusion PDE, and it would be good to begin by simply verifying this fact, as requested in the first exercise.

Exercise 1. Show that the one-dimensional diffusion equation is a solution to the one-dimensional diffusion PDE. (Hint: this computation may be somewhat complex so organize your work carefully to help you keep track of various terms.)

Exercise 2. Can you find any other solutions to the one-dimensional diffusion PDE? Begin by looking for the simplest possible functions that work when substituted into the equation, and then gradually try to find some more complicated ones. Don't assume that they all have to have the very complicated form of our one-dimensional diffusion equation.

Exercise 3. Suppose you have found a number of solutions to the one-dimensional diffusion PDE. Can you think of any way to modify or combine them in order to find yet more solutions?

Exercise 4. [For readers with a background in linear algebra.] Considering the previous exercises, especially Exercise 3, can you describe in the language of linear algebra a key observation about the set of all solutions to the one-dimensional diffusion PDE?

Exercise 5. [For readers with a background both in linear algebra and in differential equations.] Does your conclusion in Exercise 4, above, apply to the set of solutions to any differential equation, either ordinary or partial?

Now we proceed to the development of the one-dimensional diffusion PDE as the basic mathematical model for the diffusion process. This derivation will be very similar to our discussion of flux in Section 6.1, especially the discussion connected with Figure 6-3. We begin

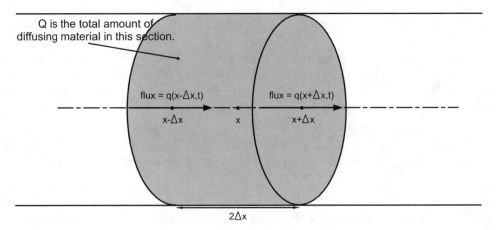

FIGURE 6-15
Short portion of the one-dimensional diffusion tube illustrating mass changes affecting a given segment

with the basic one-dimensional diffusion tube situation, as shown in Figure 6-15, and with the basic diffusion process assumption that the transport of diffusing material takes place at a rate that is proportional to its concentration gradient.

We are working in a short section of the tube centered at a fixed location x and extending a length Δx both to the left and to the right. Let Q represent the total mass of diffusing material in this particular segment of the tube. Remember M is the amount of mass distributed through the entire tube, so Q is only a portion of this mass. Naturally, Q depends on Δx.

As long Δx is relatively small, the average concentration of material within the tube segment should be well approximated by its value at the midpoint. In other words,

$$\frac{Q}{2\Delta x} \approx C(x,t).$$

We have used the approximate equality sign ($\approx$) to indicate that the above two quantities are not exactly equal, but they are approximately equal for small values of Δx; and, in fact, they would be equal if we were to take the limit as Δx approaches 0. We can rewrite the above equation to give us an expression for Q itself

$$Q \approx C(x,t) \cdot 2\Delta x.$$

We will now look at two different ways to represent the rate of change of the mass located within the given segment. The first way to represent this change is simply by the derivative $\partial Q/\partial t$. That is,

$$\text{Rate of change of mass in segment} \approx \frac{\partial Q}{\partial t}$$

$$\approx \frac{\partial}{\partial t}\big[C(x,t) \cdot 2\Delta x\big]$$

$$\approx C_t \cdot 2\Delta x$$

Note that we used the partial derivative expression in taking the time derivative of Q, since x had been fixed by the original discussion. A totally different way to calculate the rate of change of mass in the given segment would be to determine the flux into the segment, which itself can be done by taking the flux going into the segment from the left side and subtracting from it the flux leaving the segment at the right side. That is,

Rate of change of mass in segment = flux in at left − flux out at right

$$= -DC_x(x - \Delta x, t) + DC_x(x + \Delta x, t)$$

We now equate these two expressions for the rate of change of mass within the segment, move the $2\Delta x$ factor to the other side, and take the limit as Δx approaches 0:

$$C_t \approx D\frac{C_x(x + \Delta x, t) - C_x(x - \Delta x, t)}{2\Delta x}$$

$$C_t = D \cdot \lim_{\Delta x \to 0} \frac{C_x(x + \Delta x, t) - C_x(x - \Delta x, t)}{2\Delta x}$$

$$C_t = D \cdot \frac{\partial}{\partial x} C_x(x, t)$$

$$C_t = DC_{xx}$$

This is the one-dimensional diffusion PDE! The kind of reasoning used in this derivation is typical of the kind of reasoning used to solve many physical and environmental problems.

As a mathematical technicality, we should take note of the fact that the above derivation was based on the implicit assumption that the diffusion process was already "under way" and that a continuous and differentiable concentration profile had been established. This does not apply to our basic diffusion problem at the initial instant $t = 0$, and so the PDE does not apply to such initial or boundary points. Nevertheless, we would expect to be seeking a solution to the PDE that would be continuous at such points and thus to approach the initial condition of 0 concentration at all points of the form $(x, 0)$, except for the point $(0, 0)$ where the initial concentration is not even defined.

With respect to the concept of flux as used in this derivation, note that we have considered it to be simply the flow rate of material along the one-dimensional axis of the diffusion tube, similar to the way we have treated it earlier in this chapter. This is consistent with the first of the two ways to think of one-dimensional diffusion that were discussed in Chapter 3 (see Figure 3-5). If we were to adopt the three-dimensional viewpoint, but with diffusion in only one direction, then the flux would be the flow rate per unit of cross-sectional area perpendicular to the diffusion direction. In this framework, the cross-sectional area of the diffusion tube would show up as a constant factor across our mass balance equations, and it would cancel out. Alternatively, one could think of the diffusion tube as having a unit cross-sectional area.

The following exercises provide some further practice with the line of reasoning used in this section.

Exercise 6. Explain why in the final sequence of steps just above, the limit expression actually does equal the indicated second derivative.

Exercise 7. Compare the derivation given above with the derivation presented in connection with Figure 6-3. Where do the derivations diverge from each other, and what are the precise objectives of the two different derivations?

Exercise 8. The following equation was derived in Section 6.1, just prior to Exercise 15 of that section:

$$-\frac{\partial q}{\partial x} = \frac{\partial C}{\partial t}.$$

Show how the one-dimensional diffusion PDE could have been derived almost instantly from this equation.

Exercise 9. Describe precisely how in the derivation given in this section we have used the fundamental property that characterizes diffusion processes.

Naturally, the one-dimensional diffusion PDE has its counterparts in higher dimensions. For the most common case, when the diffusion constant is the same for all directions, these are the following:

$$C_t = D(C_{xx} + C_{yy}) \qquad \text{(two-dimensional diffusion PDE)}$$

$$C_t = D(C_{xx} + C_{yy} + C_{zz}) \qquad \text{(three-dimensional diffusion PDE)}.$$

The reader is asked to investigate some two- and three-dimensional diffusion aspects in the following exercises.

Exercise 10. Draw a clear diagram and use it to derive the two-dimensional diffusion PDE for the case where the diffusion constant D is the same for all directions. Be sure to state the basic diffusion property precisely so that it is available at the key point in your derivation.

Exercise 11. In the context of the previous problem, but assuming that different diffusion coefficients apply to the two diffusion directions, what should be the partial differential equation satisfied by the concentration?

Exercise 12. Sometimes a multidimensional problem has sufficient symmetry to it so that it is "essentially" one-dimensional. For example, in a two-dimensional diffusion situation with the same diffusion constant for both directions and with a single initial mass M injected at one point, the concentration will obviously be the same all the way around every circle centered at the point of mass injection. Thus the only spatial variable that really should enter into the answer is the radius r. Considering C as a function of r in this symmetric case, draw a good diagram that enables you to derive directly from the mass balance the differential equation for C as a function of the single spatial variable r, as well as t.

Exercise 13. As suggested in the previous exercise, the variables x and y are not the only variables with which you might describe the two-dimensional domain for diffusion. For example, since the polar coordinates r and θ can also be used to describe the two-dimensional domain, you would expect that the two-dimensional diffusion situation could be described in terms of these two new variables. In this exercise you are asked to develop the corresponding version of the two-dimensional diffusion PDE (that is, considering C as a function of r and θ) using two distinct lines of reasoning, namely:

(a) by using the standard relations between the Cartesian coordinates x and y and the polar coordinates r and θ, in combination with the chain rule for partial derivatives, in order to transfer our original version of the two-dimensional diffusion PDE into these new coordinates;

(b) by beginning with a basic diagram of the physical situation in which you represent a spatial element whose geometry is appropriate to polar coordinates, and then you perform a mass balance on this element similar to the way it was done in the text for Cartesian coordinates.

Exercise 14. Beginning with a clear diagram, derive the three-dimensional diffusion PDE for the case where the diffusion constant D is the same in all directions.

For the moment we shall restrict our discussion to the one-dimensional diffusion situation. What has been shown above is that every single one-dimensional diffusion situation must satisfy the one-dimensional diffusion PDE, meaning that at every point and every time after the beginning of the process, the concentration function $C(x, t)$ must satisfy the PDE. This goes well beyond the scope of our original one-dimensional diffusion equation, namely:

$$C = \frac{M}{\sqrt{4\pi Dt}} e^{-\frac{x^2}{4Dt}}.$$

This equation does *not* apply to every single one-dimensional diffusion situation, but only to the particular situation that begins with an initial mass M injected at a particular point, $x = 0$, at the beginning of the experiment. Certainly, one can think of many different diffusion situations that do not begin in this particular way. For example, you could begin a diffusion experiment by injecting various amounts of mass at various points along the diffusion tube. In fact, some examples of such cases have previously been considered. To be even more radical, you could begin a diffusion experiment without the injection of any specified amount of mass at all. For example, you could have a diffusion tube in which the *concentration* is maintained at some specific value in a given portion of the tube, and as it diffuses out through the ends of that portion of the tube, just the right amount of additional mass is brought in to replace it in that segment so as to keep the concentration there at its constant value. Or you could have a diffusion experiment in which the diffusion tube, rather than being our idealized infinitely long version, actually has finite length. The one-dimensional diffusion equation does not apply to this situation, but the one-dimensional diffusion PDE certainly does.

The auxiliary conditions that distinguish one problem from another are generally called *boundary conditions*. Sometimes boundary conditions that apply only at the initial instant of the experiment are more precisely called *initial conditions*. Boundary conditions can take a wide variety of forms, both in terms of their physical description and their corresponding mathematical characterization. For example, a diffusion tube of finite length would be characterized by a boundary condition, among others, that would require that there be no-flow through either end point. Since this is a diffusion process, that would mean that the one-sided the derivative C_x at each endpoint would need to be 0. Hence this would be the mathematical expression of the physical constraint. One of the things that makes the subject of partial differential equations a very complex subject is not the fact that the equations themselves are necessarily complex for, as has been seen, the one-dimensional diffusion PDE is relatively simple, but rather that there can be a very wide range in the kinds of boundary conditions needed to treat real physical problems with these partial differential equations.

But for now let us focus on the original one-dimensional diffusion problem, wherein a hypothetical mass M is initially injected at the point $x = 0$ in an infinitely long diffusion tube. We know that the solution function $C(x, t)$ must satisfy the one-dimensional diffusion PDE for every value of x at every time $t > 0$. But how could you find this solution just starting with this basic information? A great deal can be learned simply from trying on your own to solve this one problem, as suggested in the following pair of exercises.

Exercise 15. State mathematically any and all boundary conditions associated with this basic problem.

Exercise 16. [This is a very valuable but quite challenging problem. Some previous experience in solving differential equations, even if only ordinary differential equations, would be helpful.] It was verified in an earlier exercise that the one-dimensional diffusion equation is indeed a solution to the one-dimensional diffusion PDE. It has also been verified that this equation does satisfy the boundary condition described for our one-dimensional diffusion problem. However, it is one thing to be given a solution to a PDE and be asked simply to verify, by plugging it in, that it is a solution, and quite another thing to derive the solution in the first place. In this exercise you are asked to "forget" that you happen to know the solution to the one-dimensional diffusion problem, and you are asked to begin simply with the one-dimensional diffusion PDE and the corresponding boundary condition and show that that combination leads necessarily to a solution which turns out to be identical to the one-dimensional diffusion equation. (Hint: think about how the solution must depend on the units of the input values.)

For some problems involving more complicated boundary conditions, it is possible to represent the solutions as the sum of solutions of problems characterized by simpler boundary conditions. This has been discussed earlier when the concept of *superposition* of solutions was introduced. For example, if a one-dimensional diffusion problem begins with mass being initially injected at two different points, the molecules originating at one of those points are generally unaffected by the behavior of the molecules being injected at the other point, so in the end, in order to calculate the total concentration at any point, you could calculate the portion of the concentration originating from each of the two sources and then add them. The following problems are amenable to this technique, although there are also other ways to solve them.

Exercise 17. Consider a one-dimensional diffusion situation for a "semi-infinite" tube, meaning one that is infinite to the right but has a closed left end. Suppose that by some ingenious apparatus, we can always maintain the concentration of diffusing material at the constant value C_0 just at the left end point. Furthermore, let us suppose that we begin the experiment with a concentration 0 everywhere else. What would happen is that material would gradually flow from the left end point towards the right through the tube and we would have to keep replenishing it at the left end point to keep the concentration in the immediate "infinitesimal" neighborhood of that point at C_0. So we would constantly be injecting mass, and mass would continue to be moving from left to right throughout the tube. Find the concentration function $C(x,t)$ and the rate at which we would have to be supplying mass to maintain this situation.

Exercise 18. Consider the original infinitely long one-dimensional diffusion tube where the initial mass M, rather than being injected at the point $x = 0$, is initially distributed uniformly over the interval from $x = -1$ to $x = 1$. (That is, the initial concentration in this particular interval of the tube would be $M/2$.) Assuming that the initial concentration is 0 elsewhere, find the concentration function $C(x,t)$ that would describe the concentration function at all times $t > 0$.

Readers who have had some past experience with differential equations have probably encountered considerable emphasis on the uniqueness of solutions to given kinds of problems. Since we have not considered a wide range of equations or boundary conditions, this has not been an important theme for us. Even for physical problems, where the uniqueness of a solution is apparent from the physical situation, and our job is to try to find it, the issue of uniqueness is important. For example, sometimes when one encounters difficulties in a solution process, such as an apparently non-unique solution, that serves to indicate that the mathematical model for the physical situation is not complete. Perhaps some aspect of function behavior or some boundary condition has been omitted from the formulation. While it is beyond our current scope to investigate these issues, there is one very important and interesting concept that must be mentioned.

Diffusion problems satisfy a *maximum principle,* to the rough effect that they always assume their maximum value, if they have one, on the boundary of the region in x, t-space on which they are defined. For example, for the infinite diffusion tube, this is the upper half plane (with the vertical axis being the t-axis), and for a tube closed at one end, it would be a quadrant. This principle is easy to see when you think of a diffusion process as an averaging out or spreading out type of process, one where material always migrates from areas of higher concentrations to lower concentrations. You would hardly expect to start a diffusion experiment with the maximum controlled initial or boundary value being 10 grams per centimeter, and then all of a sudden find that the concentration later on at some point had reached 20! Given this maximum principle, if you had two solutions to a given problem, just look at their difference. It came up in an earlier exercise that the difference would also satisfy the diffusion PDE. Furthermore, the difference would satisfy a 0 initial and boundary condition on every relevant boundary in x, t-space. But then the maximum principle would imply that the maximum value of the difference of the functions would be 0 everywhere, and thus the two solution functions would have to be exactly the same function, showing uniqueness. This logic, which applies equally well to Laplace's equation, is very useful in deriving additional properties of the solutions to such equations.

For readers who have studied the previous chapter on ground-water modeling, considerable attention was given there to Laplace's equation. In closing this chapter on diffusion processes, it might be useful to compare the diffusion PDEs and Laplace's equation, as follows:

	Diffusion	Laplace
1-D	$C_t = DC_{xx}$	$0 = h_{xx}$
2-D	$C_t = D(C_{xx} + C_{yy})$	$0 = h_{xx} + h_{yy}$
3-D	$C_t = D(C_{xx} + C_{yy} + C_{zz})$	$0 = h_{xx} + h_{yy} + h_{zz}$

Looking at the diffusion equations, note that if in any such diffusion situation we were to reach a "steady state," meaning a situation in which the concentration at a point ceased to change in time, this would make the term on the left, C_t, equal to 0, at which point we could divide through by D and we would find that the concentration function actually satisfied Laplace's equation. In other words, one might think of Laplace's equation as the limiting form of a diffusion process that essentially reaches a steady state distribution in the variable of interest (which has been concentration C in the foregoing discussion)!

To turn the situation around, recall that in the chapters on ground water, all the problems we treated were steady-state problems. There was always a steady flow, and there were no auxiliary conditions that would cause it to vary with time. But not all ground-water problems are of this limited type. For example, if you have a well that is sitting idle and then all of a sudden you start to pump water out of it, the water table in the vicinity of the well starts to fall, or, if it is a pressurized confined aquifer, the head value at the well will drop, and therefore the head value at points in the vicinity of the well will start to change, making them functions of both spatial location and time after the onset of pumping. Nor will these changes be instantaneous, for as the head drops, water will actually come out of storage in some of the pore spaces, so it is even more complicated than a basically steady-state system responding instantaneously to new boundary conditions.

We did not model problems like this in the ground-water chapters, but we might well have done so. In fact, building on the analogies discussed earlier between Darcy's law and other diffusion processes, we should expect immediately that the hydraulic head will in general satisfy an equation of the form of the diffusion PDE, and only in the special case of steady-state processes will it reduce to Laplace's equation. This is actually one of two important ways in which the diffusion equation enters into ground-water problems. The other way is perhaps more obvious. Dissolved material in ground water not only moves along with the ground water itself (called advective transport), but it also tends to diffuse out through the ground water in any and all directions that might be available to it. Consideration of this additional process for the movement of dissolved materials in ground water was also beyond the scope of our earlier treatment, but the reader can well imagine at this point how the diffusion equations treated in this chapter could be applied to this other analogous situation.

Exercise 19. Describe a physical ground-water problem that has the following two characteristics: a) the hydraulic head values are a function not only of spatial location but also of time, and b) the hydraulic head profile actually does approach a steady state value for large values of t.

Exercise 20. With reference to the previous exercise, give an example of a real nontrivial physical ground-water situation that satisfies condition a in the exercise but not condition b.

6.5 Guide to Further Information

The key material for this section is related to probability, statistics, and partial differential equations. There are many good references on these topics, and interested readers can do well simply choosing from whatever is readily available to them. With the background gained in this chapter, it may be useful to reread the discussion at the end of Chapter 3, for more of the air pollution literature should now be readily accessible.

7

Additional Topics in Hazardous Materials Modeling

The purpose of this section is to investigate the detailed structure of some of the submodels that are typically used in hazmat modeling packages. A complete investigation of all the submodels for a given package would take us beyond the scope of this book, as it would require the development of extensive background information from several fields, including chemistry, fluid mechanics, combustion processes, meteorology, and others. Our objective is primarily to introduce the reader to the types of reasoning and data typically used by model developers, and for this purpose we have chosen several relatively simple modeling topics that represent a range of both theoretical and empirical lines of development. These pertain to the subject of flammable and toxic vapor hazard modeling, which is the principal area of applications to which the reader was exposed in Chapter 4. Our discussion will follow the sequence represented by the top and right side of Figure 7-1. A footnote at the beginning of each section will give an indication of the type of background recommended for that section.

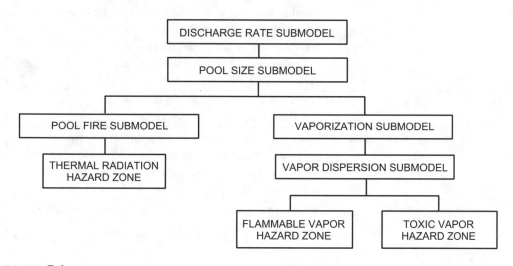

FIGURE 7-1

Typical submodel structure for liquid pool incidents

7.1 Discharge Submodels*

In investigating leaks from ruptured tanks, pipes, or hoses, we have seen that the material discharge rate is of key importance and is one of the first quantities typically estimated by a modeling package. For example, the discharge rate for liquids determines how fast the pool is forming and thus can affect how large it gets and how fast evaporation will take place.

As you would expect, the input values to determine discharge rate will include quantities that describe the size of the opening as well as the forces pushing liquid out through that opening. The discharge rate may, of course, decrease with time, for, as the tank empties, there may be less gravitational (or other) force on the remaining liquid inside. Table 7-1 summarizes some of the typical input information usually requested, although of course this may vary slightly from one modeling package to another.

TABLE 7-1
Typical input information required by most discharge rate submodels

Input information/parameter	Comments
Shape and dimensions of tank	Vertical and horizontal cylinders most common, but may be spheres or other shapes. May need to approximate to closest shape covered by model.
Nature of opening	Flow may be affected by roughness of opening, or whether it is in the tank itself or in a section of pipe.
Dimensions of opening	Model may just request area and use equivalent circular opening, or opening geometry may be requested.
Whether tank is pressurized	For materials with low boiling points, tank will either have to be pressurized or refrigerated to maintain material in liquid form. Overpressure due to vapor pressure of many non-boiling materials or to inert gas blankets usually not significant compared to gravitational forces. Pressurization due to immersion in fire environment may be significant, and could even lead to a BLEVE.
Degree to which tank is filled	May be requested as a per cent, volume or mass of material, or as height of liquid.
Specific gravity of chemical	Relevant to weight, and hence gravitational force, pushing material out of discharge hole (which cancels out in some models) and also to mass discharge rate associated with a given volumetric discharge rate.
Vapor pressure of chemical	Relevant to internal pressure, which can contribute to driving force for discharge.

---

* This section makes use of elementary concepts of physics such as potential and kinetic energy, as well as simple derivatives and integrals from calculus.

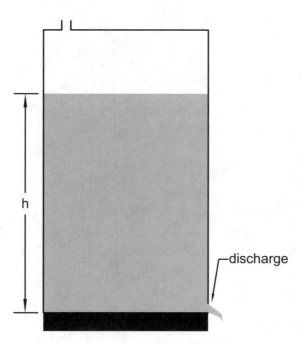

FIGURE 7-2
Simple discharge framework

In this section, we will begin by investigating the discharge of a liquid from a hole at the bottom of a vertical cylindrical tank, where the dominant force is provided by the weight of the liquid in the tank. This situation is shown schematically in Figure 7-2.

This problem was studied in detail at least as far back as 1644 by Torricelli, who showed that the theoretical velocity of discharge is given by:

$$v = \sqrt{2gh}.$$

Here g is the gravitational acceleration constant. This is actually a very interesting result, and you may recognize this expression from your earlier study of velocity problems in physics or mathematics. In particular, we shall investigate this situation in the light of Figure 7-3, which shows a progression of hypothetical situations from left to right.

Case A in Figure 7-3 depicts the case of a mass that begins at rest at a height h above a base level and is allowed to fall under the influence of gravity. The question is what will its velocity be when it finally falls through distance h and reaches the base level. There are a number of common ways to do this, and you are invited to use any with which you are familiar in connection with the first in the following set of exercises.

Exercise 1. For the situation shown in Case A of Figure 7-3, determine the final velocity of the mass as it reaches the base level. Express your answer in terms of the height h and the gravitational constant g. (No time values should show up in your final answer.) Describe the relationship between this answer and the equation of Torricelli given above for a somewhat different situation.

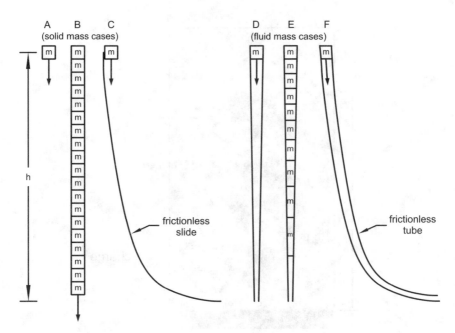

FIGURE 7-3
Progressive cases of solid and fluid behavior under gravity

Exercise 2. Consider the situation shown in Case B of Figure 7-3. Here a whole stack of equal masses are allowed to fall freely, beginning from rest. What will be the velocity of the topmost mass at the time it reaches the base level?

Exercise 3. If you tie a weight onto a piece of rope and swing it around over your head, the weight is constantly accelerating because its velocity is changing. (The velocity is not changing in magnitude, but it is changing in direction.) This is, of course, consistent with Newton's law, $F = ma$, for there is a real force that you can feel your hand and arm exerting as you swing the weight around. Calculate the amount of work you are performing on the weight. You may assume that the weight has mass m, the rope is 3 feet long, and the weight is rotating at one revolution per second.

There are a number of ways in which you might have analyzed the previous exercises, especially the first. Perhaps the most common would be to use Newton's law, $F = ma$, to analyze the motion of the falling mass, leading first to the time of arrival at the base level and from there to the associated velocity at that time. Newton's law is not the only approach to mechanics problems such as this, however, and perhaps you have already seen others. In particular, a very useful tool for analyzing this kind of situation is the *principle of conservation of energy,* which for purposes here and later may be formulated along the following lines, which are probably familiar.

Energy is the capacity of a body to perform work. *Work* is the process of applying a force through some distance. The total energy (or total mechanical energy) of a body subject to the Earth's gravity is the sum of its kinetic energy and its potential energy due to gravity. *Kinetic energy* is the energy of motion, given by

$$KE = \tfrac{1}{2}mv^2.$$

(We are using the word "velocity" here to refer to the speed or the magnitude of the velocity vector.) The *potential energy* is the energy stored in the body by virtue of its location above the base level. It is given by

$$PE = mgh.$$

Here g is the gravitational acceleration constant (32 ft/sec^2 in one set of units), and h is the height above the base level. Thus the total energy E is given by

$$E = \tfrac{1}{2}mv^2 + mgh.$$

The *principle of conservation of energy* says that if this body moves along from one position and velocity condition to another, the *net change in total energy* between the two endpoints does not depend at all on the particular path or intermediate conditions, but only on the conditions at beginning and end; and, in fact, the actual value of this net change in energy *must be equal to the total work done on the body by any forces other than gravity.*

Using this principle, for example, which you yourself might well have used earlier in doing the exercises, we would observe that no work was done on the mass in Case A of the figure, other than that done by gravity, and so there could be no net change in total energy. Therefore we can equate total energy values at top and bottom of the motion and solve for velocity, as follows:

$$E \text{ at top} = E \text{ at bottom}$$

$$\tfrac{1}{2}m(0)^2 + mgh = \tfrac{1}{2}mv^2 + mg \cdot 0$$

$$mgh = \tfrac{1}{2}mv^2$$

$$v = \sqrt{2gh}.$$

Case B of the figure has the same final answer, the key issue being that the masses below the top one will all accelerate at the same rate, and therefore none will exert any force to hold up the previous one. (This latter would happen only if one of the lower ones were slowed down, in which case there would be a "traffic jam," if not a chain reaction of collisions.)

What about Case C, where the mass is deflected to the right by a frictionless slide? Since the slide is frictionless, the only force it can exert on the mass is perpendicular or normal to the slide, and hence normal to the direction of motion. Even though it is a force, it does not move through any distance, and hence it does not do any work on the mass. Therefore the energy conservation principle leads to exactly the same equations as above, and hence to the same final velocity, except that it is pointed off to the right instead of straight down. (See the following two exercises for further investigation of certain subtleties in this line of argument.)

Exercise 4. Clarify the logic in the previous paragraph concerning the issue of whether the slide performs any work on the mass. In particular, formulate a precise mathematical expression for work and use it to explain your argument clearly. Relate this situation to that given in Exercise 3.

Exercise 5. In Case C, the mass is actually rotated as it slides down. Where does this fit into the application of the conservation of energy principle? What reasonable assumptions could you make to avoid this issue?

Now we proceed to the right side of Figure 7-3, where the mass is actually an element of fluid. For the sake of this discussion, we will limit ourselves to liquids, which can be assumed to be incompressible. Cases D and E are analogous to Cases A and B, the only difference being the tapering effect as the fluid falls. You have probably noticed that if you turn on a faucet and the water comes out smoothly, the stream gradually narrows as it falls. This is even more noticeable if you smoothly pour a liquid from a container into another receptacle. The reason for this phenomenon is primarily the fact that if the flow is steady, then there is no such thing as a completely stationary element at the top. For, if we look at the top element in Case D or E, the bottom portion of it will already have started to fall by the time the top part is poured, and so the bottom part will already be starting to travel faster. This difference leads to a gradual "stretching" of the element as it falls; and since the total amount is fixed, the diameter of the stream must decrease at the same time.

The conservation of energy principle, Newton's law of motion, and the other concepts from physics that we have been using do not depend on whether our mass is a solid, liquid, or gas. Therefore, the same analysis presented earlier applies to the calculation of the velocity of the fluid elements as they also reach the base level, and hence the final velocity is again calculated to be

$$v = \sqrt{2gh}.$$

However, the situation covered by Torricelli's equation is still more complicated than that treated by Figure 7-3. The next exercise asks you to try to extend the above arguments to the general situation.

Exercise 6. Torricelli's equation applies to the general situation shown in Figure 7-2. Explain the key physical differences in fluid flow between this situation and that shown in Figure 7-3, Case F. Can you work through these differences, either heuristically or mathematically, to develop a convincing argument of why Torricelli's equation is a plausible or reasonable answer to the original problem?

Exercise 7. What would Torricelli's equation predict as the discharge velocity from a half-inch diameter hole in the side of a vertical cylindrical tank of water, given that the water level in the tank is 25 feet high and the hole in the side is 3 feet high? What would be the effect on discharge velocity of doubling the diameter of the hole?

Exercise 8. How would you expect Torricelli's equation to fit into the determination of actual volumetric or mass discharge rates from holes in tanks?

Exercise 9. Consider two vertical tanks of water, each 40 feet high and each filled to the top. Tank A has a diameter of 80 feet. Tank B has a diameter of 20 feet. Let P_A and P_B represent the respective pressures inside the two tanks right at a point where the sides and bottoms meet. Determine the relation between P_A and P_B.

Just as the conservation of energy principle has been seen to be a useful tool for solving mechanics problems, letting one avoid some of the "nitty-gritty" that might be encountered in applying Newton's law, a version of this principle for fluid motion problems is equally valuable for a wide range of common fluid questions. It also provides a valuable introduction to the methodology of the field of fluid mechanics, and as such provides a good basis for further study of issues that arise in environmental problems in many other areas as well, including groundwater, air pollution, ocean circulation, river flow, and others. Therefore, we shall develop this basic principle, called *Bernoulli's equation,* and then use it to quickly and deftly derive both Torricelli's equation and some important variations that are typically used in hazmat modeling.

To begin, we need certain basic terminology and simplifying assumptions from fluid mechanics, namely:

- Our fluid is an *ideal fluid,* meaning that it is *incompressible* and that it exhibits *no internal friction* or *viscosity.* Thus the density is constant, and layers of fluid can slide by each other without creating any shear effects, even if the velocities of the layers are different.
- Our flow is steady or stationary, meaning that it has achieved a steady pattern such that at any given fixed point in space, the velocity (including both magnitude and direction) of the flow stays the same as time goes on. Of course, the velocity may vary from one location to another, which is a different issue. Closely related to this is the assumption that our flow is laminar rather than turbulent; that is, the flow lines followed by individual "particles" of fluid always maintain the same pattern.

In this context we can talk about *flow lines* or *streamlines,* referring to the paths traced out in space by the particles of fluid. We can also then make reference to a *flow tube,* which is a region in the flow that is completely bounded by streamlines. Figure 7-4 shows a typical portion of a flow tube with a few of the streamlines sketched in. Note that no material can flow through the boundaries of a flow tube, because if there were such a path, it would intersect the boundary of the flow tube in some point P. But since the flow is steady, laminar flow, there is only one path for fluid through P, and that is the streamline through P that forms part of the boundary of the flow tube. Therefore, we can think of a flow tube as a pipe (but with variable cross section) that constrains a fluid element as it moves along.

This idea is represented in Figure 7-5, which shows an element of mass in two different positions as it moves along such a flow tube. We have actually exaggerated the thickness of the flow tube to give room for the labels and symbols, but for this discussion you should think of it as quite narrow. On the basis of this assumption, key physical quantities such as velocity and fluid pressure may be assumed to have the same value throughout any cross section of the flow tube. Thus they would be functions of only the downstream distance, which we shall call s.

Notice how the mass element increases in length when it reaches the narrower portion of the tube. This is due simply to conservation of mass. Since the fluid is incompressible, the only

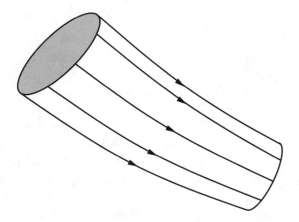

FIGURE 7-4
Portion of a flow tube

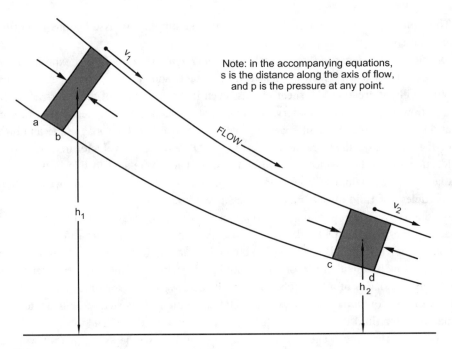

FIGURE 7-5

Side view of flow tube showing movement of a mass element

direction in which it can stretch out to compensate for the smaller cross section is along the axis of flow. In fact, to make this idea quantitative, let us denote by A_1 and A_2 the respective areas (perpendicular to the flow direction) of the two representations of the mass element. Then, since there can be no net gain or loss of total mass in the section of the flow tube between these two locations over any interval of time, we must have:

$$\text{mass in through left} = \text{mass out through right}$$

$$\rho A_1 v_1 \Delta t = \rho A_2 v_2 \Delta t$$

$$A_1 v_1 = A_2 v_2.$$

Here, ρ represents the density of the fluid, and Δt is any arbitrary interval of time, long or short. This is called the *equation of continuity,* and another way to state it would be to say that the product Av must stay constant as the mass element moves along the flow tube. Thus, in terms of Figure 7-5, this confirms our intuition: the smaller cross-sectional area portion to the right is associated with a higher velocity. We will denote this changing value of area by $A(s)$.

But now that it is clear that velocity must change, which means that there is acceleration, there must be some force bringing about that acceleration. The only two forces on our fluid element are those due to gravity and to the pressure exerted on its upstream and downstream faces by the surrounding fluid. The gravitational force is given by mg, where m is the mass of the element, and the pressure force, always in a direction normal to the cross-sectional area, can be denoted by $p(s)$ to emphasize its dependence on location along the flow tube. Recalling the energy conservation principle, we need to consider only this latter force in calculating the work done on the element as it moves from the left position to the right position. In particular,

Total energy at right − total energy at left = work done on element by other than gravity

$$[\tfrac{1}{2}mv_2^2 + mgy_2] - [\tfrac{1}{2}mv_1^2 + mgy_1] = \int_a^c p(s)A(s)\,ds - \int_b^d p(s)A(s)\,ds$$

$$[\tfrac{1}{2}mv_2^2 + mgy_2] - [\tfrac{1}{2}mv_1^2 + mgy_1] = \int_a^b p(s)A(s)\,ds - \int_c^d p(s)A(s)\,ds$$

where in the last step the portion of the integrals between b and c have canceled out. Note also that the work done on the left face of the fluid element as it moves along is taken as positive because this is the direction of positive distance s. Using the subscripts 1 and 2 to denote values of A and p at the left and right positions, we may approximate the integrals as products, thus obtaining:

$$[\tfrac{1}{2}mv_2^2 + mgy_2] - [\tfrac{1}{2}mv_1^2 + mgy_1] \cong p_1 A_1(b-a) - p_2 A_2(d-c)$$

where we have used an approximately equal sign as a reminder. But the quantities $A_1(b-a)$ and $A_2(d-c)$ are both equal to the same thing, namely, the volume of the fluid element, which is the same at both ends! Furthermore, this volume can be written in what may initially look like an awkward form as

$$V = \frac{m}{\rho}$$

but which has the advantage of using variables that have already been introduced. Therefore, our previous equations reduce to:

$$[\tfrac{1}{2}mv_2^2 + mgy_2] - [\tfrac{1}{2}mv_1^2 + mgy_1] \cong p_1 V - p_2 V$$

$$[\tfrac{1}{2}mv_2^2 + mgy_2] - [\tfrac{1}{2}mv_1^2 + mgy_1] \cong p_1 \frac{m}{\rho} - p_2 \frac{m}{\rho}$$

$$[\tfrac{1}{2}mv_2^2 + mgy_2] + p_2 \frac{m}{\rho} \cong [\tfrac{1}{2}mv_1^2 + mgy_1] + p_1 \frac{m}{\rho}.$$

As the last step, we divide through by m, multiply through by ρ, and take the limit as the thickness of the fluid element along the flow axis approaches 0, and thus we have the equality:

$$\tfrac{1}{2}\rho v_2^2 + \rho g y_2 + p_2 = \tfrac{1}{2}\rho v_1^2 + \rho g y_1 + p_1.$$

This is called *Bernoulli's equation,* and was first derived by Daniel Bernoulli in about 1738. Recalling our initial assumption that the flow tube itself was narrow, to be precise we should also say that we are taking the limit as the flow tube closes down on the streamline that runs down its center. Therefore, another useful way to state the principle of Bernoulli's equation is to say that the quantity

$$\tfrac{1}{2}\rho v^2 + \rho g y + p$$

must remain constant along any streamline.

You have to be very careful with units in working with Bernoulli's equation because of the common confusion that exists between mass units and force units, especially in the engineering and English systems of units. Table 7-2 provides a reminder of various consistent systems. However, if you are like many people and really insist on using pounds to represent both mass and force, then none of these four systems will work for you. For example, if you are dealing

TABLE 7-2

Common systems of units for force, mass, and acceleration

System	Force/weight	Mass	Acceleration	Gravitational acceleration, g
mks	newton	kilogram	m/sec^2	9.8 m/sec^2
cgs	dyne	gram	cm/sec^2	980 cm/sec^2
engineering	pound	slug	ft/sec^2	32 ft/sec^2
English	poundal	pound	ft/sec^2	32 ft/sec^2

with the flow of oil, say, and you are given that the density is 55 lbs/ft^3 and, further, that the pressure is expressed in pounds per unit area, then the straightforward application of Bernoulli's equation with the given numerical values would be absolutely wrong! This is so important to understand that you should not go on until you have worked out the following exercises.

Exercise 10.   Consider the situation represented in Figure 7-6, which shows a large-diameter cylinder through which a piston is moving to squeeze oil out through a small nozzle at the right end. The force on the piston is such that it maintains a constant pressure of 60 psi (pounds per square *inch*) within the large cylinder, and the diameter of the cylinder is so much larger than the nozzle that the velocity of the piston (or the fluid in the large cylinder) can be treated as negligible. At the tip of the nozzle, the pressure may be assumed to be atmospheric (14.7 psi). Find the velocity as the oil exits the nozzle. Hint: consider the cylinder/nozzle combination as a flow tube and study the streamline right along the center using Bernoulli's equation. Assume that the oil weighs 50 lb/ft^3. If your answer turns out less than 20 ft/sec, then you have made a mistake somewhere.)

Exercise 11.   Repeat the previous exercise assuming that the axis of the cylinder is now pointing up to the right at an angle of $45°$ and that the cylinder is 3 feet long and the nozzle 3 inches long. (Assume, to be precise, that the 60-psi constant pressure is measured right at the center of the face of the piston.)

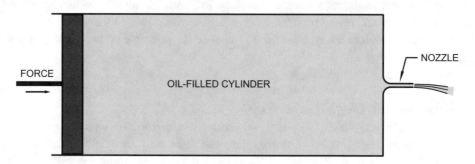

FIGURE 7-6

Piston pushing oil out nozzle at end of cylinder

Exercise 12. Suppose that you are going to be working a lot of problems in the same units as the previous two exercises, or that you are programming a computer model for which you would like to allow the user to input these units. That is, both force and mass will be given in pounds, and pressure will be given in psi. Distance units will still be in feet and time in seconds. Modify Bernoulli's equation so that it applies directly to this situation. (Keep your result handy for future reference.)

The previous exercises should have given you a working knowledge of Bernoulli's equation. Now let us see how easily it leads to Torricelli's equation, which was discussed earlier in this section. The situation is that shown in Figure 7-7, which is the same as an earlier figure except that some streamlines have been drawn in from the top of the liquid level to the discharge point. In other words, we can think of the system as a nonuniform pipe or a flow tube, and then apply Bernoulli's equation along a typical streamline. Therefore, focus your attention on any streamline in the figure. At its top end, the velocity is (essentially) 0, since the area is so much bigger than that at the discharge point. (Continuity equation.) Also at the top, the pressure is atmospheric, p_a, since the tank is connected to the atmosphere by the vent in the upper left. And the height at the top is just h. Now, moving to the bottom, the velocity there is the unknown to be solved for, the pressure is again atmospheric, and the height is 0. Putting all these values into Bernouilli's equation, we obtain:

$$\tfrac{1}{2}\rho 0^2 + \rho g h + p_a = \tfrac{1}{2}\rho v^2 + \rho g 0 + p_a.$$

The p_a terms cancel; then the ρ's can be divided out, and we easily solve for v as:

$$v = \sqrt{2gh}$$

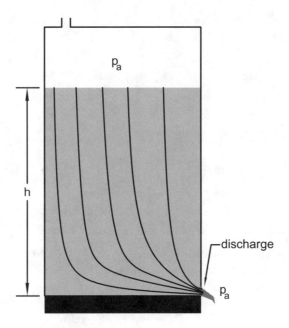

FIGURE 7-7
Tank flow to discharge point

which is Torricelli's equation!

The following exercises ask you to investigate some more general cases of tank discharge and also to look back again on some of the assumptions that we made at the outset of this section.

Exercise 13. Consider the situation shown in Figure 7-7 except for the following modification. The vent on the tank has been sealed, and, instead, the contents are kept under a constant pressure p_b. Find an expression for the discharge velocity. (Note: the contents might be kept under pressure either by the vapor pressure of the material itself, especially if it is a material with a low boiling point, or by an externally supplied pressurized gas source. The latter is generally accomplished with an inert gas like nitrogen so as to keep oxygen out of the space above the liquid and thus minimize fire and explosion potential.)

Exercise 14. Ammonia is a compound that has a low boiling point ($-28°$F), so that at ambient temperatures it will produce vapor pressures well in excess of atmospheric pressure. In particular, at $80°$F its vapor pressure is about 150 psi, or about ten times atmospheric pressure. Suppose you have a vertical tank of ammonia that is 6 feet in diameter and 20 feet high, and that the internal pressure is its vapor pressure at the ambient temperature of $80°$F. Assume that it is half full. If a small gauge fitting at the bottom of the tank breaks off due to undetected corrosion, what would our model predict the velocity of discharge to be? The specific gravity of ammonia is about 0.68.

Exercise 15. Repeat the previous exercise except for the following single change: the tank is now a horizontal tank of the same dimensions. (This would be more common for an

Exercise 16. Torricelli's equation may be thought of as a special case of your solution to Exercise 13, above, when the pressure forces cancel out and do not affect the discharge rate, so that the gravitational force is the driving force for discharge. At the other end of the spectrum is the case when the pressures inside and outside the tank are so different that they far overshadow the effects of gravity. In this case, how would your solution to Exercise 13 simplify?

Exercise 17. Apply your solution to Exercise 16 to the problem given in Exercise 14, and relate the result to your answers to both Exercises 14 and 15.

Exercise 18. A number of assumptions were made in the derivation of Bernoulli's equation. These involved both the physical basis for the problem as well as the mathematical framework for analyzing it. For each of the assumptions listed below, clearly identify how it fits into the derivation:

   a) Incompressibility of the fluid.

   b) Zero viscosity of the fluid.

   c) Small (eventually "infinitesimal") cross section of the flow tube.

   d) Small (eventually "infinitesimal") length of the fluid element.

Exercise 19. [Involves theoretical aspects of calculus.] In the derivation of Bernoulli's equation, one step involved the approximation

$$\int_a^b p(s)A(s)\,ds \cong p_1 A_1 (b - a)$$

as well as another similar approximation. Although it is convenient to speak in terms of such approximations, it is good to know how to test them rigorously by more precise mathematical analysis. Use the mean value theorem for integrals to simplify the expression on the left, and then show how the result would fit into the subsequent part of the derivation of Bernoulli's equation.

Exercise 20.   Give an intuitive explanation of why the density of the liquid does not enter into Torricelli's equation. Wouldn't you expect that a heavier liquid would exert more pressure and hence flow through the discharge opening faster?

Now you have lots of experience in calculating *discharge velocities,* but naturally what we really want is to calculate *discharge rates,* meaning flow rates in terms of volume per unit time (e.g., gallons per minute) or mass per unit time (e.g., pounds per minute). Exercise 8 asked you to think about the connection between these quantities, and the most logical answer would be that if you multiply the discharge velocity by the area of the opening, then you should get the discharge rate. Right? No, wrong! Even though it sounds so logical, it's wrong most of the time. To see this, consider a magnified view of a discharge point, such as in Figure 7-8. Because of the necessarily smooth nature of the streamlines (due to momentum effects in the fluid), the discharge jet does not occupy the entire area of the opening as it moves through. Rather, it generally continues to contract for a short distance. This distance and the extent of contraction vary with the geometry of the opening. For example, they may be minimal for a smoothly curved nozzle type opening, as seen earlier in Figure 7-7, but for a sharp opening, not intended to facilitate discharge, the distance is about half the diameter of the opening and its corresponding area is about two-thirds of the area of the opening itself. The coefficient of contraction C_c is the parameter used to represent this fractional reduction to a smaller effective discharge opening. The point at which contraction ends and the fluid begins to be more subject to external forces is called the *vena contracta.* This is actually the effective discharge point for the purposes of calculation.

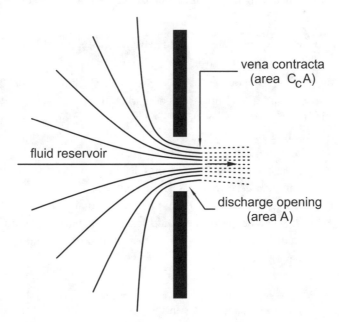

FIGURE 7-8
Magnified view of streamline pattern near discharge point

There is one additional effect that is also occasionally taken into account by discharge models, and this is the fact that there may be a frictional force on the fluid as it passes through a hole in a tank or other non-engineered opening serving as a point of discharge. This would reduce the velocity term itself to some fractional amount, with the fraction generally denoted by C_v, called the *coefficient of velocity*. Although you might think that this would have a significant effect, it rarely does, as the value of C_v is usually greater than 0.95.

Putting these two new factors together with Torricelli's equation, the volumetric discharge rate Q through an opening of area A at a height h below the top of the liquid in a tank at atmospheric pressure would be given by:

$$Q = \text{effective velocity} \times \text{effective cross-sectional area}$$
$$Q = C_v \sqrt{2gh} \times C_c A$$
$$Q = C_d \times \sqrt{2gh} \times A$$

where C_d is called the *coefficient of discharge* and is just the product of the previous two coefficients. For most sharp or rough openings its value is in the vicinity of 0.62, and only for very natural or well designed openings, such as pipe ends or special nozzles, does it approach 1.

This completes our treatment of discharge modeling itself. Naturally, your modeling package may cover additional cases that do not fall within the simplifying assumptions that we made at the outset, but the cases treated here are both important themselves and also serve to introduce the basic fluid mechanics point of view that can also be extended to other cases.

Naturally, given the discharge rate from a tank, even if this changes as the level in the tank decreases, we should be able to calculate how long it will take for the tank to empty, or, alternatively, how much will have leaked out by any given time. This is pursued in the following exercises.

Exercise 21. If you were to calculate the volumetric discharge rate Q, how would you then calculate the mass discharge rate?

Exercise 22. Consider a vertical cylindrical storage tank that is 40 ft high and 20 ft in diameter, vented to the atmosphere, and filled to the 38-foot mark with diesel fuel (specific gravity about 0.85). Assume that a three-inch diameter hole develops at the bottom of the tank wall where a pipe breaks off right at the tank wall.

a) What will be the initial discharge rate Q?

b) Letting F represent the total volume of diesel fuel in the tank at any moment, what is the relation between Q and the derivative $\frac{dF}{dt}$? Be careful.

c) Develop the relationships between F, h, Q, and t so that you can finally express F as a function of t. (Here, h is the height of the liquid level in the tank at any time t.)

d) If the leak continues unabated, how long would it take for the tank to reach the point where it is essentially empty?

e) Assuming the liquid is discharging into some kind of pool, find an expression for the total pool volume as a function of time.

f) Find an expression for the total mass in the pool as a function of time.

Exercise 23. Apply your computerized modeling package to the situation in the previous exercise and compare the result with your calculations in part d, above.

Exercise 24. Suppose that the tank in Exercise 22 is reoriented so that it now is horizontal. Assume that it has the same initial amount of diesel fuel in it as in the earlier situation and that the same size opening occurs at the bottom. How long will it now take to empty?

Exercise 25. Test your result on the previous problem against the corresponding result obtained with your modeling package.

7.2 Pool Size Submodels*

For most hazmat scenarios involving liquid spills, the next modeling step after determining a discharge rate is to estimate the size of the liquid pool. (Refer back, for example, to Figure 7-1 to review how the various submodels typically fit together.) Naturally, if the pool is large, then there is a larger area available for evaporation or boiling, which is likely to increase the hazard from the vapor cloud, which will then reach higher concentrations.

But before investigating the size of such pools, it is important to keep in mind that not all liquid releases actually result in liquid pools. Some liquids that have boiling points well below the ambient temperature are likely to "flash" immediately to the vapor state when they are released from pressurized tanks. Flashing is essentially extremely rapid boiling, and this process can be sufficiently agitated that even though not all the liquid turns immediately into vapor, the liquid part itself is "exploded" into tiny droplets, called aerosols, that can move along with the air, eventually either "raining out" on the ground or subsequently changing to actual vapor themselves.

These low-boiling-point materials are very important in our economy and also often extremely hazardous, and much of the early work on hazard modeling was motivated by their handling and transportation in large quantities. (Obviously, transportation in liquid form uses much less space.) Common examples include liquefied natural gas (LNG), propane, ammonia, chlorine, ethylene oxide, and many more. While small leaks are likely to completely flash upon release, larger discharges will generally reach the ground and begin to form a pool.

Assuming the liquid will indeed form a pool, we clearly need to estimate how large that pool will be. For the case of tank leaks within fixed facilities (i.e., plants that use or manufacture the chemicals), there is almost always some form of "secondary containment," at least around larger tanks. Most common is the use of dikes to surround the chemical storage tanks, as was discussed earlier in Chapter 4. For multiple tank storage, there can even be a main dike around an entire set of tanks, as well as "subdikes" to contain smaller spills to the vicinity of their individual tanks. See Figure 7-9 for a sample layout. This figure also calls attention to other possible spill sources in the area, such as broken pipelines and leaking pumps, valves, or connections.

Dikes or other engineered barriers (e.g., curbs, ditches, controlled drainage systems) provide a well defined pool area, except for relatively small spills that would not even generate pools as large as the diked area. (One would not be modeling these latter in most cases.) Some modeling packages might use the full dimensions of the diked area, whereas others might subtract out

---

* This short section does not use any calculus-based methods.

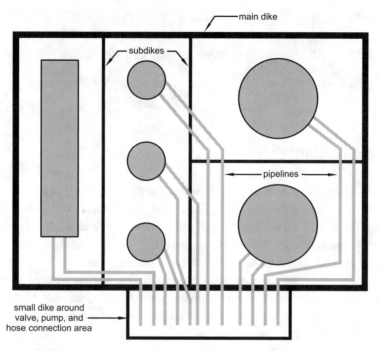

FIGURE 7-9
Tank storage area showing dikes and pipelines

the footprint of the tank or tanks within that area. The difference is not significant considering the far greater levels of uncertainly built into other stages of the model calculations anyway.

The more difficult case is when the spill occurs outside a diked area, such as along a highway or rail line as a result of a transportation accident. In such cases there may still be some kind of physical or topographic structure that might serve to contain the spill in a well defined location. If this is the case, then one simply needs to estimate its area and use that in the model. However, if no such natural containment structure exists, then one has to estimate how large the pool will become as it spreads. The basic physical processes involved are shown in Figure 7-10, and the size of the pool really depends on how these processes "balance out." For example, a high rate of seepage into the ground or a relatively low spread rate (controlled by the chemical's interaction with the ground and by its viscosity) would tend to inhibit the rapid development of a large pool. The difficulty in modeling these processes is that they are so highly dependent on specific conditions and on the chemical involved that the data input requirements are very demanding and the results can vary widely, depending on the assumptions made. Since the models are most often used for planning purposes, and since these kinds of accidents are likely to be transportation accidents, where the terrain could be almost anything, many of the models that have been used or that are incorporated into the modeling packages are based on relatively simple and general assumptions, usually intended to yield approximate answers and ones that, if anything, will tend to err on the conservative side.

Unlike the fluid mechanics calculations used in the previous section to model discharge rates, these models tend to be based on empirical evidence or simple assumptions. Three of these simplified approaches that have been used by investigators in the past include:

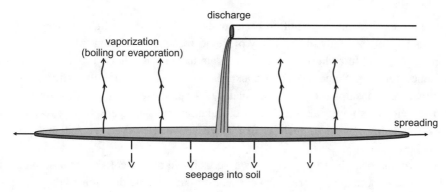

FIGURE 7-10

Processes affecting pool size in absence of physical constraints

- equilibration between discharge and vaporization rates;
- extrapolation from empirical data on spill sizes;
- assuming a fixed pool depth.

All three of these approaches generally ignore seepage into the ground, although for certain site-specific cases, there have been efforts to take this additional process into account. Following the more common practice, we will also ignore seepage in what follows.

The first approach is based on the following logic. If the discharge rate is greater than the vaporization rate at any given moment, then the pool should keep expanding. On the other hand, if the vaporization rate is larger than the discharge rate, then the pool is likely to be shrinking. Assuming a circular pool geometry with diameter d and a constant depth throughout its extent, it should be possible to find where these two processes balance each other.

In particular, let Q represent the discharge rate in units of volume per time. It may in principle be a function of time, but let us assume we can represent it by a constant value. It is common in modeling packages to use the average discharge rate in this way. Furthermore, let us denote by V the unit vaporization rate. Since we might expect the total amount of vaporization to be proportional to the total surface area, as discussed earlier, V will stand for the rate of vaporization in terms of volume units per time, per unit of surface area. So, for example, this might be cubic feet per minute per square foot, which actually reduces to the simpler units of feet per minute, or, more generally, length per time. Thus we must have:

$$\text{vaporization} = \text{discharge}$$

$$V \times \pi \left(\frac{d}{2}\right)^2 = Q$$

$$d = \sqrt{\frac{4Q}{\pi V}}.$$

This last equation would let us calculate the pool size as long as we had an independent way of estimating the vaporization rate. This latter aspect will be left to the next section because it is a subject in itself. In any case, this would be one way to calculate pool size.

The second approach listed above for calculating pool size involves the use of empirical data. As you can well imagine, one would not be encouraged to conduct a large number of experiments that would involve spilling large quantities of hazardous materials on the ground.

But some such experiments have indeed been conducted from time to time in isolated locations, generally as part of the study of the safety of liquid rocket fuels during the Cold War or early in the space program. In addition, some further data have been collected by investigators based on unplanned releases. Some modeling packages do indeed offer the user the opportunity to use empirical relationships (i.e., size vs. quantity curves) based on these data for their model calculations, either as the primary mode of calculation or to compare against results from an alternative modeling approach, such as that given above. However, we will not go into the details of these data.

The third approach listed previously is a very simple one. It consists of assuming that the depth (or thickness) of the pool remains constant throughout the duration of the spill process. For a viscous material like crude oil, one might assume a thickness of a foot or more, depending on temperature. For an inviscous or "thin" material like acetone, one might assume a thickness of some fraction of an inch. The underlying assumption is that the viscosity of the material and the nature of its interaction with the surface is such that there is only some equilibrium thickness that can be supported, and anything more added to the pool just causes it to spread out further at the edges. This is also a reasonable approach, and the thickness to use can be estimated using empirical data or fluid mechanics equations. However, investigating this approach in detail would take us too far afield, and so we shall not pursue it further.

In summary, three reasonable approaches have been identified for calculating pool sizes, and the first has been carried to the point where the reader could actually implement it by hand if only the vaporization rate were known. This will be the subject of the next section.

Exercise 1. Discuss any possible limitations you can identify in the logic underlying the first modeling approach given in this section for calculating pool size.

Exercise 2. It was shown in the text that the units of V are length per time. Give a direct physical interpretation of the V value based on these units, explaining your reasoning.

Exercise 3. The text referred to the assumption that the total vaporization rate should be proportional to the area of the pool. Explain why this is a reasonable assumption, but then also describe physical factors that might cause it to be only an approximation.

Exercise 4. Based on the discussion in this section, identify a feedback loop that is inherent in Figure 7-1 but that is not explicitly identified there.

7.3 Evaporation/Vaporization Submodels*

In this section we focus on the problem of determining how fast a pool of boiling or volatile material will be converted from liquid to vapor form. If its temperature is higher than its boiling point, then boiling will actually take place. On the other hand, if its temperature is below its boiling point, then evaporation will take place. These two cases actually turn out to be quite different in terms of the processes and parameters that control the vaporization.

---

* Much of this discussion in independent of calculus, but as there is some variation, footnotes concerning required background will be included at the beginning of each individual subsection.

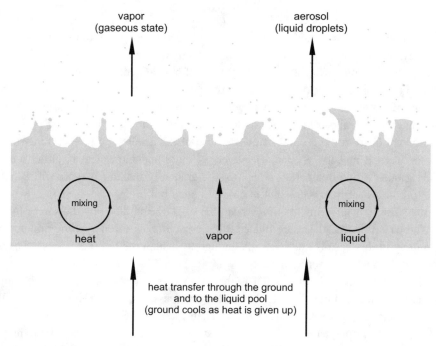

FIGURE 7-11
Portion of turbulent boiling pool showing physical processes involved

First, we consider the case when the pool actually boils. That is, some chemical with a low boiling point has spilled in sufficient quantity onto the ground so that a pool has formed. A portion of such a pool is shown in Figure 7-11, illustrating the various processes at work. It is not much different from boiling water on a stove; it's just that for this chemical the normal ground "feels" as hot as a stove and is enough to make it boil! The liquid level itself is rather turbulent, as vapor bubbles form and rise to the top, where they break through the surface. This turbulent action at the surface not only releases vapors, but also small droplets of the chemical, some of which are small and buoyant enough to rise into the air. Some will rain back down on or near the pool, but others will subsequently vaporize themselves and add to the mass in the vapor form. The turbulence also serves as a mechanism for mixing and hence distributing the applied heat throughout.

To understand what controls this process, recall that whenever a liquid is converted into vapor, the process requires a certain amount of heat energy per unit mass, called the *heat of vaporization*. This is true whether the vaporization takes place through boiling or through simple evaporation. While there is a slight temperature dependence, for our purposes we can treat it as a constant, denoted H. In simplified terms, a liquid with a low boiling point really "wants" to move into the vapor state, and as soon as you give it the energy required by the heat of vaporization, it makes the transformation. The boiling process is a very rapid process of vaporization, and so it requires a lot of heat to keep it going. When the liquid first lands on the ground, it primarily uses its own internal stored energy to make this process go forward. (Using its own internal energy basically involves lowering the temperature of the remaining liquid, hence freeing up energy, to supply energy for some its molecules to move up into the vapor phase.) This process is known as *flashing* or *flash boiling*. But since the heat of vaporization is considerable, even for materials that boil at low temperatures, the pool uses up its own heat supply rapidly, and its temperature drops from its original storage temperature (often ambient temperature) down to approximately its boiling point. (For example, you may recall from physics that it takes 540 calories to vaporize one gram of water, but only one calorie is made available when a gram of water drops $1°C$. So a large amount of liquid temperature drop is needed to provide the energy for vaporization.) As its temperature falls toward the boiling point, it starts to pull more heat in from the ground below because of the developing temperature gradient or difference between it and the ground. However, as more and more heat is pulled out of the ground, the ground's own temperature also drops, gradually making it more difficult for the pool to get the heat energy it needs if it is going to keep boiling.

You may also be wondering about the heat that might be supplied by the sun or by contact with the warmer air. While these are indeed additional sources of energy, the amount of heat that they supply in this case is not generally enough to sustain the boiling process. The ground is a much more productive source of heat.

The conclusion from this discussion is that the vaporization process for such a material goes through two stages. First there is the flashing process, which is primarily controlled by the availability of heat within the liquid itself. Then there is the more usual boiling process, where heat transfer through the ground is almost always the single most important process controlling the rate of vaporization. Therefore, a vaporization submodel for this process would generally be based on individual models for each of these stages. These two processes, as well as evaporation from liquids that are below their boiling points, will be treated in the three subsequent subsections.

7.3.1 Flash Boiling Submodels*

The simplest models for flashing generally calculate the excess heat energy available in the liquid, that is, the stored heat represented by the mass's being a certain number of degrees above the boiling point. This corresponds to the concept of *heat content,* which is somewhat analogous to the concept of potential energy that we have used earlier. In particular, it is simply the heat energy that would be released if the mass were to be lowered in temperature to some baseline level. We can measure the heat content with respect to any baseline that is appropriate to our purposes. In this case, we will measure it with respect to the boiling point of the liquid because the only energy available for flashing is from material that is at a temperature higher than the boiling point. In later cases, where we will be tracking only the *change* in heat content, we use the $0°$ value on whatever temperature system we are working in. So, for example, if we are using Celsius temperature units, then our assumed baseline energy level is that corresponding to $0°C$, and similarly for Fahrenheit or even absolute or Kelvin units.

The heat content is an amount of energy, and it can be calculated by the following relationship:

$$\text{heat content} = \text{mass} \times \text{specific heat} \times \text{degrees above baseline}$$

---

* The discussion here is based on concepts from elementary physics, which are introduced at a basic level. There is no use of calculus in this section. Work with a computerized modeling package is included in some of the exercises.

The specific heat, generally written c, is an experimentally determined quantity whose units would be energy per mass per degree. For example, the specific heat of water is approximately 1 Btu per pound of water per degree Fahrenheit, meaning that 1 Btu of energy is needed to raise the temperature of 1 pound of water by 1°F. In other units, it is 1 calorie per gram of water per degree Celsius. To work an example with the latter units, if 500 grams of water were superheated (in a pressurized container) to 120°C, above the normal boiling point, and then released to atmospheric pressure, then the energy available for immediate flashing would be:

$$\text{available heat energy} = \text{heat content above baseline}$$

$$= 500 \text{ g} \times 1 \text{ cal/g/°C} \times 20°C$$

$$= 10,000 \text{ cal}$$

(The specific heat value for water used in this calculation is actually the average heat energy necessary to raise one gram of water by one degree Celsius within the range from 0 to 100 degrees.) How much of the original mass of water would be vaporized by this release of heat? It would be:

$$\text{mass boiled} = \frac{\text{total energy available}}{\text{heat of vaporization}} = \frac{10,000 \text{ cal}}{540 \text{ cal/g}} = 18.5 \text{ g}$$

This is only about 4% of the total mass (500g) we started with.

Now to repeat this same logic more generally, let us suppose that we have a mass m of liquid at a temperature u somewhat higher than the boiling point u_b. Its specific heat is given by c and its heat of vaporization by H. The energy given off by a temperature fall from u to u_b is enough to vaporize the following amount of mass:

$$\text{mass vaporized} = \frac{\text{energy available}}{\text{heat of vaporization}}$$

$$= \frac{mc(u - u_b)}{H}.$$

It is even more common to talk about the fraction that could be vaporized by the flash process, and this would then be:

$$\text{flash fraction} = \frac{c(u - u_b)}{H}.$$

Obviously, therefore, one can calculate the flash fraction without knowing the total mass involved.

Exercise 1. Use the above approach to calculate the flash fraction for a spill of liquid propane onto the ground at an ambient temperature of 25°C. (Data for propane are: boiling point at atmospheric pressure = −40°C; average specific heat over the temperature range from ambient down to the boiling point = 0.59 cal/g; and heat of vaporization at the boiling point = 102.5 cal/g.)

The calculations shown above have a sound theoretical basis, but they have the unfortunate characteristic that the results are often not borne out by experiment! This raises the interesting question frequently faced by modelers as to whether it is better to use theoretical or empirical models. In fact, we are not talking about small differences in this case. Experimental results show that even for predicted flash fractions of around 20%, sometimes, under certain kinds of conditions, the actual fraction is in the range of 70% to 100%.

Sometimes in the past we have ignored issues like this in the name of conservatism. That is, if leaving out the further complication would show only an increased risk, one might try the calculation and see if the result still lies in an acceptable range. If so, there is no need to refine the model. Unfortunately, since the flash fraction is the amount of mass immediately injected into the vapor space, from where it can be transported with the air, to underestimate this fraction would be to underestimate the risk, which is nonconservative.

A principal reason for the model's shortcoming has to do with the formation of aerosols, mentioned earlier. The flash process is so violent and turbulent that a significant amount of liquid in the form of small droplets can be entrained in the escaping vapors. Once carried into the vapor space, these can also boil or evaporate (depending on temperature) and move along with the vapor cloud, although some may also rain back down out of the cloud almost immediately and reenter the pool or land nearby on the ground.

Several approaches exist to deal with this issue:

1. Skip the pool formation and flash calculations and assume that the liquid flashes instantaneously as it is discharged. This would be the most conservative approach. It is probably most reasonable for relatively small discharges.

2. Modify the flash fraction, as calculated above, by some multiplier intended to include the "entrainment fraction." For example, an entrainment fraction assumed equal to the flash fraction has been used by some modelers. Some models may give the user the option of inputting such a fraction, based on experimental data for situations near the conditions for which the calculations are being carried out.

3. Include detailed physical models for the various processes involved in entrainment, such as droplet size distribution, buoyancy and frictional forces, fall rate, evaporation, and others. The complexity of these models is beyond the scope of this book, but if millions of dollars of investment and the safety of persons and property are at stake from some large proposed project involving such materials, it is not unusual to invest appropriately in the development of highly detailed models that are tailored to the specific situation.

In fact, some modeling packages provide several different options for the user to choose from, depending on the objectives of the analysis and the availability of data.

Exercise 2. Read the documentation for your modeling package and summarize the approach taken to deal with the flashing phenomenon.

Exercise 3. When aerosols evaporate or boil, where does the required heat of vaporization come from?

Exercise 4. Discuss the temperature conditions you would expect to find in the air just over a pool of liquid boiling rapidly at ambient temperature. Would this have any implications for the spread of the vapors?

Exercise 5. Describe a physical storage situation involving a low-boiling-point liquid in which a catastrophic tank failure and spill to the ground would not be expected to yield a significant flash fraction.

7.3.2 Normal Boiling Submodels*

As discussed earlier, the "normal" boiling process continues even after all the energy available for flash boiling has been used up. Now the boiling rate is limited by the flow of heat through the ground and into the pool, and such heat flow naturally depends on the ground's characteristics. This process is actually very similar in mathematical terms to the diffusion equation, which was treated in Chapters 3 and 6.

In particular, heat flow (by conduction, which is the case here) is governed by *Fourier's law*, which says that the heat flux at any point is proportional to the temperature gradient. This should sound very similar to other principles used in earlier chapters, such as Darcy's law for groundwater movement and Fick's law for diffusion. (In fact, see Table 5-2 in Chapter 5 for a comparison.) Our case is essentially one-dimensional, as the heat is generally moving straight up through the ground under the pool. The only place that there would be some sideways movement of heat would be out near the edges of the pool, where some additional heat could be pulled in from outside the radius of the pool; but hazmat modelers would generally choose to ignore this minor factor in the interests of a simpler model. Therefore, even though we are working in three-dimensional space, we can use a one-dimensional analog to analyze the flow of heat. This is just like what we did in Chapter 3 with mass diffusion, and you may wish to review Figure 3-5 there and its discussion before proceeding further.

Letting the variable q stand for the heat flux, u for the temperature, and K for the constant of proportionality, Fourier's law may be stated mathematically as:

$$q = -K\frac{\partial u}{\partial x}.$$

Here we should keep in mind that both q and u are functions of both position and time, and thus they can be written $q(x,t)$ and $u(x,t)$ when useful for emphasis or clarification. The constant K is called the *thermal conductivity,* and, of course, it looks just like the hydraulic conductivity from Darcy's law and the diffusion coefficient from Fick's law or the diffusion principle. Based on this governing equation and other fundamental properties from elementary physics, we want to derive a version of the one-dimensional diffusion partial differential equation applicable to this problem. We will then solve this equation (under somewhat different boundary conditions from the mass diffusion case) in order to determine the temperature distribution and heat flux. Note that we use the same term, "diffusion," for this problem as we did in Chapter 3 for diffusion of air pollutants. The only difference is that in this case *heat energy* is diffusing through a *solid,* whereas in the earlier case, *matter* was diffusing through a *gas.*

To preserve this analogy, consider Figure 7-12, which is almost identical to an earlier mass diffusion figure, Figure 6-3, except for minor changes to make it apply to the heat flow situation instead. A long thin solid rod is the analog in heat conduction to our earlier one-dimensional diffusion tube. The two differences here are that there is no concentrated source (spray can in the diffusion case) and the temperature function $u(x,t)$ has replaced the concentration function $C(x,t)$. The basic conservation principle here is the following:

net flow rate of heat into this section = time rate of change of heat stored in section.

---

* This section uses partial derivatives and other calculus concepts. It will be best understood if the reader has studied Chapter 6, especially Section 6.4, which treated the one-dimensional diffusion partial differential equation. For readers with the appropriate background, there is some mention of linear algebra.

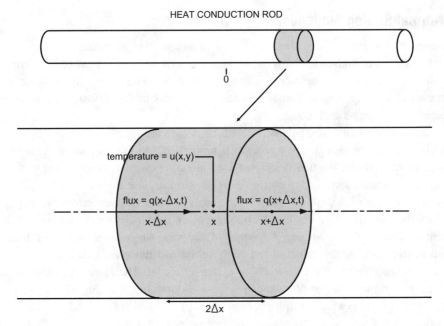

FIGURE 7-12
Selected short portion of a one-dimensional heat conduction rod

The quantity on the left is just the heat flux in minus the heat flux out, or:

$$\text{net flow rate of heat into this section} = q(x - \Delta x, t) - q(x + \Delta x, t).$$

The quantity of heat stored in the section is, as earlier, the product:

$$\text{heat content of section} = \text{mass of section} \times \text{specific heat of material} \times \text{temperature}.$$

As earlier, this heat content is measured with respect to some baseline, and for purposes here that baseline will be simply the $0°$ level on our temperature scale. Furthermore, the mass of that section of the rod should be simply its density ρ (measured in this one-dimensional case as mass per length) times the length of the section, which is $2\Delta x$.

Now we can combine the above expressions and then take the limit as Δx approaches 0, as follows:

$$q(x - \Delta x, t) - q(x + \Delta x, t) = \frac{\partial}{\partial t}[(\rho \cdot 2\Delta x) \times c \times u(x, t)]$$

$$\frac{q(x - \Delta x, t) - q(x + \Delta x, t)}{2\Delta x} = \rho c \frac{\partial}{\partial t} u(x, t)$$

$$-\frac{\partial q}{\partial x} = \rho c \frac{\partial u}{\partial t}.$$

This is very similar to the diffusion situation except that the factor ρc had no previous analog there. Now recalling that we have an expression for q from Fourier's law,

$$q = -K \frac{\partial u}{\partial x},$$

we can differentiate this with respect to x to obtain

$$K\frac{\partial^2 u}{\partial x^2} = \rho c \frac{\partial u}{\partial t}$$

and thus

$$\left(\frac{K}{\rho c}\right)\frac{\partial^2 u}{\partial x^2} = \frac{\partial u}{\partial t}.$$

The quantity in parentheses in this last equation is called the *thermal diffusivity,* and it is denoted by the Greek letter, κ. Thus, for reference,

$$\kappa = \frac{K}{\rho c}.$$

This gives the final form of our partial differential equation (PDE):

$$\kappa \frac{\partial^2 u}{\partial x^2} = \frac{\partial u}{\partial t},$$

which is exactly analogous to the diffusion PDE treated in the last chapter (Section 6.4), with the diffusivity constant κ taking the place of the earlier diffusion coefficient D.

We are now at the point where we know that that the heat flow and temperature distribution are always controlled by this PDE, whatever the starting conditions or other constraints on the system. So the question now is what specific starting conditions or constraints exist on the heat flow problem for heat moving up through the ground to a pool of boiling liquid. For convenience, we will consider the spatial variable x to be distance measured down into the ground from the surface. There are two constraints on this system.

1. The temperature in the boiling pool remains constant at the boiling point of the liquid throughout the process. (This ignores the brief initial period during which the pool temperature falls from ambient.) Letting this temperature be denoted u_b, this constraint has the mathematical form:

$$u(0,t) = u_b \qquad \text{for all times } t.$$

2. The initial temperature in the ground is constant for all depths, and has the value u_g. Since we are dealing with only that part of the underground that can yield heat to this process, namely, a few meters or tens of meters, this is a very reasonable assumption. This is represented mathematically by the equation:

$$u(x,0) = u_g \qquad \text{for all } x > 0.$$

Condition 1 would be called a boundary condition, and condition 2 would be called an initial condition, although together they are generally referred to as *boundary conditions* for the PDE.

You may remember from the previous chapter that the solution to the diffusion equation is not straightforward. In fact, at first you were simply given the solution and asked to plug it in and verify that it worked, and that was a lot of work in itself! A later and very difficult exercise did ask you to try to develop a solution from first principles.

Therefore, because of the very close relationship that exists between this problem and our earlier problem involving mass diffusion, we want to build on the earlier work to solve

TABLE 7-3

Comparison of mass diffusion and heat flow equations

	Mass diffusion	Heat flow
Partial differential equation	$D\dfrac{\partial^2 u}{\partial x^2} = \dfrac{\partial u}{\partial t}$	$\kappa\dfrac{\partial^2 u}{\partial x^2} = \dfrac{\partial u}{\partial t}$
Boundary conditions	1. Mass injection M at $x = 0, t = 0$ 2. $C(x, 0) = 0$ for $x \neq 0$	1. $u(0, t) = u_b$ for all t 2. $u(x, 0) = u_g$ for all $x > 0$
Interval of x values	$-\infty < x < +\infty$	$0 \leq x < +\infty$
Flux equation	$q = -D\dfrac{\partial u}{\partial x}$	$q = -K\dfrac{\partial u}{\partial x}$
Notes	Same constant in PDE and in flux equation	$\kappa = \dfrac{K}{\rho c}$

the new problem. To begin, Table 7-3 gives a side-by-side comparison of the two problems. Furthermore, recall that the solution to the mass diffusion problem was given by:

$$C = \frac{M}{\sqrt{4\pi D t}} e^{-\frac{x^2}{4Dt}}.$$

Therefore, by analogy and taking the simple case $M = 1$, you already know one solution to the heat flow equation, namely:

$$u = \frac{1}{\sqrt{4\pi\kappa t}} e^{-\frac{x^2}{4\kappa t}}.$$

Here is our surprising result: *the integral of this solution with respect to x, with minor modifications, turns out to be the solution to our basic heat flow problem, including both the PDE and the correct boundary conditions!*

This no doubt sounds strange, so let's define exactly what we mean and verify that it is true. (Issues related to the correct dimensions for this and subsequent equations will be investigated further in the exercises.) To begin, given the function $u(x, t)$ defined above, let us define a new function $U(x, t)$ by taking, for each fixed t value, $t > 0$,

$$U(x, t) = \int_0^x u(x, t)\, dx.$$

This means that for each individual fixed moment in time t, we look at u as a function of one variable x, and we compute the area under its graph all the way from 0 to the x value for which we are defining U. The variable x inside the integral is really a "dummy variable," and we could just as well have written

$$U(x, t) = \int_0^x u(s, t)\, ds$$

to emphasize this fact (in fact we will continue to use this second system of notation).

The above definition does not apply when $t = 0$ (since the formula for $u(x, t)$ is undefined at such points), so we will define U at these boundary points by the limiting condition

$$U(x, 0) = \lim_{t \to 0} \int_0^x u(x, t)\, dx$$

which is shown in one of the exercises to have an interesting nonzero value that does not even depend on t! The reason we use a limit for this definition is that we expect our final temperature function to be continuous even at $t = 0$, except at the point where both x and t are 0, where we do not even try to define U because its initial values and boundary values do not approach the same limit at this point.

You know from elementary calculus how to differentiate U with respect to x, and thus we can write its first and second derivatives as:

$$U_x = u(x, t)$$

$$U_{xx} = u_x(x, t)$$

where we are now shifting to the subscript notation for partial derivatives because it will make our computations simpler. We can also take the partial derivative of U with respect to t, as follows:

$$U_t = \frac{\partial}{\partial t} \int_0^x u(x, t)\, dx = \int_0^x \frac{\partial}{\partial t} u(x, t)\, dx = \int_0^x u_t(x, t)\, dx.$$

Now, using the actual equation for $u(x, t)$ given above, you should be able to use its x and t derivatives to verify that $U(x, t)$ does indeed satisfy the appropriate PDE for all points with $t > 0$, as is pursued in the exercises.

Exercise 1. Verify that $U(x, t)$ satisfies the heat flow PDE for all points with $t > 0$. Furthermore, determine the correct dimensions for κ, u, and U, as used in the calculations above.

Exercise 2. Determine the boundary conditions satisfied by $U(x, t)$ at $x = 0$ for all $t > 0$ and at $t = 0$ for all $x > 0$.

Exercise 3. Based on your results in the previous two exercises, find a function $U^*(x, t)$ that completely solves the heat flow problem summarized in Table 7-3.

Exercise 4. Discuss the validity of the step in the equations in the text where the partial derivative with respect to t was moved inside the integral sign. (You may wish to consult a calculus book as a reference for this.)

Exercise 5. If $C(x, t)$ is the solution to the standard mass diffusion problem (with mass M initially injected at the origin), give a physical interpretation of the integral

$$m(x, t) = \int_0^x C(s, t)\, ds.$$

Using this interpretation together with the diffusion principle, show that m must satisfy the one-dimensional diffusion PDE. (Hint: this is quite simple.)

Exercise 6. Give two distinct physical interpretations of the function $U(x, t)$ encountered in this section, or of some constant multiple of $U(x, t)$.

Exercise 7. For this problem, you are given that u, u_1, and u_2 are arbitrary solutions to the heat flow PDE. They do not necessarily have the form of u above, nor do they necessarily satisfy the boundary conditions for our problem. Furthermore, a and b are constants. Determine which of the following related expressions are also necessarily solutions to the heat flow PDE:

a) au

b) $u + b$

c) $au_1 + bu_2$

d) $u + bx$

e) u_x

f) u_t

g) u_{xxt}

h) $\int_0^x u(s,t)\,ds$

i) $\int_0^1 u(x,t)\,dx$

j) $\int_{-x}^x u(s,t)\,ds$

Figure 7-13 may cast further light on this interesting way to develop U. For this part of the discussion, we may assume that $u(x,t)$ represents any temperature distribution in the rod, where the temperature scale itself is also arbitrary. (So, for example, $u = 0$ could represent any actual "baseline" temperature.) This figure looks something like the previous figure, but in this case we are looking at the entire section of the rod from 0 to x, rather than a small incremental portion. The energy conservation principle must, of course, still apply to this portion. Since we are using x to refer to the right endpoint, let us use s to represent the general horizontal distance. The mass of any small subinterval would be just the linear density times the length, or $\rho \Delta s$. Thus the "excess heat content" (i.e., above the 0 baseline on our arbitrary temperature

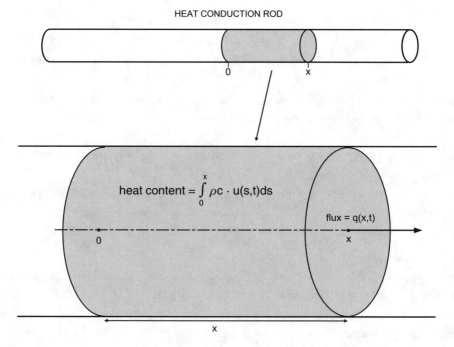

FIGURE 7-13

Framework for applying the energy conservation principle to a finite length of a rod scale) of that short length would be this mass times the specific heat (c) times the temperature, or:

$$\text{excess heat content of short length} = \rho \Delta s \cdot c \cdot u(s, t).$$

Therefore the total excess heat content of the entire length from 0 to x is just the limit of the sum of such terms, which, with slight rearrangement, is the definite integral:

$$\text{total excess heat content} = \int_0^x \rho c \cdot u(s, t) \, ds.$$

The only way this heat content can change with time is by the flow of heat through one of the ends. Thus its rate of change with respect to time is just the difference in the flux values at the two ends:

$$\frac{\partial}{\partial t} \int_0^x \rho c \cdot u(s, t) \, ds = q(0, t) - q(x, t).$$

Now using Fourier's law for the flux at x and simplifying, we obtain:

$$\frac{\partial}{\partial t} \int_0^x \rho c \cdot u(s, t) \, ds = q(0, t) + K u_x(x, t)$$

$$\rho c \frac{\partial}{\partial t} \int_0^x u(s, t) \, ds = K u_x(x, t) + q(0, t)$$

$$\rho c U_t = K U_{xx}(x, t) + q(0, t)$$

$$U_t = \frac{K}{\rho c} U_{xx}(x, t) + \frac{1}{\rho c} q(0, t)$$

$$U_t = \kappa U_{xx}(x, t) + \frac{1}{\rho c} q(0, t).$$

This is practically the original heat flow equation, except for the second term on the right side. Since this term is a constant times the heat flux at $x = 0$, it simply tells us that our original integral function $U(x, t)$, defined by

$$U(x, t) = \int_0^x u(s, t) \, ds$$

for any given temperature function u, will also satisfy the heat flow PDE *if and only if* the heat flux at $x = 0$, namely, $q(0, t)$, is always 0! This bears some further thought, as in the following exercises.

Exercise 8. For our original problem involving a pool of boiling liquid, is this last condition of zero heat flux at $x = 0$ satisfied?

Exercise 9. For the function u defined by

$$u = \frac{1}{\sqrt{4\pi \kappa t}} e^{-\frac{x^2}{4\kappa t}},$$

is the zero heat flux condition met at $x = 0$? (Hint: this should be very easy to see.) Relate your result here to your solution to Exercise 1.

Exercise 10. Other than for the situations investigated in the previous exercises, describe a physical situation of one-dimensional heat flow in which there would be a zero-flux boundary.

The discussion above provides an alternative confirmation of the fact that the function $U(x, t)$ defined by

$$U(x, t) = \frac{1}{\sqrt{4\pi\kappa t}} \int_0^x e^{-\frac{x^2}{4\kappa t}} ds$$

is indeed a solution to the same PDE as $u(x, t)$. It would be nice if this function were actually the solution to the entire problem of PDE plus boundary conditions given in Table 7-3, but it is not quite that good. However, it is close enough for some final adjustments. In particular, the boundary conditions for our original problems apply to $U(0, t)$ and $U(x, 0)$. For this function U, these values are:

$$U(0, t) = \frac{1}{\sqrt{4\pi\kappa t}} \int_0^0 e^{-\frac{s^2}{4\kappa t}} ds = 0, \qquad \text{instead of the desired value } u_b;$$

$$U(x, 0) = \lim_{t \to 0} \frac{1}{\sqrt{4\pi\kappa t}} \int_0^x e^{-\frac{s^2}{4\kappa t}} ds = \frac{1}{2}, \qquad \text{instead of } u_g.$$

It would be well to think through this second limit, as called for below.

Exercise 11.   Verify the limit of $\frac{1}{2}$ in the previous equation. (Hint: you should be able to do this practically by inspection based on the work in the previous chapter or on any previous experience you may have had in statistics.)

Exercise 12.   As an alternative strategy to the previous problem, you might argue that since the integral is taken with respect to s, terms involving t can be moved in or out of the integral to suit your convenience, just like constants. Therefore you could take the factor in front of the integral inside, and then also the limit. For this line of argument, either justify it based on the rules of calculus and use it to get the correct answer, or determine precisely where it violates a rule of calculus for taking constants and limits inside an integral.

The encouraging thing about the above limit calculations is that the limiting values are constants; they don't depend on the remaining variable, t in the first case and x in the second. Since the desired boundary conditions are also of this form, it suggests that some change of scale might be all the adjustment that is needed. In particular, let us see if we can find constants A and B such that the new function $U^*(x, t)$ defined by

$$U^*(x, t) = AU(x, t) + B$$

might work for the original problem. Based on the results of Exercise 7 in this section, this new function will also be a solution to the PDE, a fact that will also be verified directly in the exercises below. What are its boundary conditions? Well,

$$U^*(0, t) = AU(0, t) + B = A \cdot 0 + B = B;$$
$$U^*(x, 0) = AU(x, 0) + B = A \cdot \frac{1}{2} + B = \frac{A}{2} + B.$$

We want to make the first of these equal to u_b and the second equal to u_g, which involves solving two simple simultaneous equations in the two variables A and B. The final result, to be verified in Exercise 13, is that the new function

$$U^*(x, t) = (2u_g - 2u_b)U(x, t) + u_b$$

TABLE 7-4

Parameters used in heat flow and boiling calculations

Parameter	Description	Source
ρ	density of the ground	reference handbook, depending on ground type; may be built into model
c	specific heat of the ground	reference handbook, depending on ground type; may be built into model
κ	thermal diffusivity of ground	reference handbook, depending on ground type; may be built into model
K	heat conduction constant for ground	calculated from three previous parameters (equation in Table 7-3)
u_b	liquid boiling point	chemical database
u_g	ground initial temperature	estimate based on climate and region
H	heat of vaporization	chemical database

is a solution to our specific problem as summarized in Table 7-3, including both the PDE and the boundary conditions!

Exercise 13. Set up and solve the two simultaneous equations referred to in the previous paragraph.

Exercise 14. Show that U^* satisfies the heat flow PDE, based on the fact that U does.

Now finally we can return to our original problem and calculate the vaporization rate from the boiling pool. The basic data we would need for this calculation are summarized in Table 7-4. The only one requiring special comment is the density, since the database value would be in units of mass per volume, and yet all our derivations for the one-dimensional heat flow problem make use of "linear density," which means mass per unit length. In fact, since we are going to be calculating the heat flow into the pool for each unit area of the ground, its linear density and normal volumetric density have the same value. For example, if the density is 200 lb per cubic foot, then if we restrict our attention to one square foot of the ground, the density can also be expressed as a linear density with the value 200 lb. per foot of depth. These parameter values would generally be used as follows in the vaporization submodel. The heat flow up through the ground to the pool would be given by Fourier's law, namely:

$$q = -K \left. \frac{\partial U^*}{\partial x} \right|_{x=0}.$$

The value of K is determined as described in Table 7-3 and 7-4. The partial derivative of U^* is just

$$\frac{\partial U^*}{\partial x} = \frac{\partial}{\partial x} \left[(2u_g - 2u_b)U(x,t) + u_b \right]$$

$$= (2u_g - 2u_b)\frac{\partial}{\partial x}U(x,t)$$

$$= (2u_g - 2u_b)\frac{\partial}{\partial x}\int_0^x \frac{1}{\sqrt{4\pi\kappa t}}e^{-\frac{s^2}{4\kappa t}}\,ds$$

$$= (2u_g - 2u_b)\frac{1}{\sqrt{4\pi\kappa t}}e^{-\frac{x^2}{4\kappa t}},$$

so the flux value at $x = 0$ is given by

$$q = -K\left.\frac{\partial U^*}{\partial x}\right|_{x=0} = \frac{-K(2u_g - 2u_b)}{\sqrt{4\pi\kappa t}} = \frac{-K(u_g - u_b)}{\sqrt{\pi\kappa t}}.$$

The negative sign simply indicates that it is an upward flux, since the positive x-direction is downward. This would yield units of energy per unit time per unit area of the pool, and since this heat would be used to vaporize liquid, which is already at the boiling point, the corresponding boiling or vaporization rate would be:

$$\text{boiling rate} = \frac{|q|}{H} = \frac{K(u_g - u_b)}{H\sqrt{\pi\kappa t}} \quad \text{(units of mass per time per unit area of pool)}.$$

This rate naturally depends on time, for, as discussed earlier, as the nearby heat in the ground is used up, the rate of heat transfer into the pool gradually diminishes. The total amount of material that vaporizes (per unit area of the pool) over any time interval would be given by the integral of this instantaneous value, namely:

total mass vaporized between times t_1 and t_2 per unit pool area

$$= \frac{K(u_g - u_b)}{H\sqrt{\pi\kappa}}\int_{t_1}^{t_2} t^{-1/2}dt$$

$$= \frac{K(u_g - u_b)}{H\sqrt{\pi\kappa}}(2\sqrt{t_2} - 2\sqrt{t_1}).$$

This completes the problem of how to model the boiling of a liquid spill under the simplifying assumptions made early in the section.

Exercise 15. Draw a series of three graphs, corresponding to progressively later times, on one set of axes to represent qualitatively the expected temperature profile going down into the ground under a liquid pool boiling at ambient temperature. Then, on a set of axes just below and lined up with the first, draw graphs of the temperature gradients for the original three graphs. Explain how you can use these graphs to illustrate some of the concepts from this section concerning the boiling rate as a function of time.

Exercise 16. Suppose that a refrigerated tank containing 1 million kilograms of LNG (liquefied natural gas) fails catastrophically and that the liquid forms a pool in the diked area, which contains 20,000 ft². How long would it take for all of the mass to enter the vapor cloud? What would be the flash fraction? (You may assume that the initial temperature of the contents of the tank is $-162°C$, the boiling point of the material. The heat of vaporization of LNG at this temperature is 119 cal/g. For other parameters, use the following typical values: ambient temperature $25°C$; initial ground temperature $18°C$; ground specific heat 0.2 calories per gram per degree Celsius; ground thermal diffusivity 0.0046 cm² per second; and ground density 2.5 grams per cubic centimeter.)

Exercise 17. Compare your results in Exercise 16 with the same results obtained using your modeling package. Can you explain any differences?

Exercise 18. Suppose you were to repeat Exercise 16 with the only change in input data being an increase in the ground density to 4 grams per cubic centimeter. Qualitatively speaking, how should this affect the results, if at all? Explain.

Exercise 19. Suppose you were building a modeling package to apply to liquids with low boiling points. Draw a general logic diagram or flowchart to illustrate how you would combine the following five submodels: discharge, pool formation, flash boiling, normal boiling, and vapor dispersion.

Exercise 20. Consider spills of low-boiling-point liquids onto bodies of water instead of onto land, such as if a tank barge carrying such material were rammed by another vessel, breaching its tanks while it is moored in a harbor. What would you expect to be the differences in the evolution of the incident compared to a spill on land? You may assume that the material is lighter than water and largely floats as a pool on the surface.

Although the above analysis of the boiling process is probably the typical approach taken in most modeling packages, there are both refinements and alternatives. Refinements might, for example, model the ground as a multilayered assemblage of different materials, each with different thermal properties. These properties may even change as the temperature falls, especially because of the formation of ice within the affected zones. More radical alternatives to the heat conduction model include empirical models based on experimental spills and data from past accidents.

The following brief additional discussion of the development of U (and hence U^*) is not essential to the solution of the heat flow and boiling problems, but it may give additional mathematical insights and facilitate a deeper understanding of the methods that have been used above. There is the very natural question of why someone might even think of investigating the function

$$U(x,t) = \int_0^x u(s,t)\, ds = \int_0^x \frac{1}{\sqrt{4\pi\kappa t}} e^{-\frac{s^2}{4\kappa t}}\, ds = \frac{1}{\sqrt{4\pi\kappa t}} \int_0^x e^{-\frac{s^2}{4\kappa t}}\, ds$$

as a possible solution to the heat flow or diffusion PDE. (Recall that this was the key to our whole solution, and it may have struck you as a "trick" that you would never have thought of.) Our analysis of Figure 7-13 did later lead to some physical interpretations of this idea, but a more practical and direct strategy might be worth mentioning. The solution of mathematical equations, especially differential equations with various boundary conditions, is usually a very hard problem. Therefore, once you find one solution to the differential equation, you usually try very hard to see if you can use either the solution itself or the same logic by which you found it in order to find additional solutions. Simple strategies to try are multiplying by constants, adding different solutions, etc. In fact, if you can show that any linear combination of solutions will be a solution, then you know that the set of solutions forms a special mathematical structure called a *vector space,* which you may have studied if you have had linear algebra. Two observations flow out of this. The first is that once you know you are dealing with a vector space of solutions, all you have to do is find a so-called *basis* for the vector space, a set of specific solutions, sometimes having a common and relatively simple form, such that every solution will be a linear combination of some of these. Then for your particular boundary value problem you just have to find the right coefficients to make up this linear combination.

The second observation is that many linear operations, or *linear operators,* when applied to a solution, will give you another solution. For example, letting D denote the operation of taking the partial derivative with respect to x, and letting u represent an arbitrary solution to the heat flow PDE, we have:

$$\kappa \frac{\partial^2 u}{\partial x^2} - \frac{\partial u}{\partial t} = 0$$

$$D\left(\kappa \frac{\partial^2 u}{\partial x^2} - \frac{\partial u}{\partial t}\right) = D \ (0 \ \text{function})$$

$$\kappa D \frac{\partial^2 u}{\partial x^2} - D \frac{\partial u}{\partial t} = 0$$

$$\kappa \frac{\partial^3 u}{\partial x^3} - \frac{\partial^2 u}{\partial x \partial t} = 0$$

$$\kappa \frac{\partial^2 (Du)}{\partial x^2} - \frac{\partial (Du)}{\partial t} = 0.$$

The final line in this sequence shows that Du is a solution to the PDE, and the two key steps were the linearity of the differentiation process (second to third step), or of the "differentiation operator," and the interchange of order of differentiation with respect to t and x, which is a valid operation whenever the function has continuous second-order partials.

A natural line of thought now suggests looking at integrals, or antiderivatives, because if solution u_2 is a derivative of solution u_1, then solution u_1 is an antiderivative of solution u_2. So since *some* antiderivatives are clearly solutions, it is reasonable to try this antidifferentiation process to generate potential solutions and see by substitution if they work. (We found earlier in the text that this worked for our case.)

Another look still at the defining equation for U, namely,

$$U(x,t) = \int_0^x u(s,t)\,ds = \int_0^x \frac{1}{\sqrt{4\pi\kappa t}} e^{-\frac{s^2}{4\kappa t}}\,ds = \frac{1}{\sqrt{4\pi\kappa t}} \int_0^x e^{-\frac{s^2}{4\kappa t}}\,ds$$

will bring out another important point. While U is written as a function of two independent variables, x and t, it really can be written as a function of a single composite variable. In particular, if you were to try to solve the integral by creating a new variable of integration z so that the exponent of e would become simpler, then a logical variable change would be:

$$z = \frac{s}{\sqrt{4\kappa t}},$$

and the integral would become

$$U(x,t) = \frac{1}{\sqrt{4\pi\kappa t}} \int_0^x e^{-\frac{s^2}{4\kappa t}}\,ds = \frac{1}{\sqrt{\pi}} \int_0^{x/\sqrt{4\kappa t}} e^{-z^2}\,dz = \frac{1}{2} erf\left(\frac{x}{\sqrt{4\kappa t}}\right),$$

where the last term is the well-known *error function* defined by the equation

$$erf(Z) = \frac{2}{\sqrt{\pi}} \int_0^Z e^{-z^2}\,dz.$$

This integral cannot be evaluated in closed form, but there are tabulated values of it available in many books on statistics, probability, and applied mathematics, and it is also available on some calculators and mathematical computer packages. However, the most important single observation here is that the function U, and hence the function U^*, representing the ground temperature under the pool and defined earlier in terms of U, depends on only this single composite value Z defined by:

$$Z = \frac{x}{\sqrt{4\kappa t}}.$$

Exercise 21. Work out the details of the change of variables from s to z in the above calculations, including consideration of the bounds of integration.

Exercise 22. For the problem given in Exercise 16 of this section, use a reference source of error function values ($erf\ z$), such as a table or standard computer program, to draw graphs of the underground temperature profile within the top 20 inches of the soil at points 1 hour and 2 days into the process. (Hint: If you have trouble getting started, begin by simply finding the underground temperature at a single time and location, say 1 hour into the process at a location 2 inches underground. This should help clarify how to do the required calculations.)

7.3.3 Evaporation Submodels*

We return in this section to the situation where we have a pool of spilled liquid that is below its boiling point. Therefore, the flashing and boiling processes discussed in the previous sections are not applicable, and the governing process is simple evaporation, which is a slower vaporization process than those that have just been discussed. Nevertheless, for relatively volatile materials or for those that are dangerous even at quite low concentrations, it is certainly possible to generate a vapor cloud that presents a hazard to people in the surrounding area. In fact, a number of our earlier scenarios in Chapter 4, such as those involving acetone and acrylonitrile, were precisely of this type.

Recall from the discussion of vapor pressure and related topics in Chapter 4 that molecules of a volatile liquid are always moving in both directions between the top of the liquid layer and the vapor space above it. When the amount of molecules in the vapor space is low, such as when evaporation first starts, there is a serious imbalance in the rates of these two transfer processes, and hence the net transfer to the vapor space is high. Gradually, if the vapor is somewhat confined so that the concentration begins to build up there, the transfer rates start to cancel each other more and the net transfer rate decreases. Finally, if the maximum vapor pressure for the material is ever reached in the vapor space, the rates are exactly balanced and there is no net additional transfer. The two principal factors governing these processes in a given material are the following:

- The **transfer from liquid to gas** is controlled by the temperature of the liquid, which determines what fraction of molecules per unit time have sufficient energy to break the bonds of attraction holding them into the liquid mass.
- The **transfer from gas to liquid** is primarily controlled by the vapor pressure of the material in the vapor space just above the liquid, as this determines the total number of molecules from which a certain fraction (i.e., those moving downward with sufficient energy) will reenter the liquid.

The temperature of the liquid may be a relatively simple issue, although if the pool is other than extremely thin, there may be important temperature variations vertically resulting from the

---

* This section does not depend on the use of calculus or any advanced mathematics.

removal of heat energy by its use (in the form of heat of vaporization) to vaporize whatever amount has already moved to the vapor state. This would then involve the consideration of heat transfer processes within the pool itself. The vapor pressure of the material just above the pool is more problematic, since this is not a simple closed container moving towards equilibrium. Just as fast as molecules enter the vapor space, some of them may be swept away by the wind or otherwise diffuse through the air to more distant locations.

Figure 7-14 suggests the complexity of the processes that may be relevant to the net evaporation rate from a liquid pool. Recognizing that simple heat conduction from the ground, as encountered in the last section, can already be somewhat complex in its modeling, one can easily imagine the complexity of models needed to treat in detail most or all of the phenomena shown in this figure.

Focusing on the mass transport components, keep in mind that the net vaporization rate will be increased whenever some mechanism helps to remove vapor molecules from the neighborhood of the surface of the liquid. (This corresponds to keeping the chemical's vapor pressure just above the liquid very low.) The two principal processes for such transport away from the surface are diffusion and advection. As discussed in Chapters 3 and 6, diffusion occurs at a rate proportional to the concentration gradient, and the proportionality constant itself depends on the wind speed and other atmospheric factors. Advection is simply the process whereby material is carried off with the wind. Just as with the heat transfer processes, these could also be modeled

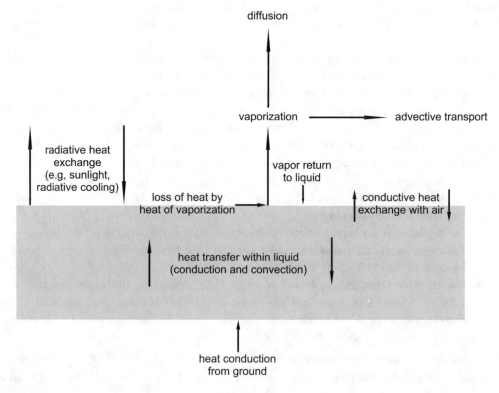

FIGURE 7-14
Processes potentially affecting evaporation rate from pool

in some detail for the region just over the pool, but it would also be quite complex to do so and the data requirements might be very demanding.

Therefore, modelers have also sought other ways to deal with the evaporating pool issue, and the intention here is only to give a typical approach. (The "best" choice, if there is one, would depend on the purposes of the modeling exercise, the availability of data, the resources available to devote to the modeling, and similar factors.) As was mentioned early in this section, it is not safe or convenient to conducts lots of experiments involving large open pools of many different highly hazardous materials. But a number of years ago the US Air Force did indeed conduct a series of experiments in the desert involving spills of hydrazine. This is a chemical that is used as a rocket fuel (as well as for other purposes) and significant quantities of it are shipped around the country in tank containers. There had been questions raised concerning the safety of such shipments, and the experiments were planned to obtain directly relevant data and to aid in the calibration of models that had been developed to treat the associated phenomena, such as evaporation from a pool on the ground. (Hydrazine does not boil at ambient temperature.) These experiments helped to sort out the key factors controlling the rate of such evaporation and led to the calibration of some rather complex models for the process. However, secondarily, modelers then began to approximate the results of the experiments and the output of the complex models by fitting them with curves that depended only on what appeared to be the dominant variables. This led, for example, to the following approximate equation for hydrazine evaporation as a function of pool temperature and wind speed:

$$E = 1.4 \times 10^{-4} u^{0.75} (1 + 4.3 \times 10^{-3} T^2)$$

where:

E is the evaporation rate in pounds per minute per square foot of pool;

u is the wind speed in miles per hour; and

T is the temperature in degrees Celsius.

(The unit conversions are included in the constant at the beginning of the equation, a fact that might horrify a scientific purist. But the emphasis here is on a user-friendly model equation, and this equation does indeed use units that most practitioners would be comfortable with.) You can see in the above that the parameters u and T are the determinants of evaporation rate in this simplified model, which reinforces our earlier observations about why they should be significant variables. However, the particular functional forms in which they are included do not have a simple intuitive explanation and should best be viewed as the result of the curve-fitting calculations leading to this equation. (We will not review these calculations.)

Not only did the investigators develop an equation for hydrazine evaporation, they studied the scaling of the results to other chemicals. How should we expect such a scaling process to be structured? First, since the evaporation calculations probably involve the *number of molecules* moving in and out of the liquid, we might need to scale the results by molecular weight. That is, if acetone molecules (molecular weight = 58.08) were being released into the vapor space *at the same rate* as predicted for hydrazine (molecular weight = 32.05), then the rate at which acetone *mass* would be being released would exceed that of hydrazine by a factor of

$$\frac{58.08}{32.05} = 1.81,$$

which is almost twice as much mass per unit time. Another factor by which we might expect to have to scale is vapor pressure. For example, the vapor pressure of acetone at, say, 20°C is 180 mmHg, whereas that for hydrazine is about 10 mmHg at the same temperature. This indicates that acetone is much more volatile; in loose terms we might say 18 times as much. To a first approximation then, one might expect that the evaporation rate for acetone might be scaled up by a factor of 18 over that of hydrazine.

This kind of logic has been borne out reasonably well by the more complex models, and the result is the following evaporation rate equation that has been incorporated in some modeling packages:

$$E = 1.49 \times 10^{-4} u^{0.75} (1 + 4.3 \times 10^{-3} T^2) \cdot \frac{M_w}{32.05} \cdot \frac{P}{P_h}.$$

This simplifies through the combination of the two multiplicative constants to

$$E = 4.65 \times 10^{-6} u^{0.75} (1 + 4.3 \times 10^{-3} T^2) \cdot M_w \cdot \frac{P}{P_h}.$$

Here, E and T are just as above, M_w is the molecular weight of the chemical in the pool, and P and P_h are the respective vapor pressures of the chemical and of hydrazine at the temperature of the pool (generally assumed to be ambient). The above equation requires some modification for $T < 0$, a case that we need not investigate here. Note that to apply this equation, one would need the vapor pressures of both hydrazine and the chemical of interest. The standard MSDS reported values are generally at 20°C, and for many calculations this would be a reasonable ambient temperature assumption. For other cases, the corresponding values could be determined from chemical reference material or from internal vapor pressure functions included in some modeling packages.

One of the most interesting aspects of the above model is what is not included in it, and the fact that it need not be included tells us something about the dominant controls on the process of evaporation. This is pursued in one of the exercises.

Exercise 1. What would be the estimated evaporation rate of acetone from a spilled pool at 20°C? Assume a wind speed of 5 mph. Find the corresponding rate for acrylonitrile, a chemical also treated in earlier examples.

Exercise 2. The heat of vaporization is not included in the equation presented above for the evaporation rate, and yet it was quite important in the previous case, which involved boiling. Can you suggest any possible basis for this?

7.4 Vapor Dispersion Submodels*

In previous sections, we have treated the sequence of events that begins with a discharge of some hazardous volatile liquid. We calculated the rate at which it would leak out through an opening, the formation of a pool, and the vaporization of that pool either through boiling or evaporation processes. The final logical step in this sequence is the modeling of the dispersion of the vapor in the air, and especially the determination of zones where the concentration of

---

* This section assumes that the reader is familiar with most of the material in Chapters 3 and 6, involving the modeling of the dispersion of air pollutants, although there is a brief review of the key results here. Some of the problems in this section involve derivatives and integrals.

the chemical vapors may pose either a flammable or toxic risk. In many cases this dispersion process is essentially the same as the dispersion of air pollution sources when released from a stack, a topic to which Chapter 3 was devoted; and so we will build upon that earlier work in the current situation.

First we begin with a quick review of the essential material to be used from Chapter 3. The primary problem treated there was the modeling of a continuous plume of material being injected into the air at a steady rate of Q, representing mass per unit time, and from an effective height H above the ground. (H may be 0, as it will be for most cases of an evaporating pool.) The air is moving at a windspeed of u, which of course is the factor that causes the release to form a long plume extending in the downwind direction. The concentration, C, is of course a function of the three spatial variables: x, the horizontal direction along the axis of the wind; y, the horizontal direction perpendicular to the wind; and z, the vertical direction. The formula we developed for C was called the Gaussian plume equation, and it had the following form:

$$ C = \frac{Q}{2\pi\sigma_y\sigma_z u} \left[e^{-\frac{y^2}{2\sigma_y^2}} \right] \left[e^{-\frac{(z-H)^2}{2\sigma_z^2}} + e^{-\frac{(z+H)^2}{2\sigma_z^2}} \right]. $$

In this expression the degree of dispersion in the transverse (y) and vertical (z) directions is accounted for by the dispersivities σ_y and σ_z, which are themselves functions of the downwind distance x. Graphs and equations for these latter functions were given in Chapter 3. Note that this equation does not include any dispersion in the x-direction.

Before moving on, it will be valuable to refresh your acquaintance with this equation by doing the following exercises.

Exercise 1.   A certain pool of acetone is estimated to be evaporating at a rate of 50 lb/min into the air. The conditions are bright sun with a wind at 8 mph. What would the ground-level concentration be at a distance of 350 feet downwind along the direct axis of the wind? Express your final answer in units of grams per cubic meter in order to facilitate comparisons with the next exercise. (Do this problem using your air dispersion tools from Chapter 3, not your hazmat modeling package.)

Exercise 2.   Repeat the above problem using your modeling package. How do the results compare?

Exercise 3.   Write a simplified version of the Gaussian plume equation for the case of a ground-level release and a ground-level receptor. (This will be our most common application.)

Exercise 4.   Discuss the last bracketed factor, involving the z-direction, in the Gaussian plume equation, explaining why it accounts for a no-flow boundary condition at the surface of the ground. Does this apply even when $H = 0$, or does it double the correct value in this situation?

Let us make a slightly more detailed list than in Chapter 3 of the assumptions inherent in the Gaussian plume model:

- *The vapors are neutrally buoyant,* meaning that the part of the air mass containing the vapors is neither lighter nor heavier than the original air itself. This might not be true, for example, if the vapor molecules are much heavier or lighter than the air molecules, especially if their concentration is relatively high, or if temperature changes above the pool, such as in flash boiling, cause the air to cool and become more dense.
- *The movement of the air is not affected by the entry of the vapors into it.* Since the model is based on dispersion processes controlled by the atmospheric stability class, any large

turbulent modification of those processes would completely disrupt the situation. This could happen in the turbulent zone above a rapidly boiling pool, or as a result of factors mentioned in the previous assumption.

- *The release can be treated as a "point source,"* meaning that the entire release is modeled as occurring at a fixed point, rather than being spread out over the physical extent of a pool or other source. This point is usually taken as the center of the pool, although for some conservative calculations, it is modeled as the source point closest to a receptor of particular interest.

- *The release occurs at a constant rate over time, and the extent of that time is relatively long, so that the plume reaches a steady state.*

- *Horizontal dispersion along the axis of the wind can be ignored.* You can see this because there is no σ_x term in the equation.

- *The dispersion coefficients σ_y and σ_z apply to the conditions being modeled.* For example, they are usually reported to apply to ten-minute average concentration values, and different sets of values apply to different kinds of terrain. Depending on their values, they also usually apply within a specified distance range from the source, usually omitting very near and very distant receptors. The published range for the standard values is 0.1 to 10 km, although these limits are often stretched for lack of anything better and with recognition of the poorer approximations expected in such applications.

The following exercises ask you to think briefly about some of these assumptions.

Exercise 5. Discuss the rationale for excluding dispersion in the x-direction in the Gaussian plume model. Is this a conservative assumption in terms of calculated values of maximum downwind concentrations?

Exercise 6. If the combination of air and vapor is heavier than the air itself, would it be conservative to ignore this difference and apply the Gaussian plume model?

Exercise 7. How does the assumption cited above relating to a long time of release relate to any of the other assumptions on the list? Describe qualitatively the effect on concentration values of dropping this assumption in favor of a constant, but finite-duration release.

Exercise 8. Suppose your σ_y and σ_z values do indeed pertain to ten-minute averages, but you interpret the model results as instantaneous concentration values. Have you overestimated or underestimated the concentrations you wanted? What if you interpreted your results as half-hour averages of concentration values?

Exercise 9. How might you adapt the Gaussian plume model to a situation where the release rate varies with time? Does your approach apply to the situation where the release rate includes a fairly steep rise? If so, explain how. If not, can you find any additional way to include this situation?

The discussion above should make you more alert to the limitations of the normal Gaussian plume model, probably more so than when it was first introduced in Chapter 3. In fact, these limitations are far more important in the current context, where the basic assumptions of the model are often violated, at least during some part of the vapor dispersion scenario.

One approach to dealing with some important variations was included in Chapter 6, where a three-dimensional puff model was discussed for application to the case of an instantaneous release of a mass M of chemical vapor from a point source. The resulting equation takes the form:

$$C = \frac{M}{(2\pi)^{3/2}\sigma_x\sigma_y\sigma_z} \left[e^{-\frac{(x-ut)^2}{2\sigma_x^2}} \right] \left[e^{-\frac{y^2}{2\sigma_y^2}} \right] \left[e^{-\frac{(z-H)^2}{2\sigma_z^2}} + e^{-\frac{(z+H)^2}{2\sigma_z^2}} \right].$$

Based on your familiarity with the Gaussian plume equation and with three-dimensional diffusion, the terms in this equation should look like what you might expect. It would be worthwhile to think through some of the details, however, as is requested in the exercises below.

Exercise 10. Give a convincing heuristic explanation for the puff equation presented above. Devote particular attention to a simple explanation for the first bracketed term, involving the x-direction, and to the determination of the composite term in front of the bracketed ones. You may use any results from earlier chapters in your explanation.

Exercise 11. Use the puff equation to model the concentration of dangerous ethylene oxide vapor resulting from an instantaneous nighttime release of 50 lb of such vapor from a vent that is 30 feet above ground level. For a ground-level receptor at a distance of 400 feet downwind, directly along the axis of the wind, find both the maximum concentration and the concentration at a time 30 seconds after the release. Assume that the wind is blowing at 10 mph and that it is a clear night. Also assume open terrain, as usual, and use the dispersion coefficients from Chapter 3 (even though they are really intended for time-averaged concentrations). If you need any additional input parameters, make reasonable assumptions for their value(s), explaining your rationale.

Exercise 12. Apply your modeling package to the previous problem and compare the results with your previous calculations. Discuss any observations.

Exercise 13. Suppose you tried to model the situation in Exercise 11 by using the standard Gaussian plume model, assuming a release rate of 50 pounds per second. (That is, you use a brief time period to approximate an instantaneous release, so that now you have a discharge rate.) Find the predicted concentration at the same point as above, and discuss your results.

Exercise 14. In previous exercises you have had to make assumptions about σ_x and other parameters. Modelers need to examine the sensitivity of their results to such assumptions, especially when there is considerable uncertainty about them. Returning to the situation of Exercise 11, consider the answer to the second question there (i.e., the concentration at a fixed point at $t = 30$) as a function $f(\sigma_x)$. You should be able to calculate its value for a range of σ_x values nearby (although it will be easier if you program the puff equation into a programmable calculator, spreadsheet, or other computer program). Draw the graph of the function $f(\sigma_x)$ for a reasonable domain of values for σ_x. What does this graph tell you about the sensitivity of the puff calculations to this parameter? (E.g., is it large or small?)

The previous exercises should have raised some questions for you concerning the choice of dispersivity values for the puff model. First, there is the question about what to use for σ_x, which does not even have a counterpart in the plume equation. A reasonable choice here might be σ_y, since it also applies to horizontal dispersion. (This would have been your most reasonable choice in Exercise 11.) But even more generally, might it be the case that the σ's for the puff model are completely different from those for the plume situation? Unfortunately, the answer is that for some conditions there may be considerable differences in the appropriate values. However, we shall not investigate these cases and shall assume in what follows that our problems are for conditions in which it is acceptable to use the same dispersivities for both puffs and plumes, a common practice in many modeling packages.

As you can see from the previous discussion and exercises, the plume and puff equations presented here are essentially two extreme cases, one assumed to persist forever and the other to occur instantaneously. In order to obtain more accurate results, the developers of model packages have derived many variations. The objective is to be able to deal with finite duration releases and releases that vary in intensity over time.

One common approach is to consider a finite duration plume as a sequence of individual puffs. For example, a half-hour release might be thought of as a sequence of 6 five-minute releases, or 30 one-minute releases, etc., with these successive approximations looking more accurate but more imposing in terms of calculation. To take the latter case, if we wanted the concentration at a fixed point at, say, 45 minutes after the onset of the release, we would use the puff model for each of 30 one-minute releases and then just total up what would pass by our observation point at this single time, regardless of which of the 30 "mini-puffs" it started out with. So, for example, we might look at the first release 45 minutes later, the second release 44 minutes later, all the way to the 30th release 16 minutes later. (In case you're wondering, yes, it would be even better to make the zero time-point the first half-minute mark, so that each puff is centered on its associated time.)

You might also be wondering about how we would combine the concentrations from the puffs. We would just use the principle of superposition that was introduced in Chapter 6. As a review, look at it this way. As all the molecules in the first puff head into the air, paint them green. Then for the second puff, paint those molecules red, and continue for all the little puffs. The red molecules don't care what the green ones are doing, and, more generally, they all behave independently, an observation we previously used in Chapter 6 when considering diffusion from multiple sources. So each puff travels oblivious to its counterparts, even though dispersion in the x-direction will certainly cause them to overlap and mix. At the end, the observer, at the correct time, simply totals up the vapor molecules of all colors in order to get a concentration.

In fact, the individual mini-puffs could even be of different magnitudes! We never made the assumption in this strategy that they have to be the same. Thus, not only would this approach give us a method to deal with finite duration plumes, but it would even apply if the source strength were to vary over that period.

The following exercises are based on this concept. The first is a straightforward numerical problem that should help you make sure that you really understand the strategy. The second asks you to carry this idea to full fruition by deriving a general formula for these kinds of releases. The final two ask you to apply the general results and compare the answers against your modeling package.

Exercise 15.   Consider a vapor release from a pool of liquid acrylonitrile with a windspeed of 5 mph and atmospheric stability class D. Suppose the pool is growing in such a way that the rate at which material enters the vapor state increases linearly from an initial rate of 100 lb/min to a rate of 600 lb/min 30 minutes later. At that point the fire department covers the pool with foam, thereby stopping evaporation. You are interested in knowing the concentration at this point in time (i.e., when the foam cover is finally established) at a location that is 350 feet downwind and 40 feet to the side of the axis of the wind through the pool center. Express your answer in grams per cubic meter to facilitate later comparisons. Do this problem along the lines just discussed in the text. In particular:

a) Model the release as 6 five-minute puffs.

b) Model the release as 60 half-minute puffs.

c) Based on this experience, construct what you believe to be a reasonably good approximation to the answer.

d) How would your solution method change if you wanted the concentration at the 20- or 40-minute marks? (You do not need to work out the actual numerical values for these cases.) [Hint: in order to make this problem computationally tractable, you will need to use a spreadsheet program (probably the best for this), other computer program, or programmable calculator.]

Exercise 16. The approach of the text and the previous problem clearly suggests a limiting process. That is, think of a finite duration plume as a sequence of mini-puffs, each representing a time interval of $\Delta t = (b - a)/n$, where a and b are start and stop times for the entire process, and n is the number of individual mini-puffs. The addition process at the observation point as n becomes large is really akin then to an integration process. Use this idea to develop a general formula for the concentration $C(x, y, z, T)$ resulting from a finite duration release on the time interval from a to b whose rate is described by a function $m(t)$. You may assume for simplicity that $T \geq b$. (Hint: this is not as imposing as it sounds if you keep good track of terms, just as in the numerical example above.) You may leave your answer in the form of an integral.)

Exercise 17. Use your integral representation from the previous exercise, together with a numerical integration routine in a standard math package, to solve Exercise 15.

Exercise 18. Solve the problem in Exercise 15 using your modeling package, and compare with the results obtained in that exercise and in Exercise 17.

Exercise 19. Discuss conceptually the use of the mini-puff concept for dealing with a "distributed source," meaning one whose spatial extent is such that it would be too inaccurate to model as a point source.

These exercises complete our discussion of Gaussian dispersion processes, whether for a plume or a puff. What remains for us to consider are those dispersion situations where the most fundamental Gaussian assumptions fail, namely, where the air/vapor mass is not neutrally buoyant, and where its formation so disturbs the air flow pattern over the pool that transport is dominated by quite different considerations. In particular, we will focus on the *heavy gas* situation, where the vapor/air mixture is strongly influenced by gravity, which pulls it down towards the surface and keeps it there. Obviously this is potentially a more dangerous situation because we would expect the ground-level concentrations to stay high for a longer period of time, at least near the release point. Materials that produce heavy gases are quite common in everyday commerce. Some have high molecular weights, and others form heavy mixtures with air because they flash boil if released and form cold vapor that is therefore more dense.

The modeling of heavy gases is a rather specialized and complex process, most models consisting of a combination of theoretical and empirical components. A number of spill experiments involving such gases have been conducted, and the model that fits one set of data best may not be the one that best fits another set. One needs a great deal of background in chemistry, chemical engineering, fluid mechanics, and related fields in order to achieve good results. Fortunately, much of this experience is built into some of the modeling packages; but, unfortunately, it is somewhat beyond the scope of this book.

Nevertheless, it is important to understand the basic physical processes involved, especially if you are going to apply a modeling package that may automatically switch to a heavy gas model under certain circumstances. Therefore, consider the top part of Figure 7-15, which depicts the vapor cloud emanating from a spilled pool, particularly in the case of the relatively rapid vaporization you would have during boiling or evaporation of a very volatile substance. When the cloud starts to form over the pool, there is likely to be a great deal of turbulent mixing within it. It addition, recalling that at the outset it has no horizontal velocity, the wind drives into it from the left side, although the bulk of its mass also causes some deflection of the wind up and over it. As the air moves through and over the cloud, there is considerable mixing. The air gives up some of its heat to the cloud, further evaporating any aerosols, and the resulting air/vapor mixture is still cooler and denser than plain air itself. As a result of this "negative buoyancy," there is a gravitational slumping process, and the heavy cloud mass tries to slide out along the ground under the lighter air surrounding it.

The bottom part of Figure 7-15 shows a later point in the evolution of this mass, which flattens, spreads, moves downwind, picks up speed, and grows in volume. Gradually, of course, the mass picks up enough heat from the air and from the ground, and it gets sufficiently diluted by the air, that it begins to lose its distinct negative buoyancy and starts to disperse more like a neutrally buoyant mass.

It is not immediately apparent whether the heavy gas phenomenon should result in longer or shorter distances to a specified concentration threshold, compared to treating the release as a neutrally buoyant gas. Since the cloud itself stays intact longer near the source, one might imagine that at higher concentration levels, such as those of concern in flammability risk, the risk might be dominated by this stage of the cloud's evolution. On the other hand, for toxic risk, which usually involves a lower threshold concentration in the ppm range, transport beyond the

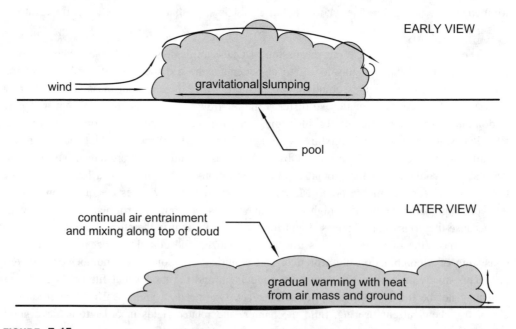

FIGURE 7-15

Processes controlling the evolution of a heavy gas vapor cloud heavy cloud zone, more in line with Gaussian dispersion, might be most representative. There are so many variables that it is difficult to generalize, but the reader is asked to explore this issue for a standard modeling package in the exercises.

How do models and modeling packages actually treat this phenomenon? Two typical approaches that roughly span the range from simple to complex are the following:

1. "Box" Models. These models treat the cloud as a simple geometrical structure, such as a vertical cylinder. This is the "box" (like a round hatbox). The cylinder is imagined to be getting shorter in height, larger in both radius and overall volume (due to air entrainment), and moving downwind. Models for these various individual processes are developed, often in the form of ordinary differential equations, based on both empirical data and simplified physical conceptualizations. Once the vapor cloud is calculated to have lost most of its negative buoyancy, the model switches to a Gaussian model whose source is the current calculated state of the vapor cloud.

2. Continuum Models. These models are based on a complex set of partial differential equations for fluid flow (since the cloud is a fluid), taking into account some or all of the following physical processes or principles:

- momentum transfer from the air to the vapors
- air entrainment at the boundaries of the cloud mass
- vapor dispersion and mixing processes within the cloud
- heat transfer to the vapors from the air and ground
- gravitational slumping (force balance on individual cloud elements)
- entrainment and vaporization of aerosols
- fluid properties of the cloud.

In addition, there may be other factors included, such as the roughness of the terrain over which the cloud is moving.

In the context of the above discussion, it may be interesting to see how your own modeling package deals with this issue, as suggested below.

Exercise 20. Review the documentation for your modeling package and summarize concisely, in the framework described above, how it deals with clouds of heavy gases.

Exercise 21. Consider the instantaneous release of 5,000 gallons of liquefied propane from a small pressurized storage tank ruptured by a runaway truck. Assuming that ignition does not take place as a result of impact, determine the flammable hazard zone (distance to $\frac{1}{2}$ LFL). Assume that it is a clear night with a windspeed of 6 mph. To keep the focus on the heavy gas issue, assume that the entire mass flashes upon release. If your modeling package chooses a modeling approach for the vapor cloud automatically, what choice does it make? If you have the option of specifying whether to apply either heavy gas or Gaussian models, apply both and compare the results.

7.5 General Comments and Guide to Further Information

Many of the comments made in the last section of Chapter 4 pertain to both that chapter and this one, and it would be good to review that section once again at this time. For further detailed

discussion of possible modeling approaches to various physical phenomena, the references mentioned in Chapter 4 are the best sources for both summary information and guidance concerning
the research literature. Another good source of information is the documentation for specific
model packages, which often discusses alternative approaches as part of its rationale for the
approach it has actually implemented.

For further information on heat conduction, especially the complete solutions for various
geometries, the classic reference is: *Conduction of Heat in Solids,* by H. S. Carslaw and J. C.
Jaeger, now also reprinted in paperback.

Our treatment of evaporation modeling was based largely on the approach in the ARCHIE
model, but work continues to be done on this issue and there is quite a range of methods. Good
references are the following:

P. W. M. Brighton, "Evaporation from a plane liquid surface into a turbulent boundary
layer," *Journal of Fluid Mechanics,* Volume 159 (1985), pp. 323–345.

Peter I. Kawamura and Donald MacKay, "The evaporation of volatile liquids," *Journal of
Hazardous Materials,* Volume 15 (1987), pp. 343–364.

P. W. M. Brighton, "Further verification of a theory for mass and heat transfer from
evaporating pools," *Journal of Hazardous Materials,* Volume 23 (1990), pp. 215–234.

With respect to the heavy gas issue, the following references give a range of approaches:

Proceedings of the Third Symposium on Heavy Gas and Risk Assessment, Bonn, Germany,
1984, ed. Sylvius Hartwig, D. Reidel Publishing Company, Holland, 1986.

Stably Stratified Flow and Dense Gas Dispersion, ed. J. S. Puttock, Clarendon Press,
Oxford, England, 1988.

Both of these are conference proceedings that may be quite interesting to peruse because they
give a good perspective on who is doing what and why. They also provide more details on
heavy gas models, which we treated only briefly. Of special note is the extensive discussion
of large scale field experiments (complete with photographs) and how these have been used to
test and refine various models.

Aside from the modeling literature, a good basic physical chemistry reference would be
valuable for gaining a deeper understanding of some of the modeling issues. There are many
introductory physical chemistry texts, and one can choose on the basis of ready availability
and general readability. Because physical chemistry covers a variety of more or less distinct
phenomena, it is generally not necessary to read such texts in a linear fashion. One can move
rather quickly to the sections dealing with the phenomena of interest.

In addition to physical chemistry, fluid mechanics is another field that underlies much
of the modeling in this book, as should be apparent. There are also many good introductory
textbooks on this subject.

Parallel to the risk-modeling strategies discussed in this book are an entire set of additional
models that focus on the release of radioactive materials, such as from nuclear power plants
that may experience severe accidents. While such events are relatively rare, their consequences
can be considerable, with radioactive fallout occurring even in distant countries (as happened,
for example, with the Chernobyl accident in the Ukraine in 1986). One encounters in these
problems many distinct and challenging issues, such as: source term values (e.g., how much
actually gets out of the plant, as opposed to 'plating out' on the walls and structures); air
transport; chemical and physical processes that change the species during transport; possible

surface and ground water transport as well; food chain transport; and calculations involving ultimate human exposure and health effects. The Nuclear Regulatory Commission has published many guidance documents on these aspects; they generally take the form of government reports rather than hard-bound texts. Such documents can be readily located by using government on-line document listings.

At the conclusion of this book, the author would make one more exhortation. Go out and see models in action; don't just think about them as textbook or academic exercises. By visiting plants, talking with operational personnel about past incidents or concerns, meeting emergency responders to learn their viewpoints, and hearing from modelers who work with real world problems, one will develop a much richer understanding of this interesting and important area of endeavor.

Index